World Encyclopaedia of
MODERN AIR WEAPONS

World Encyclopaedia of MODERN AIR WEAPONS

Christopher Chant

PSL

Patrick Stephens

First published in 1988

British Library Cataloguing in Publication Data

Chant, Christopher, 1945-
Encyclopaedia of modern air weapons.
1. Air operations. Military equipment.
Weapons
I. Title
355.8,2

ISBN 1-85260-028-4

Patrick Stephens Limited is part of the Thorsons Publishing Group, Wellingborough, Northamptonshire, NN8 2RQ, England

Printed in Great Britain at The Bath Press, Avon
Typeset by MJL Limited, Hitchin Hertfordshire

10 9 8 7 6 5 4 3 2 1

Contents

Introduction

At all levels of modern warfare, the combat aircraft can and often does exert a decisive influence on the outcome of events on the ground and at sea. This fact has received ample confirmation from conflicts in recent years and from the huge quantity of published material about modern combat aircraft. Yet in all this interest, both general and specialized, there appear to be two important failings: specifically which weapons any particular aircraft can carry and the nature of those weapons.

A companion volume, *Modern Aircraft Armament*, published by Patrick Stephens Ltd, is addressed to the first of these limitations, while *Modern Aircraft Weapons* is designed to overcome the second shortcoming. That there are failings in these two fields can readily be checked: all too many modern reference books about aircraft state baldly that the type in question 'can carry 8,000 lb of ordnance on six pylons, including two air-to-air missiles', or words to the same effect. But the nub of the matter to those genuinely interested in the capabilities rather than just the performance of modern combat aircraft should be firstly the types, numbers and dispositions of the whole range of weapons that can be delivered by the aircraft in question, and secondly the nature of the weapons themselves. Reference works often seem to forget that combat aircraft are, in themselves, only the platforms for the delivery of weapons, and that it is the latter which really count: performance and agility may enable the aircraft to reach its target and undertake its primary mission before departing, but the effect on the target is caused by the weapons whether they be simple and comparatively cheap 'iron' bombs, a whole range of guns and rockets or complex and increasingly expensive guided missiles. Yet this is an aspect of air operations that has hitherto, in general, not received the attention that it deserves.

Modern Aircraft Weapons is designed to go a fair way to satisfying this requirement. The selection of weapons is intended to reflect the whole range of weapon types currently available for combat operations, and within each entry emphasis has been placed textually on current variants of the particular weapon and its applications. The accompanying photographs have been judiciously selected and their captions carefully written to highlight the salient features of the weapons.

The author cannot claim that the treatment is exhaustive, for the pace of development is extraordinarily rapid, and the space available within the book limited. Nevertheless he feels that the most important weapons have been covered, and a fair indication of future developments given. Missiles, more than most weapons have been generally well covered in the current literature. Here, however, they are included partly for the sake of completeness, but chiefly for the overriding importance of this class of weapons in many of the roles undertaken by combat aircraft. The best known variety of guided missile is probably the air-to-air variety, which is used for offensive or defensive purposes. The three guidance modes used in such weapons are passive infra-red homing (in which the missile locks onto a heat source such as the target's engine exhaust), semi-active radar homing (in which the missile locks onto electro-magnetic radiation reflected from the target) and active radar homing (in which the missile has it own active radar to search for and lock onto the target). Infra-red homing is generally used for short-range (often dogfighting) missiles, semi-active radar homing for medium-range missiles, and active radar homing for long-range missiles.

Designed for offensive use is the air-to-surface missile. As befits its primary task, such a missile generally carries a more substantial warhead than its air-to-air counterpart. Air-to-surface missiles also use a wider spectrum of guidance modes, including

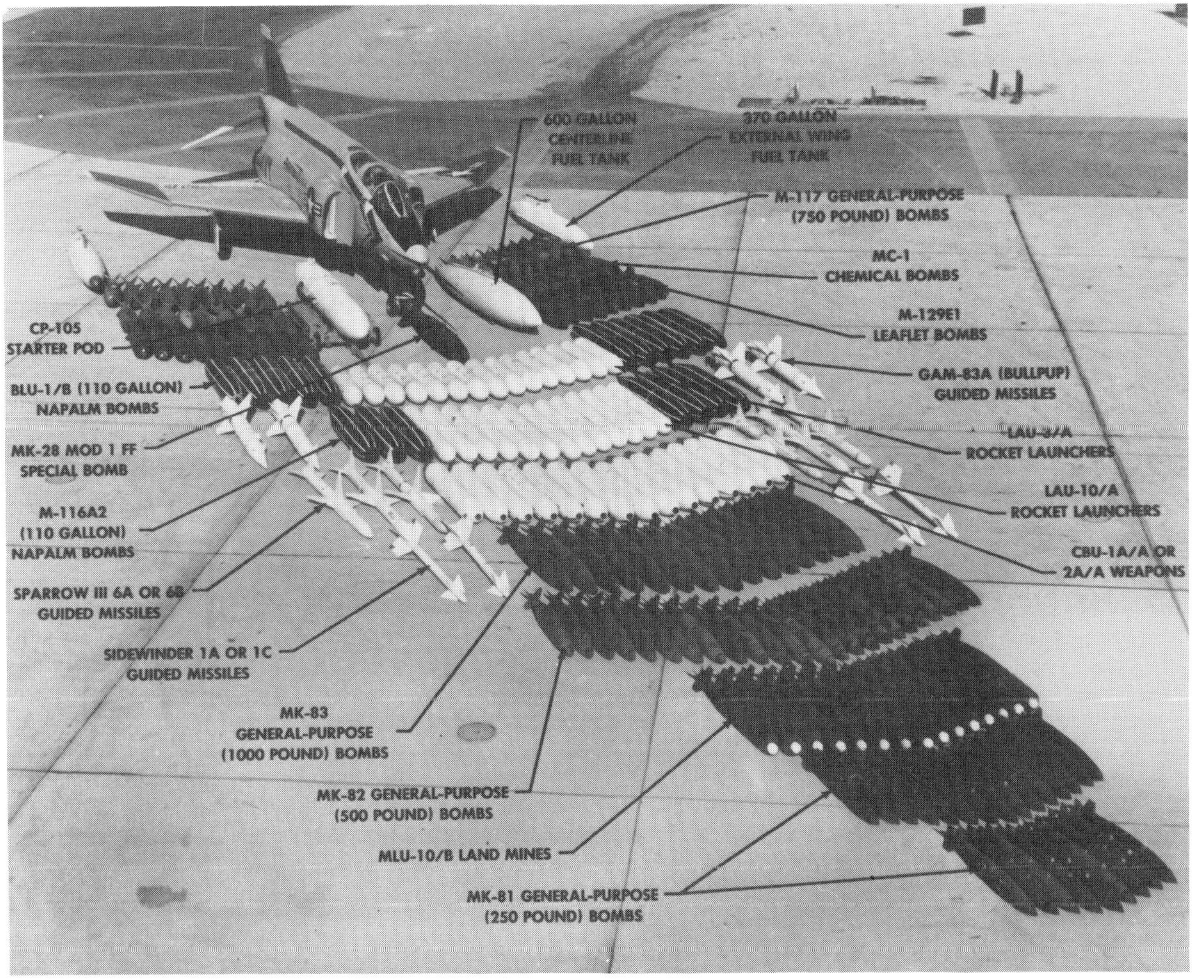

Though elderly (the aircraft being a McDonnell F-4B Phantom II of the US Navy), this pre-1962 annotated layout photograph of armament capabilities reveals some of the more important guided and unguided weapon options available at the time. And though the guided missiles have been replaced by more modern equivalents, the unguided ordnance is still in the US inventory.

electro-optical and laser homing as well as command guidance by radio or microwave. A specialized subvariant of the air-to-surface missile is the anti-radiation missile, an extremely important type designed to allow attack aircraft to 'take out' the enemy's air-defence radars with missiles that home on the surface-based emitter. The most modern anti-radiation missiles can also choose between various potential targets, selecting the type which offers the greatest threat to the attacking force and homing on its location even if the emitter shuts down.

Obviously related to the basic air-to-surface missile are the specialized anti-tank missiles, which mostly rely on manual or automatic command to line-of-sight guidance, though the most modern missiles use semi-active laser homing. Anti-tank missiles are generally of larger diameter and slower than pure air-to-surface missiles, these factors being necessary to provide the special hollow-charge warhead with the right radius for good armour penetration, and to ease the problem of detonating the warhead at exactly the correct stand-off distance from the target's armour for the hollow-charge warhead's gas/vaporized metal jet to be focussed right on the armour.

Just about every air-to-surface missile can be used in the anti-ship role, but there is also a role-dedicated group of missiles for this task. These missiles tend to be massive weapons characterized by long range (often as a result of using air-breathing rather than rocket propulsion), an exceptionally powerful warhead and, in most cases, a special guidance system designed to allow the missile to approach its target in sea-skimming flight before adopting one of several homing and attacking modes.

In this sense the air-launched torpedo can also be regarded as a guided missile, the latest weapons of this variety being incredibly sophisticated devices using the most advanced cybernetic and propulsion technologies. Like anti-ship missiles, torpedoes carry a powerful warhead, and weapons such as the American Mk 50, the British Sting Ray and the French Murène are capable of deep diving and repeated attacks against nuclear-powered submarines.

But other weapon categories have been largely ignored by all other than the highly specialized literature. In a way this belies the importance of these less 'glamourous' weapons which, when accurately delivered against the right target, can be just as effective as vastly more expensive guided missiles. To take but one example, imagine the effect on a target of a 500-lb (227-kg) bomb and an air-to-surface missile such as the AGM-65 Maverick: both weapons deliver roughly equal blows (the bomb having more explosive but less kinetic energy, and the missile having less explosive but more kinetic energy), but the bomb is very considerably cheaper for a comparable result. Of course it cannot be denied that the hit probability of a single missile is considerably greater than that of a single bomb, and that the launch aircraft must approach the target more closely (and thus more vulnerably) to deliver a bomb than a missile, but modern weapon-delivery systems have improved the accuracy of bomb delivery to a very marked degree. The last twenty years have witnessed the development of a hybrid weapon type in which an 'iron' bomb can be fitted with small flying surfaces and a guidance package plus control surfaces. This development, epitomized by the American Paveway series (using semi-active laser guidance) and the Israeli Opher (using passive infra-red guidance) gives the 'iron' bomb the accuracy of the missile and a fair measure of that weapon's stand-off range, but at only a fraction of the missile's cost.

The combination of accurate delivery and launch aircraft escape is also offered by low-level attack using retarded bombs: the low-level attack enhances accuracy and the bomb-retarding system (using drag vanes, parachutes or 'ballute' inflatable retarders) allows the launch aircraft to clear the area before the bomb detonates. Parachute retarders are also used for specialized bombs such as the anti-tank and anti-runway varieties. In these cases the retarder serves the dual functions of allowing the launch aircraft to escape and causing the bomb to drop towards the optimum vertical nose-down position for warhead penetration. In the case of the anti-runway bomb the retarder is discarded as soon as the nose has pitched down, and a rocket motor ignited to boost the weapon and so increase its kinetic energy.

Similar developments are evident in the field of unguided rockets. Much developed and used since World War 2, rockets have usually been considered area-attack weapons because of their relative inaccuracy. However, recent developments in propulsion and ballistics have allowed the development of rockets with much greater intrinsic accuracy, while the evolving capabilities of warhead design have enabled even small rockets to be produced with submunition or sub-calibre flechette warheads, the latter relying on kinetic energy and small diameter for very significant capability against light and medium armoured vehicles. The most notable of these weapons are the 68- and 100-mm (2.68- and 3.94-in) Thomson Brandt Multi Dart types, the 2.75-in (69.85-mm) BEI Hydra 70 and comparable Forges de Zeebrugge FZ-100 types, and the 2.75-in (69.85-mm) Bristol Aerospace CRV7 types.

Submunitions have also gained great favour for area attacks and area denial in the form of payloads carried by cluster bombs and dispenser weapons. Both types are designed to strew a carpet of anti-armour and/or anti-personnel munitions, the former from a casing dropped from the launch aircraft and the latter from a larger unit attached under the fuselage or wings of the launch aircraft. This evolutionary process has also taken the further step along the road to autonomous operations with the first stand-off dispenser weapons: glider-like devices released some distance from the target to cruise into the attack area before dispensing their submunitions. The process of giving these stand-off weapons their own propulsion and 'smart' guidance systems is already under way, and 'smart' submunitions are already appearing. This all helps to distance the launch aircraft and its crew from the increasingly effective anti-aircraft defences of the modern land and sea battlefields, and also enhances the capabilities of manned aircraft against high-value targets such as armour concentrations. Submunition weapons are one of the most important growth fields in modern weapon design, and great things are expected of the stand-off varieties, especially as these adopt 'stealth' technology for low observability and are loaded with 'smart' submunitions such as the Avco Skeet anti-tank device.

Mines fall into the same basic category, though clearly the sea mine has, in its latest forms, much of the torpedo's cybernetic modernity, being designed to identify and attack only the right specific target, and to resist the enemy's best endeavours to 'sweep' it.

Another weapon type that has made a strong comeback in recent years is the gun. During the 1950s it was thought that while machine-guns and 20-mm cannon might retain a limited utility as podded weapons for the ground-attack role, the gun was basically obsolete as an air-to-air weapon. The close-range limitations of the air-to-air missile became all too evident in Vietnam and other wars of the late 1960s, however, and the fixed cannon has re-emerged as an essential component of the combat aircraft's armament. Thus a new generation of weapons has been developed to provide aircraft with greater tactical flexibility through the availability of inbuilt barrelled weapons possessing great reliability and firing rate, and using high-velocity rounds with an increasingly sophisticated array of projectiles. The same basic technology has also been pressed into service to produce devastating point- and area-attack air-to-surface weapons of the barrelled variety for fixed- and rotary-wing aircraft. Indeed, so important has the helicopter become on the tactical battlefield that a readily discernible trend is that towards the combat helicopter, fitted with a fixed or turreted heavy-calibre cannon plus short-range air-to-air missiles to attack the enemy's battlefield helicopters and to escort friendly helicopters carrying anti-tank missiles, air-to-surface rockets and flexible single- or multi-barrel machine-gun and light cannon.

As always, there is great difficulty in providing accurate data about Soviet weapons, which in this context are more difficult to assess even than Soviet aircraft, themselves the subject of much dispute in the West. But whereas Soviet aircraft (and to a lesser extent missiles) are eventually seen and photographed, this is often not the case (at least in the published literature) with Soviet free-fall weapons and rockets. Thus the subject remains one of conjecture rather than hard fact, a situation exacerbated by the reluctance of Western intelligence sources to reveal the information they inevitably possess. Even when such weapons are photographed, there is comparatively little of a precise nature that can be deduced from so simple a structure as a bomb casing.

Readers are reminded that within each chapter each manufacturing country appears alphabetically by its full name in English (thus USSR precedes UK and USA). Individual weapons' entries within each country are listed in alphabetical and series number order.

The author would like to take this opportunity to thank the professional colleagues and manufacturers who have aided so willingly and extensively in the preparation of *Modern Aircraft Weapons*. Any errors are, of course, entirely the responsibility of the author.

Grantham, Lincolnshire. *August 1988.*

GUNS

Fabrique Nationale (Browning) M3P

Belgium

Type: aircraft machine-gun (air-to-air and air-to-surface)
Calibre: 12.7 mm (0.5 in)
Operating method: recoil operation
Feed: left- or right-hand linked belt
Overall dimensions: length 1653 mm (65.0 in)
Barrel length: 1143 mm (45.0 in), or 90 calibres
Rifling: 8 grooves with right-hand twist
Weight: complete weapon 29 to 35 kg (63.9 to 77.2 lb) depending on barrel
Recoil force: data not available
Muzzle velocity: 850 to 920 m (2,789 to 3,018 ft) per second depending on ammunition type
Rate of fire: variable between 900 and 1,100 rounds per minute by a special back plate group
Effective range: 3000 m (3,280 yards)
Ammunition types: M2 AP, FN 169 APEI, M8 API, M20 API-T, M2 and M33 Ball, M1 and M23 Incendiary, and M10, M17 and M21 Tracer
Ammunition weight: complete round 120 g (4.23 oz); projectile between 39.7 and 6.8 g (1.4 and 1.65 oz) according to type
Ammunition length: complete round 138.4 mm (5.45 in)

Variants and applications
M3M: The M3 series of heavy machine-guns for aerial use was developed during World War 2 as a faster-firing version of the **Browning M2**, which was listed at a cyclic rate of between 750 and 850 rounds per minute but a practical rate of about 550 rounds per minute. The later weapon thus offered comparable safety and reliability with a significantly higher rate of fire. FN in Belgium is now one of the main suppliers of these weapons (though substantial numbers are produced elsewhere either by remanufacture of existing guns or by production to differing standards), and the M3M is the version intended for use in helicopter doors with pintle mounts such as the Belgian **FN ETNA MPS 1050** and **External Pintle Support** (EPS, a special mounting attached to the External Mounting Assembly developed jointly by FN ETNA and Bell to provide Bell helicopters with the ability to carry external loads on each side of the fuselage), and the Italian **Aerea DGP** (Door Gunner Post). The 30-kg (66.1-lb) EMA is a tripod structure attached to the fuselage by three fasteners, and the installation of such a fitting on each side of the fuselage provides considerable weight-lifting capacity for weapons. Each EMA can carry a fixed gun pod or rocket launcher, or a combined gun pod and rocket launcher, and can also support the 12-kg (26.5-lb) EPS for one M2/M3 machine-gun and its case/link collector; the total weight with weapon and ammunition is 143 kg (315.3 lb). On the EPS the machine-gun can be elevated from +5° to −70°, and traversed from 90° forward to 60° backward. The Aerea DGP weighs 70 kg (154.3 lb) with gun, and

An important weapon for light tactical aircraft and helicopters is the Aerea MTP (Multi Task Pod), which contains the useful combination of a 12.7-mm (0.5-in) machine-gun and 68-mm (2.68-in) or 2.75-in (69.85-mm) rockets.

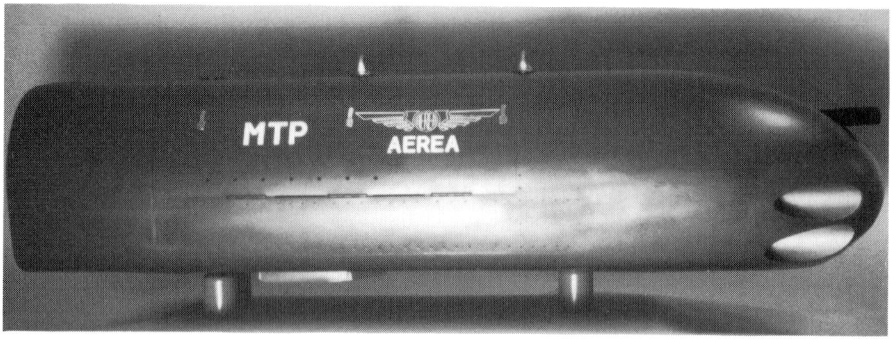

has provision for 100 rounds of ammunition on the mounting and another 400 to 1,000 rounds off it depending on the helicopter type; the gunner remains seated even when the gun is being fired at its maximum depression −70°.

M3P: This is the standard version of the M3 series for fixed installations, and incorporates such features as a firing solenoid, flash suppressor, a device to prevent 'cook offs', and different barrel options. The most recent round developed for this still-important weapon is the FN 169 Armour-Piercing Explosive Incendiary type, which possesses a complete round weight of 112 g (3.95 oz) and a projectile weight of 43 g (1.52 oz) including its heavy alloy penetrator located just behind the explosive/incendiary filling in the nose of the projectile. This latter can penetrate the side armour of most armoured personnel carriers at a range of 700 m (765 yards), and the armour of attack helicopters at 1000 m (1,095 yards).

Aerea P/N 006-3552: This Italian pod is otherwise known as the **Multi-Task Pod** and is designed for carriage on NATO standard 356-mm (14-in) suspension lugs. The MTP offers an interesting combination of firepower as it has an internal M3 machine-gun with 250 rounds as well as six launcher tubes for 2.75-in (69.85-mm) rockets. The pod is 1800 mm (70.87 in) long, 410 mm (16.14 in) wide and 410 mm (16.14 in) high, and weighs 79 kg (174.2 kg) empty and 175 kg (385.8 lb) loaded.

Avibras MS10-21: This Brazilian gun pod carries one 12.7-mm machine-gun and 350 rounds.

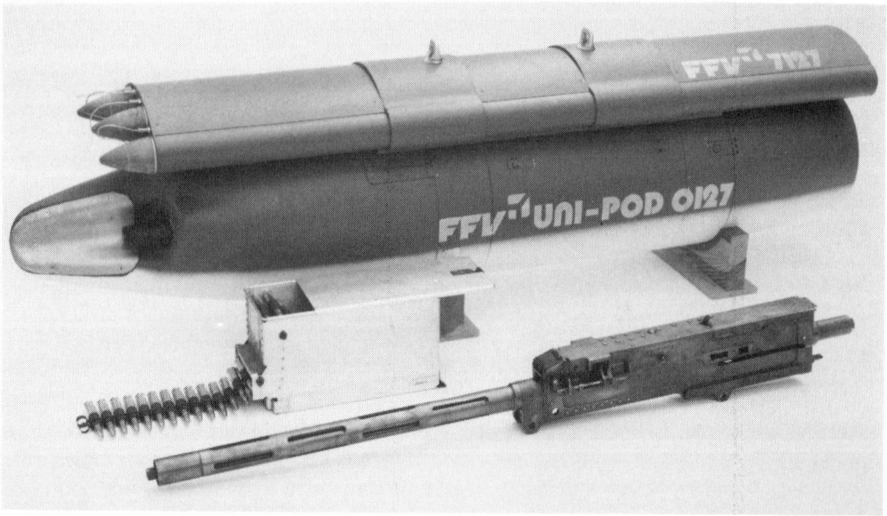

The Swedish FFV Uni-Pod 0127 carries a 12.7-mm (0.5-in) heavy machine-gun and its ammunition, and can also be boosted in power by the addition of the FFV 7127 rocket launcher adapter between the pods and the suspension lugs.

B-52 installation: The B-52G variant of the Boeing Stratofortress has a tail armament of four 0.5-in (12.7-mm) Browning M3 machine-guns each provided with 600 rounds of ammunition. This Bosch Arma MD-9 remote tail barbette is controlled by the ASG-15 fire-control system with separate target-acquisition and target-tracking radars.
CASA twin-gun pack: As part of the weapon fit for its C-101 Aviojet trainer and light attack aircraft, the Spanish manufacturer CASA has developed a special twin-gun pack for installation in the lower-fuselage weapon bay as an alternative to reconnaissance or ECM gear, or the alternative barrelled installation of one 30-mm DEFA cannon. The twin-gun pack contains two M3 machine-guns semi-exposed in the airflow beneath the aircraft, with ammunition storage above the weapons.
FFV Uni-Pod 0127: This is a Swedish offering on the market for light gun pods, and carries the standard M3 machine-gun with 200 rounds, whose spent cases are ejected. The gun is located on the left of the pod's forward section, with the magazine beside and to the right of the barrel. A useful feature is the provision of a pneumatic re-cocking system so that the weapon need not be cocked until the aircraft is safely airborne. The pod's weights are 90 kg (198.4 lb) empty and 118 kg (260.1 lb) loaded, and the overall dimensions include a length of 1916 mm (75.43 in) and a diameter of 360 mm (14.17 in). The pod can also be used in combination with the **FFV 7127** rocket-launcher adapter, which can be used singly or doubly (with four or eight rockets) between the gun pod and its NATO standard 356-mm (14.in) suspension lugs.

FN ETNA HMP: This is the basic podded installation of the M3P for light tactical aircraft with NATO standard 356-mm (14-in) suspension lugs, and flying at speeds up to Mach 0.75. The machine-gun runs virtually the full length of the pod with its muzzle and flash suppressor emerging through a port in the upper nose. Other main items are the 250-round magazine (under the barrel in the forward part of the pod), the spent case collector (to the side of the action opposite the feed), and the electro-mechanical safety and cocking device (on the left of the receiver) so that the weapon can be cocked when the aircraft is in the air. The pod's overall dimensions include a length of 1810 mm (71.26 in) and a diameter of 420 mm (16.54 in), and the loaded weight is 116 kg (255.75 lb). The HMP can also be fitted on its underside with the **MRL 70** launcher with four 2.75-in (69.85-in) rockets.
Kondor Quad Fifty: This is an American package designed to provide medium helicop-

An MBB BO105LS light helicopter shows off a useful armament kit, namely two FN ETNA HMP gun pods (each carrying a 12.7-mm/0.5-in M3P heavy machine-gun) with MRL-70 four-tube launchers for 2.75-in (69.85-mm) rockets suspended under each.

ters with a powerful mix of fixed and flexible machine-gun firepower. The package consists of a pallet (designed to fit into the payload bay of the helicopter) plus recoil-attenuating 'soft' mounts for a maximum of four M3 heavy machine-guns each with 600 rounds of ammunition. The two forward mountings are cantilevered outside the helicopter for a pair of solenoid-operated fixed guns (controlled by the pilot, who has a special sight and 'rounds remaining' counter), while the two rear mountings are pintles for a pair of flexible guns and traversing bicycle-type seats.
Lucas Helicopter Gun Turret: This is a British development using the M3 machine-gun (from any of several sources) in an extremely neat underfuselage damped mounting. The whole barbette is controlled electrically from the co-pilot/gunner's weapon-sighting system, with hydraulic power to train the turret through 240° (120° left and right of the aircraft centreline) and elevate the gun to any angle between +15° and −45°, both traverse and elevation being accomplished at the rate of 80° per second. A magazine just to the rear of the gun holds 400 rounds, and the manufacturer can provide installations tailored to specific helicopters and requirements. The turret is also offered with the new Ferranti helmet-mounted sight, allowing rapid and straightforward training and elevation of the gun under control of the operator's helmet, who has merely to acquire the target in the reticle of his helmet sight: the system automatically measures the orientation of the helmet's sensor to the reference magnetic field, and moves the Lucas turret/gun assembly accordingly.

Belgium

Fabrique Nationale MAG 60/40

Type: air machine-gun (air-to-surface)
Calibre: 7.62 mm (0.3 in)
Operating method: gas operation
Feed: left- or right-hand linked belt
Overall dimensions: length 1255 mm (49.4 in) with flash suppressor
Barrel length: 545 mm (21.46 in), or 71.52 calibres
Rifling: 4-groove right-hand twist
Weight: complete weapon 10.1 kg (22.27 lb); barrel 2.75 kg (6.06 lb)
Recoil force: data not available
Muzzle velocity: 840 m (2,756 ft) per second
Rate of fire: 1,000 rounds per minute
Effective range: 1200 m (1,315 yards
Ammunition types: NATO standard types
Ammunition weights: complete round 37.5 g (1.32 oz); projectile 14.4 g (0.51 oz)
Ammunition length: complete round 71.2 mm (2.8 in)

Variants and applications

MAG 60/40: This is the version of the ground MAG 58 rifle-calibre machine-gun with small revisions for the air role. The type can be located in helicopter doors on pintle mounts such as the Belgian **FN ETNA AFSA II** (complete with ammunition box and combined link plus case collector) for the fire suppression role, but is generally fitted in pods. The weapon can also be fitted on each of the two possible **External Pintle Supports** (see the entry on the FN M3P machine-gun). Apart from its gun and/or rocket pods, the EMA can also support the 12-kg (26.5-lb) EPS for one MAG machine-gun and its case/link collector; the total weight with weapon and ammunition is 49 kg (108 lb). On the EPS the machine-gun can be elevated from +5° to −60°, and traversed from 90° forward to 60° backward.

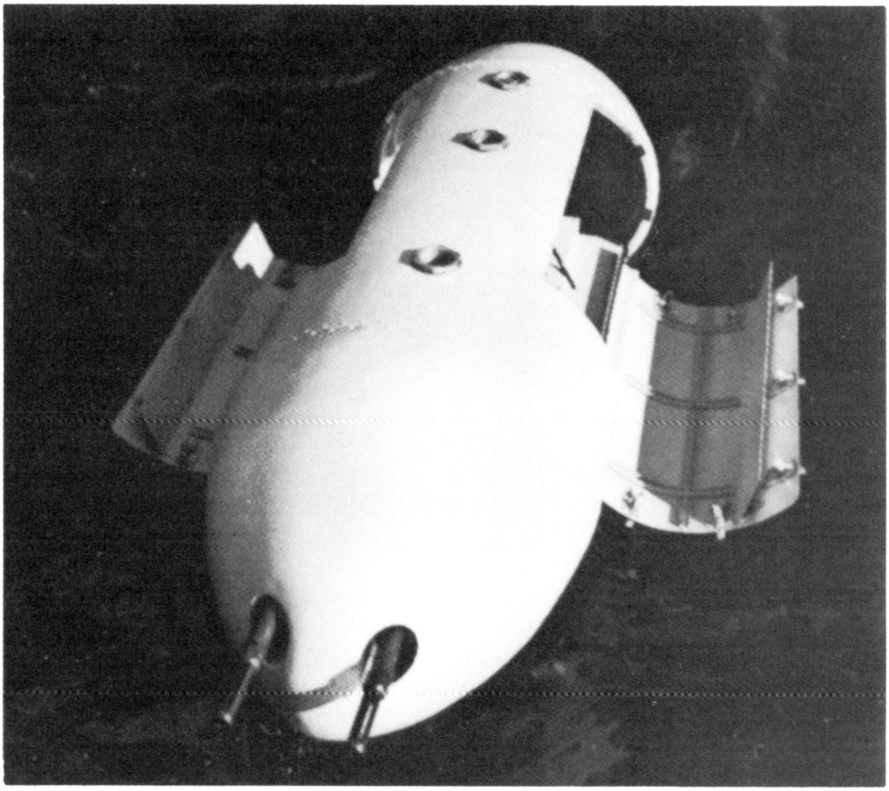

In common with other countries facing the possibility of low-intensity warfare, Brazil finds many advantages in podded rifle-calibre machine-guns, typical being this Avibras pod for two 7.62-mm (0.3-in) FN MAG 60/40 weapons.

Aerea P/N 781-586: This is an Italian pod generally equivalent to the TMP-5, with two MAG machine-guns and 1,000 rounds of ammunition. The standard pod allows the spent cases to be ejected, though an optional modification allows the collection of the cases. The pod's overall dimensions are a length of 1750 mm (68.89 in), a width of 395 mm (15.55 in) and a height of 290 mm (11.42 in). The pod weighs 55 kg (121.25 lb) without and 84 kg (185.2 lb) with ammunition.

Avibras gun pod: The Brazilian company Avibras has produced a gun pod suitable for carriage by light tactical aircraft. This pod carries a pair of FN MAG machine-guns, but other details are lacking.

FN ETNA TMP-5: This is a podded installation of two MAG 60/40 machine-guns for carriage by helicopters and light tactical aircraft. The two guns are located side-by-side in the forward upper portion of the pod, with their flash suppressors and muzzles protruding, on an elastic cradle. Outside each gun and extending to the rear of the pod are two magazines, each holding 500 rounds of linked ammunition whose spent cases are ejected inwards by each gun and collected in a central container. Like the HMP pod for FN's version of the 0.5-in (12.7-mm) Browning M3, the TMP-5 is designed for carriage on NATO standard 356-mm (14-in) suspension lugs, is controlled electrically, and features an electro-hydraulic safety and cocking system for the guns. The pod's loaded weight is 108 kg (238.1 lb), and the overall dimensions include a length of 1900 mm (74.80 in), a width of 422 mm (16.61 in) and a height of 425 mm (16.73 in).

FN ETNA CMP: This is a development of the TMP-5 with an inbuilt 35-mm camera in the deepened underside of the nose. The carriage of two identical pods by a light tactical aircraft allows stereoscopic images to be produced. Overall dimensions are iden-

A Pilatus PC-7 trainer and light attack aircraft shows off FN ETNA armament items, including two TMP-5 machine-gun pods (each carrying a pair of 7.62-mm/ 0.3-in MAG 60/40 machine-guns) and four seven-tube launchers for 2.75-in (69.85-mm) Forges de Zeebrugge unguided rockets.

tical with those of the TMP-5, but weight is increased slightly to 114 kg (251.3 lb). **HAL pod:** In India Hindustan Aeronautics Ltd produces a pod suitable for carriage by light tactical aircraft. This pod contains two 7.62-mm (0.3-in) FN machine-guns. **Royal Small Arms Factory L7A2:** This is the British version of the ground-use MAG 58 series, and differs little in essential respects from the Belgian weapon other than in its heavier barrel, designed to provide the weapon with better sustained-fire capability. The L7A2 has a maximum sustained rate of 200 rounds per minute to a range of 1,800 yards (1645 m), and is fed from the left with 100-round linked ammunition belts. The weapon weighs 24 lb (10.9 kg) and is 48.5 in (1232 mm) long, and is frequently used for the fire-suppression role when fitted on a pintle mounting in helicopters such as the Aérospatiale Puma and Westland Lynx. The basic weapon has also been developed for podded installation as the **L20A1:** this has a solenoid-operated firing mechanism, and the feed system has been adapted to take belts from either the left- or right-hand sides.

China

NORINCO Aircraft Automatic Gun Type 1

Type: aircraft cannon (air-to-air and air-to-surface)
Calibre: 30 mm
Operating method: gas operation with pyrotechnic re-cocking and electrical ignition
Feed: linked belt
Overall dimensions: data not available
Barrel length: data not available
Rifling: data not available
Weight: complete weapon 66.5 kg (146.6 lb)
Recoil force: 6000 kg (13,228 lb)
Muzzle velocity: 780 to 795 m (2,559 to 2,608 ft) per second

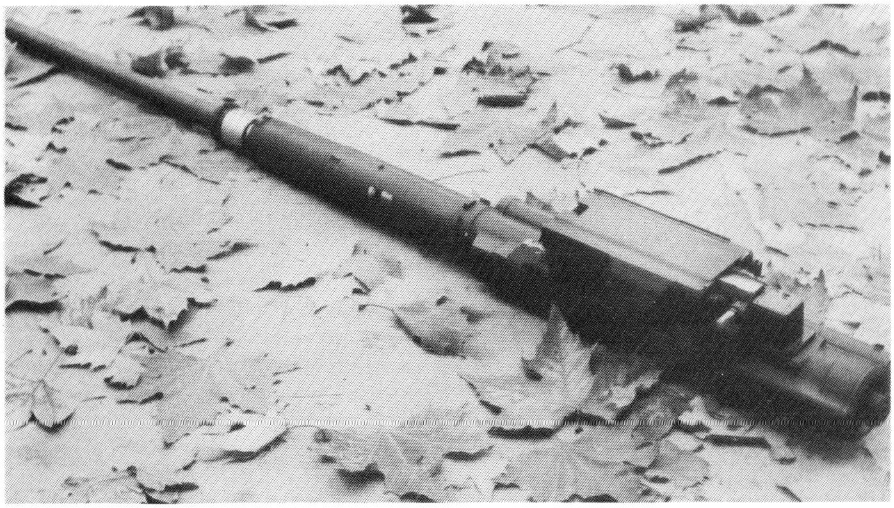

The Chinese 30-mm Aircraft Automatic Gun Type 1 is a conventional yet effective weapon firing a heavy projectile, and suitable for fixed installation in fighters and attack aircraft, or for flexible installation in bombers.

Rate of fire: 850 rounds per minute
Effective range: 800 m (875 yards)
Ammunition types: APHE and HEI
Ammunition weights: complete round 840 g (29.63 oz) for HEI; projectile 418 g (14.74 oz)
Ammunition length: complete round 266 mm (10.47 in)

Variant and application
Aircraft Automatic Gun Type 1: This is the standard Chinese heavy-calibre aircraft cannon, and better suited to the air-to-surface than air-to-air role because of its modest effective range and firing rate. The Type 1 HEI projectile fired by this heavy weapon contains 40.6 g (1.43 oz) of desensitized RDX explosive, and has considerable incendiary capability.

NORINCO Aircraft Automatic Gun Type 2K

China

Type: aircraft cannon (air-to-air and air-to-surface)
Calibre: 23 mm
Operating method: gas operation with pyrotechnic re-cocking and electrical ignition
Feed: linked belt
Overall dimensions: data not available
Barrel length: data not available
Rifling: data not available
Weight: complete weapon 43 kg (94.8 lb)
Recoil force: 2600 kg (5,732 lb)
Muzzle velocity: 705 to 720 m (2,313 to 2,363 ft) per second

Rate of fire: 1,200 rounds per minute
Effective range: 1000 m (1,095 yards)
Ammunition types: API and HEI
Ammunition weights: complete round 314 g (11.08 oz) for API; projectile 174 g (6.14 oz)
Ammunition length: complete round 199 mm (7.83 in)

Variants and application
Aircraft Automatic Gun Type 2: This is an effective air-to-air weapon, with the comparatively low muzzle velocity offset by the rate of fire and the weight of the projectile. At a complete weight of 38 kg (83.8 lb) the gun is lighter than the Type 2K, being intended for employment mainly in turreted installations with a rate of fire of 800/950 rounds per minute using Type 1 HEI or HEI-T ammunition.
Aircraft Automatic Gun Type 2H: This is a version of the Type 2K with a weight of 47 kg (103.6 lb) and possessing a rate of fire probably in the order of 1,150 rounds per minute. The variant is fitted with a large perforated muzzle brake.
Aircraft Automatic Gun Type 2K: This is the version of the Type 2 optimized for installation in fighters, and differing from the basic Type 2 in having a weight of 43 kg (94.8 lb), and firing Type 2 ammunition. The HEI projectile of this latter round has 17.8 g (0.63 oz) of explosive and a considerable incendiary effect.

China

Other Chinese cannon

The People's Republic of China also produces a number of Soviet-designed weapons (see section for the USSR), but has more recently moved towards the design of purely Chinese weapons. The most important of these is a new 23-mm cannon whose existence (but not designation) was revealed in 1986. This is a compact design working on the same type of 'Gatling' principle as the American M61 Vulcan and its derivatives. The 'Gatling' operating system offers an effective combination of fire rate and reliability within a comparatively small package, while the use of ammunition presumably derived from Soviet 23-mm types should generate very great hitting power.

France

GIAT 30/781

Type: aircraft cannon (air-to-surface and air-to-air)
Calibre: 30 mm
Operating method: externally-powered electric motor with automatic re-cocking and electrical ignition
Feed: left- or right-hand de-linked to a multi-chamber revolver cylinder
Overall dimensions: length 1870 mm (73.62 in); width 225 mm (8.86 in); height 310 mm (12.20 in)
Barrel length: data not available
Rifling: data not available
Weight: complete weapon 65 kg (143.3 lb) with fire-control electronics
Recoil force: 610 kg (1,345 lb)
Muzzle velocity: 1025 m (3,363 ft) per second
Rate of fire: 750 rounds per minute

For the new generation of French combat aircraft the national armaments concern GIAT is developing two powerful cannon: the 781B is designed for helicopter installation and fires standard DEFA ammunition, while the 791B (**overleaf**) *is designed for future fighters such as the Dassault-Breguet Rafale-B and has its own range of high-performance ammunition.*

GIAT 30 mm (781B) cannon

GIAT 30 mm (791B) cannon

Effective range: 1500 m (1,640 yards)
Ammunition types: APHEI-SD, HEI, HEI-SD, TP and TP-T
Ammunitions weights: complete round 449 g (15.84 oz) for HEI-SD, 450 g (15.87 oz) for HEI, TP and TP-T, and 472 g (16.65 oz) for APHEI-SD; projectile 243 g (8.57 oz) for HEI-SD, 244 g (8.61 oz) for HEI, TP and TP-T, and 270 g (9.52 oz) for APHEI-SD
Ammunition lengths: complete round 200 mm (7.87 in); projectile 110 mm (4.33 in)

Variant and application

30/781: This powerful weapon is designed for helicopter applications, initially the Eurocopter family of multi-role types, now being recast as a less complex series but probably still using this impressive cannon. The weapon can be fitted on a variety of mountings (door-mounted pintles or fuselage fixed installations for area fire suppression, or turrets for point attack), though fixed and turreted installations clearly offer the greatest possibilities for accurate delivery of the type's useful projectiles given the weapon's considerable bulk and recoil. The 30/781 is designed to fire single rounds or limited-number bursts.

France

GIAT 30/791B

Type: aircraft cannon (air-to-air and air-to-surface)
Calibre: 30 mm
Operating method: externally-powered electric motor with automatic re-cocking and electric ignition
Feed: left- or right-hand de-linked to a multi-chamber revolver cylinder
Overall dimensions: data not available
Barrel length: data not available
Rifling: data not available
Weight: complete weapon 110 kg (242.5 lb)
Recoil force: 2750 kg (6,063 lb) peak
Muzzle velocity: 1025 m (3,363 ft) per second
Rate of fire: selectable 300/600, 1,500 and 2,500 rounds per minute
Effective range: data not available
Ammunition types: APHEI-SD, AP-T, HEI, HEI-SD, TP and TP-T (30 × 150 types)
Ammunition weights: complete round (weight not available); projectile 275 g (9.7 oz)
Ammunition lengths: data not available

Variant and application

30/791B: This advanced revolver cannon is under development as successor to the DEFA 550 series, its initial application being as the armament of the Dassault-Breguet Rafale combat aircraft (both land- and carrier-based versions). The cannon offers a combination of considerable kinetic energy with several projectile types, a number of firing rates suiting the weapon to several applications, and the electronically-controlled choice between continuous fire, or 0.5- or 1-second bursts.

France

GIAT DEFA 553

Type: aircraft cannon (air-to-air and air-to-surface)
Calibre: 30 mm
Operating method: gas operation with pyrotechnic cocking and electrical ignition
Feed: left- or right-hand belt to a five-chamber revolver cylinder
Overall dimensions: length 1660 mm (65.35 in)
Barrel length: 1400 mm (55.1 in), or 46.67 calibres
Rifling: data not available

Weight: complete weapon 85 kg (187.4 lb), or 160 kg (352.7 lb) with feed mechanism and empty magazine
Recoil force: data not available
Muzzle velocity: 765 m (2,510 ft) per second with APHEI-SD and TP, 815 m (2,674 ft) per second with HEI, and 820 m (2,690 ft) per second with the other types
Rate of fire: 1,300 rounds per minute
Effective range: 1500 mm (1,640 yards)
Ammunition types: APHEI-SD, HEI, HEI-SD, TP and TP-T
Ammunition weights: complete round 449 g (15.84 oz) for HEI-SD, 450 g (15.87 oz) for HEI, TP and TP-T, and 472 g (16.65 oz) for APHEI-SD; projectile 243 g (8.57 oz) for HEI-SD, 244 g (8.61 oz) for HEI, TP and TP-T, and 270 g (9.52 oz) for APHEI-SD
Ammunition length: complete round 200 mm (7.87 in)

Variants and applications:
DEFA 552: This important French aircraft cannon was developed as the DEFA 551, and entered production in 1954 as the DEFA 552 (otherwise known as the **Canon 550-F2**) for installation in fighters intended for the high-altitude air-to-air role. The design is modelled closely on German revolver developments in World War 2 (notably the Mauser MG213C), and thus has many similarities to the British Aden, which was developed at much the same time from the same conceptual sources. The weight of the complete weapon is 80 kg (176.4 lb) and the rate of fire 1,250 rounds per minute.
DEFA 552A: This is the **Canon 550-F2A** development of the DEFA 552 with features of the DEFA 553 such as a chromed barrel and a forged rather than welded drum casing. This increases weight to 81 kg (178.5 lb), but boosts the rate of fire to 1,300 rounds per minute and increases barrel life to 5,000 rounds.
DEFA 553: This is a product-improved DEFA 552 whose development as the **Canon 550-F3** was initiated in 1968 to allow the start of production in 1971. The areas in which improvement were sought were electro-mechanical reliability and ease of operation, and installation was also facilitated by provision of a feed mechanism able to accept linked ammunition from either side without modification. The adoption of a nitro-chrome steel

Examples of the DEFA 552, 553 and 554 30-mm cannon family clearly reveal the close relationship between these important weapons.

Above *Visible under this CASA C-101 Aviojet trainer and light attack aircraft are Explosivos Alaveses BR free-fall general-purpose bombs and, in the shadow under the fuselage just forward of the airbrake, the semi-external installation that can accommodate one 30-mm DEFA cannon or two 12.7-mm (0.5-in) heavy machine-guns.*

Right *The second prototype Dassault-Breguet/Dornier Alpha Jet shows off one of its armament options: an under-fuselage pod containing a DEFA 30-mm cannon and 125 rounds of ammunition, and two underwing Matra F1 rocket launcher pods each containing 36 68-mm (2.68-in) SNEB rockets.*

barrel reduces barrel wear and so increases life to 4,000 rounds, and a pressure-reducing fitting can be installed at the muzzle to mitigate the problem of propellant gas ingestion when the muzzle is close to an inlet. The type has provision for continuous fire, or for 0.5- or 1-second bursts.
CASA DEFA installation: As part of the weapon fit for its C-101 Aviojet trainer and light attack aircraft, the Spanish manufacturer CASA has developed a special single-cannon pack for installation in the lower fuselage weapon bay as an alternative to reconnaissance or ECM gear, or the alternative installation of two 12.7-mm (0.5-in) M3 machine-guns. The pack contains one DEFA cannon semi-exposed in the airflow beneath the aircraft, with storage for 130 rounds of ammunition above the weapon.
Dassault-Breguet CC 420: This is the main podded installation of the DEFA 552, 552A, 553 or 554 cannon types for carriage on underfuselage or underwing installations with NATO standard 356- or 762-mm (14- or 30-in) suspension lugs. The pod comes in two basic configurations, namely the **CC 420 Type A** with a double-ended (closed-loop) feed system for 180 rounds, and the **CC 420 Type B** with single-ended (case-ejecting) system for 250 rounds. Development of the pod began in 1970, and in 1978 the system was put into production to meet French and export orders. The dimensions include a length of 4200 mm (165.35 in) and a diameter of 420 mm (16.535 in), and the empty weight is 120 kg (264.6 lb). The loaded weights of the Type A and Type B pods are 300 and 325 kg (661.4 and 716.5 lb) respectively.
Matra SA 10: This is a podded installation of the DEFA 552, 552A or 553 (or perhaps more commonly the British Aden of the same calibre) with 400 rounds. The pod can be carried at speeds up to Mach 2, and the cannon fired at speeds up to Mach 1.6.

GIAT DEFA 554 — France

Type: aircraft cannon (air-to-air and air-to-surface)
Calibre: 30 mm
Operating method: gas operation with pyrotechnic cocking and electrical ignition
Feed: left- or right-hand belt to a five-chamber revolver cylinder
Overall dimensions: length 2010 mm (79.13 in)
Barrel length: data not available
Rifling: data not available
Weight: complete weapon about 85 kg (187.4 lb), or 160 kg (352.7 lb) with feed mechanism and empty magazine
Recoil force: data not available
Muzzle velocity: 765 m (2,510 ft) per second with APHEI-SD and TP, 815 m (2,674 ft) per second with HEI, and 820 m (2,690 ft) per second with the other types
Rate of fire: selectable 1,200 or 1,800 rounds per minute
Effective range: 1500 m (1,640 yards)
Ammunition type: APHEI-SD, HEI, HEI-SD, TP and TP-T
Ammunition weights: complete round 449 g (15.84 oz) for HEI-SD, 450 g (15.87 oz) for HEI, TP and TP-T, and 472 g (16.65 oz) for APHEI-SD; projectile 243 g (8.57 oz) for HEI-SD, 244 g (8.61 oz) for HEI, TP and TP-T, and 270 g (9.52 oz) for APHEI-SD
Ammunition length: complete round 200 mm (7.87 in)

Variant and application
DEFA 554: This is a considerable development of the basic DEFA 552/553 theme, with

Ultimate development of the outstanding DEFA series of 30-mm aircraft cannon, the DEFA 554 embodies the lessons of practical experience with earlier cannon for great reliability, long barrel life and the capability for several re-cockings in the air.

greater electro-mechanical reliability and much improved rate of fire. The latter has been obtained by better use of the five-chamber revolver cylinder, in which three rather than two chambers are now used for the loading process. This boosts firing rate considerably without the need for any increase in the speed of the action's operation, and the weapon has thus been fitted with an electronic control unit to provide rates of 1,200 rounds per minute for the ground-attack role, and 1,800 rounds per minute for the air-to-air role. Whereas the earlier models were fitted with just a single pyrotechnic re-cocking cartridge, the DEFA 554 has three, allowing the pilot to arm the weapon only after take-off and still have two cartridges for inflight re-cocking.

France

GIAT M621

Type: aircraft cannon (air-to-surface and air-to-air)
Calibre: 20 mm
Operating method: gas operation with cable cocking and pyrotechnic re-cocking, and electrical ignition
Feed: left- or right-hand linked-belt
Overall dimensions: length 2228 mm (87.72 in)
Barrel length: data not available
Rifling: data not available
Weight: complete weapon 45 kg (99.2 lb); barrel 17 kg (37.5 lb)
Recoil force: 460 kg (1,014 lb)
Muzzle velocity: 980 to 1030 m (3,215 to 3,379 ft) per second depending on ammunition type
Rate of fire: single rounds, or 300 or 740 rounds per minute
Effective range: more than 1500 m (1,640 yards)
Ammunition types: API, HE and HEI

Ammunition weights: complete round 252.3 g (8.9 oz); projectile 100 g (3.53 oz) for HE and HEI, or 110 g (3.88 oz) for API
Ammunition length: complete round 168 mm (6.61 in)

Variants and application
M621: This light cannon is intended for use primarily in helicopters, and was developed between 1962 and 1973 on the basis of the AME582 prototype weapon. The M621 has in its turn been adapted for the surface-to-surface role as the M693, which is used on a number of French light armoured fighting vehicles. The weapon has been used extensively on the Aérospatiale Alouette III light helicopter after the development of a special shock-absorbed mounting and muzzle brake (the latter in place of the otherwise standard flash suppressor) to reduce the recoil force to 250 kg (551.1 lb).
M621/19A: The 19A mounting was developed to allow the installation of the M621 in the door of the Aérospatiale SA 330 Puma series of medium-lift helicopter. The mounting and gun together weigh 165 kg (363.75 lb), the mounting allowing traverse through an arc of 90° to the side of the helicopter, and elevation from +5° to −40°. Ammunition is fed from four 240-round boxes, and the overall weight of the system is 632 kg (1,393.3 lb).
M621/22A: This is the fixed installation of the M621 to provide light and medium battlefield helicopter with a point-attack weapon. The primary application of the installation is on the starboard side of the Aérospatiale SA 341F/Canon's fuselage, but it has also been adopted for other helicopter types including the military derivatives of the Aérospatiale Ecureuil and Dauphin light helicopters.
NC 20M621: This is a podded installation of the M621 cannon, intended for use by helicopters and light tactical aircraft fitted with NATO standard 356-mm (14-in) suspen-

The 20-mm GIAT M621 cannon can be attached in a fixed installation on the starboard side of the Aérospatiale SA 342M Gazelle, turning this light scouting helicopter into a powerful yet agile weapon with capabilities against other battlefield helicopters as well as troops and light vehicles on the ground.

Fixed armament is becoming increasingly important on battlefield helicopters for anti-helicopter and fire-suppression purposes, most such installations making use of the projecting beam otherwise used for the accommodation of rocket pods and/or anti-tank missiles. The weapon favoured by the French for this role is the highly reliable GIAT M621, seen here in the fit that turns the Aérospatiale SA 341F Gazelle into the SA 341F/Canon. Ammunition is fed from a magazine in the fuselage, discarded links and spent cases being ejected outboard through the chutes visible on the right of the weapon, which is electrically controlled and hydraulically buffered.

sion lugs, and designed for carriage and firing speeds up to Mach 0.7. The cannon is electrically controlled, while the pod has an overall length of 3516 mm (138.43 in) and a diameter of 398 mm (15.57 in). It weighs 93 kg (205 lb) empty and 160 kg (352.75 lb) loaded with 180 rounds. The pod has an electro-hydraulic system for unlimited inflight re-cocking, and the spent cases are collected in a container in the lower portion of the pod. Turn-round time on the ground is reduced by the fact that the ammunition box can be refilled in four minutes, or replaced with a full box in two minutes.

France

GIAT NF-1

Type: aircraft machine-gun (air-to-surface)
Calibre: 7.62 mm (0.3 in)
Operating method: delayed blowback
Feed: left-hand disintegrating linked belt
Overall dimensions: length 1080 mm (42.52 in)
Barrel length: 600 mm (23.62 in) without flash suppressor, or 78.74 calibres
Rifling: 4-groove right-hand twist
Weight: complete weapon 11.15 kg (24.58 lb) with flash suppressor
Recoil force: data not available
Muzzle velocity: 800 m (2,625 ft) per second
Rate of fire: 900 rounds per minute
Effective range: 1200 m (1,315 yards)
Ammunition types: NATO standard types
Ammunition weights: complete round 37.5 g (1.32 oz); projectile 14.4 g (0.51 oz)
Ammunition length: complete round 71.2 mm (2.8 in)

Variants and application

NF-1: This is the standard French medium machine-gun designed for ground use in the original 7.5-mm (0.295-in) **AA 52** and later NATO standard 7.62 (0.3-in) AA 7.62-mm NF-1 variants. The latter is now the more common variant and is also used on French tactical helicopters fitted with a door-mounted pintle. The weapon has the option of light and heavy barrels, the former being 100 mm (3.93 in) shorter and 1.45 kg (3.2 lb) lighter than the latter; the figures quoted above are for the weapon with a heavy barrel, which has a practical rate of between 250 and 700 rounds per minute compared with the light barrel's 150 rounds per minute.

Matra C.M.7.62 mm: This is the light podded installation of two 7.62-mm (0.3-in) NF-1 machine guns for carriage by light tactical aircraft fitted with NATO standard 356-mm (14-in) suspension lugs. The pod is similar in many respects to the FN ETNA pod with two MAG guns, being a slab-sided container with two FN-1 machine guns in the upper side of the nose, and the ammunition containers below and behind them. The pod is faired into a neat tail containing the control equipment.

Armscor GA1 Cobra

South Africa

Type: aircraft cannon (air-to-surface and air-to-air)
Calibre: 20 mm
Operating method: gas operation with electrical ignition
Feed: left- or right-hand linked belt
Overall dimensions: length 1780 mm (70.08 in); width 244 mm (9.61 in); height 190 mm (7.48 in)
Barrel length: data not available
Rifling: 8-groove twist
Weight: complete weapon 39 kg (85.98 lb); barrel 10 kg (22.05 lb)
Recoil force: 545 kg (1,201 lb) peak and 275 kg (606 lb) average
Muzzle velocity: 720 m (2,362 ft) per second
Rate of fire: 700 rounds per minute
Effective range: 1250 m (1,365 yards)
Ammunition types: APC-T, HEI, HEI-T, Practice and Practice-T
Ammunition weights: complete round 311 g (10.97 oz) for APC-T and 314 g (11.07 oz) for the other types; projectile 97 g (3.42 oz) for HEI and Practice, 98 g (3.46 oz) for HEI-T and Practice-T, and 100 g (3.53 oz) for APC-T
Ammunition lengths: complete round 212 mm (8.35 in); projectile 83 mm (3.27 for APC-T and 92 mm (3.62 in) for the other types

Variants and applications

GA1 Cobra: This is a much modernized South African development of the German MG151/20 weapon of World War 2. The type is generally used on a flexible mounting in helicopters, but has also been developed in a computer-controlled underfuselage mounting for battlefield helicopters. This latter system was pioneered on the Atlas XH-1 Alpha experimental gunship developed entirely in South Africa from the French Aérospatiale Alouette III, and is now being prepared for production on the service version of the Atlas XTP-1 Beta battlefield helicopter developed in South Africa on the basis of the Aérospatiale SA 330 Puma. In this installation the cannon is located in a servo-controlled underfuselage mounting, and fed from a 1,000-round magazine in the lower fuselage:

targets are designated by the gunner's helmet sight, the computer system using this and inputs from the nose-mounted air-data probe to generate extremely accurate aiming instructions.

Rattler: This is the servo-assisted mounting for heavier helicopters, the system being designed round a gunner's seat, cannon (or in alternative form four S77 7.62-mm/0.3-in machine-guns), ammunition and Gannet fixed reflector optical sight system. When fitted with the GA1 cannon the whole system weighs 228 kg (502.6 lb) empty and 310 kg (683.4 lb) loaded with 240 rounds of 20-mm ammunition. The mounting can be traversed from +20° to −40°, and the cannon can be elevated from 0° to −30°, the slewing rate being 50° per second in each plane. The overall dimensions of the mounting include a length of 2600 mm (102.36 in), a width of 500 mm (19.7 in) and a height of 1000 mm (39.4 in).

The alternative fitting is four 7.62-mm (0.3-in) machine-guns with 1,000 rounds per gun. In this configuration the maximum possible rate of fire is 2,000 rounds per minute, reduced to 1,000 rounds per minute when the gunner uses the facility to select just two guns (either the upper or the lower pair). This installation weighs 300 kg (661.4 lb) empty and 335 kg (738.5 lb) loaded. The azimuth and elevation limits are revised (−25° to +25° and 0° to −38° respectively) though the slewing rate remains the same at 50° per second. The machine-gun mounting is 1900 mm (74.8 in) long, 800 mm (31.5 in) wide and 1000 m (39.4 in) high.

Switzerland/ UK

Hispano-Suiza (Oerlikon-Bührle/ BMARC) Mk 5

Type: aircraft cannon (air-to-air and air-to-surface)
Calibre: 20 mm
Operating method: gas operation
Feed: linked belt
Overall dimensions: length 2052 mm (80.79 in)
Barrel length: data not available
Rifling: data not available
Weight: complete weapon with feed mechanism 42 kg (92.6 lb)
Recoil force: data not available
Muzzle velocity: 850 m (2,789 ft) per second
Rate of fire: 580 to 640 rounds per minute
Effective range: 1600 m (1,750 yards)
Ammunition types: AP, Ball, HEI and Tracer
Ammunition weights: complete round 367 g (12.95 oz); projectile 138 g (4.87 oz)
Ammunition length: data not available

Variants and applications

Hispano-Suiza Mk 2*: This is the oldest surviving model of what was once the world's most important aircraft cannon, which was developed just before and during World War 2 on the basis of the Hispano-Suiza 404. The weapon is now virtually obsolete, but survives in service on a number of older aircraft.

Hispano-Suiza Mk 5: This is the definitive version of the 'Hispano', slightly lighter and possessing a shorter barrel than the Mk 2*. The weapon has a marginally higher rate of fire, and can be fitted in lower-drag installations.

Oerlikon-Bührle Type KAA

Switzerland

Type: aircraft cannon (air-to-surface)
Calibre: 20 mm
Operating method: blowback operation with a gas-operated breech locking mechanism and electro-mechanical ignition
Feed: left- and right-hand linked belt
Overall dimensions: length 2627 mm (103.425 in); width 220 mm (8.66 in); height 241.5 mm (9.51 in)
Barrel length: 1700 mm (66.93 in) without the muzzle brake, or 85 calibres, and 1856 mm (73.07 in) with the muzzle brake, or 92.8 calibres
Rifling: 12-groove right-hand twist increasing from 0° to 6°30′
Weights: complete weapon 88 kg (194.00 lb), and 133 kg (293.2 lb) with cradle; barrel 26 kg (57.3 lb)
Recoil force: 1400 kg (3,086 lb)
Muzzle velocity: 1150 m (3,773 ft) per second for AP-T and 1100 m (3,600 ft) per second for the other types
Rate of fire: 1,100 rounds per minute
Effective range: 2000 m (2,185 yards)
Ammunition types: AP-T, HEI, HEI-T, SAPHEI, SAPHEI-T, TP and TP-T
Ammunition weights: complete round 322 g (11.36 oz) for AP-T, and 337 g (11.89 oz) for the other types; projectile 110 g (3.88 oz) for AP-T, and 125 g (4.41 oz) including 10 g (0.35 oz) of explosive for the other types
Ammunition length: complete round 203.5 mm (8.01 in)

Variant and application
Type KAA: Originally known as the **204 GK**, this powerful cannon is one of a series designed as a multi-role land and air weapon, and as the armament of helicopters can be pintle-mounted in a door or fixed in an external cradle. For its calibre the weapon is very powerful, firing a substantial projectile at high muzzle velocity: the AP-T projectile can penetrate 20 mm (0.79 in) of armour at 800 m (875 yards) at an incidence of 0°.

Oerlikon-Bührle Type KAD-B

Switzerland

Type: aircraft cannon (air-to-surface and air-to-air)
Calibre: 20 mm
Operating method: gas operation with pyrotechnic re-cocking and electro-mechanical ignition
Feed: left- or right-hand linked belt
Overall dimensions: length 2744 mm (108.03 in); width 158 mm (6.22 in); height 239 mm (9.41 in)
Barrel length: 1906 mm (75.04 in) with flash suppressor, or 95.3 calibres
Rifling: 15-groove right-hand twist of 7°
Weight: complete weapon 65.5 kg (144.4 lb) with 8.5-kg (18.74-lb) feed mechanism, and 114.5 kg (252.4 lb) with cradle; barrel 31 kg (68.3 lb)
Recoil force: 830 kg (1,830 lb)

A powerful 20-mm fixed cannon that can be fitted to the Westland Lynx is the Oerlikon-Bührle Type KAD. This trim installation, with ammunition fed from a magazine in the cabin, does not preclude the mounting of a quadruple launcher for TOW anti-tank missiles just outboard of the cannon.

Muzzle velocity: 1150 m (3,773 ft) per second for AP-T and 1100 m (3,600 ft) per second for the other types
Rate of fire: 1,000 rounds per minute
Effective range: 1500 m (1,640 yards)
Ammunition types: AP-T, HEI, HEI-T, SAPHEI, SAPHEI-T, TP and TP-T
Ammunition weights: complete round 306 to 321 g (10.79 to 11.32 oz) depending on type; projectile 125 to 144 g (4.41 to 5.08 oz) depending on type
Ammunition length: complete round 213 mm (8.39 in)

Variant and application
KAD-B: This cannon is one of the most important produced by Oerlikon-Bührle, and was originally developed and built as the **Hispano-Suiza HS 820**. The weapon has applications mainly as the ordnance for light anti-aircraft and light armoured fighting vehicle mountings. More recently, however, the weapon has been developed for fixed installation on helicopters in single or twin mountings on the side(s) of the fuselage and belt-fed with anything between 200 and 370 rounds.

Although the HBS202 under-fuselage armament kit for the MBB BO105 light helicopter was designed for the Rheinmetall MK 20 Rh202, the basic system can be adapted to accept other cannon of the same 20-mm calibre, such as the Oerlikon-Bührle Type KAD seen in this illustration.

Oerlikon-Bührle Type KBA-B

Switzerland

Type: aircraft cannon (air-to-surface and air-to-air)
Calibre: 25 mm
Operating method: gas operation with electrical re-cocking and ignition
Feed: left- and right-hand linked belt
Overall dimensions: length 3038 mm (119.6 in); width 263 mm (10.35 in); height 253 mm (9.96 in)
Barrel length: 2000 mm (78.74 in) without the muzzle brake, or 80 calibres, and 2173 mm (85.55 in) with the muzzle brake, or 86.9 calibres
Rifling: 18-groove right-hand twist with final pitch of 7°30′
Weight: complete weapon 112 kg (246.9 lb) with cocking device plus recoil mechanism and 150 kg (330.7 lb) with cradle; barrel 27 kg (59.52 lb)
Recoil force: 1630 kg (3,593 lb) average
Muzzle velocity: 1360 m (4,462 ft) per second with APDS-T, and 1100 m (3,609 ft) per second with the other types
Rate of fire: 570 rounds per minute
Effective range: 2500 m (2,735 yards)
Ammunition types: APDS-T, HEI, HEI-T, SAPHEI, SAPHEI-T, TP and TP-T
Ammunition weights: complete round 454 g (16.01 oz) for APDS-T, and 502 g (17.71 oz) for the other types; projectile 125 g (4.41 oz) for APDS-T, and 180 g (6.35 oz) for the other types
Ammunition length: complete round 223 mm (8.78 in)

Variant and application

Type KBA-B: This is an exceptionally powerful multi-role cannon generally found in anti-aircraft and naval applications, but sometimes used on helicopters. The most notable features are extremely high muzzle velocity with a powerful round, and a good rate of fire. The HE projectiles contain 27 g (0.95 oz) of explosive, and the SAPHEI projectiles 11 g (0.39 oz); the APDS-T projectile can penetrate 25 mm (0.98 in) of armour

Fixed installation of an Oerlikon-Bührle Type KBA 25-mm cannon on the starboard side of the fuselage of a Bell UH-1 helicopter, with twin ammunition feed from magazines located on the centreline in the passenger compartment.

The Oerlikon-Bührle Type KBA is a hard-hitting 25-mm cannon that can fire a wide assortment of projectile types, giving even light/medium helicopters such as this Agusta (Bell) AB.212 great offensive capability.

at 2000 m (2,185 yards) at an incidence of 30°. The flight time to this range is 1.7 seconds for the APDS-T projectile, and 3.33 seconds for the other projectiles.

Switzerland

Oerlikon-Bührle Type KCA

Type: aircraft cannon (air-to-air and air-to-ground)
Calibre: 30 mm
Operating method: gas operation with pyrotechnic re-cocking and electrical ignition
Feed: right-hand belt to a four-chamber revolver cylinder
Overall dimensions: length 2691 mm (8.83 ft); width 242.5 mm (9.5 in); height 249.5 m (9.82 in)
Barrel length: 1976 mm (77.8 in), or 75 calibres
Rifling: 21-groove right-hand twist increasing from 0° to 8°15′
Weight: complete weapon 136 kg (299.8 lb), and barrel 27 kg (59.5 lb)
Recoil force: 4800 kg (10,582 lb) peak and 4200 kg (9,259 lb) average
Muzzle velocity: 1030 m (3,379 ft) per second
Rate of fire: 1,350 rounds per minute
Effective range: 2500 m (2,735 yards)
Ammunition types: HEI, HEI-T, SAPHEI, TP and TP-T
Ammunition weights: complete round (with steel cartridge case) 890 g (31.39 oz); projectile 360 g (12.7 oz)
Ammunition length: complete round 290 mm (11.42 in)

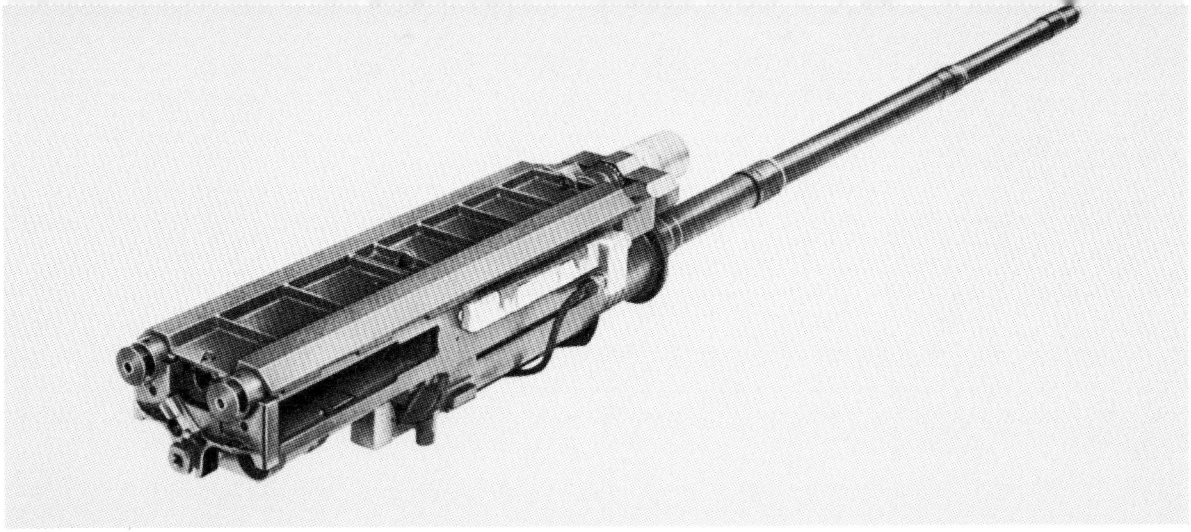

Above *The Oerlikon-Bührle Type KCA 30-mm cannon is currently the most powerful air-to-air cannon in the world, offering exceptionally high muzzle velocity with a heavy projectile for greater flight accuracy and enormous impact on the target.*

Left *The Saab JA 37 Viggen is by any criteria an exceptional interceptor, and is seen here in typical interception mode with the fixed belly pack accommodating one 30-mm Oerlikon-Bührle Type KCA cannon plus the underwing fit of two RB 71 (Sky Flash) medium-range and two RB 24 (AIM-9 Sidewinder) short-range air-to-air missiles.*

Variant and application

Type KCA: This is one of the world's most powerful cannon, and is intended for aircraft and pod installations. The weapon is currently used only on the JA 37 interceptor version of the Saab Viggen, and is characterized by high muzzle energy to provide a 1000-m (1,094-yard) flight time of 1.139 seconds. The SAPHEI round can penetrate 12 mm (0.47 in) of armour at 1000 m at an incidence of 60°.

USSR

Soviet aircraft machine-guns

The USSR still makes use of heavy-calibre machine-guns on its lighter tactical aircraft and also on a number of battlefield helicopters. The oldest weapon of this type still in service is the **Universal Beresin**, a type pioneered in 1940 as a standard 12.7-mm (0.5-in) weapon for aircraft use. The UB's typical weight is 21.5 kg (27.56 lb), rate of fire 1,000 rounds per minute and muzzle velocity 860 m (2,822 ft) per second, firing a projectile weighing 48 g (1.69 oz). The weapon is conventional by the standards of the day, and fires belted ammunition in both its unsynchronized **UBK** fixed installation (as on the Mikoyan-Gurevich MiG-15UTI 'Mongol' conversion and weapons trainer) and **UBT** flexible fire-suppression and anti-helicopter installation (as on various battlefield helicopters from the Mil design bureau).

More recently the USSR has introduced a four-barrel 12.7-mm (0.5-in) machine-gun working on the Gatling principle: this is fitted in the power-operated chin turret of the Mi-24 'Hind' helicopter, and is assumed to fire standard 12.7-mm (0.5-in) ammunition at a selectable rate up to 4,000 rounds per minute. It is probable that an advanced fire-control system is used to provide effective medium-range capability with armour-piercing ammunition against battlefield targets such as helicopters and anti-aircraft vehicles.

USSR

Soviet aircraft cannon

The USSR has long been a firm believer in the tactical efficacy of the aircraft cannon, in both the air-to-air and air-to-surface roles, and continues to use older weapons as well as developing more advanced cannon. The Soviet practice has generally been to produce weapons of slightly greater calibre than their Western counterparts, with emphasis on great strength so that a massive projectile can be fired at high muzzle velocity: the rate of fire inevitably suffers as a consequence, but the net result is more destructive power (in terms of explosive content and kinetic energy) delivered with a flatter trajectory that eases aiming problems.

An extremely powerful aircraft weapon is the gas-operated 37-mm **Nemenov-Suranov-Rikhter-Gribkov N-37** introduced in the mid-1940s, and intended as the primary armament of new Soviet jet fighters: though the type fires a very useful 735-g (25.93-oz) projectile at a muzzle velocity of 690 m (2,264 ft) per second, the 103-kg (227.1-lb) weight of the installation (together with that of a tactically significant quantity of ammunition) was a problem for aircraft designers, and the low rate of fire (400 rounds per minute) also had disadvantages except in the ground-attack role. The type is now found only on some variants of the Mikoyan-Gurevich MiG-17 'Fresco' and MiG-19 'Farmer'.

Of a similar vintage is the gas-operated 23-mm **Nudel'man-Rikhter NR-23:** this is thus a fairly old design by modern standards, but still an effective air weapon for its very considerable power and excellent 200-g (7.05 oz) projectile fired at a muzzle velocity of 690 m (2,264 ft) per second and rate of 850 rounds per minute. The basic weapon weighs 39 kg (85.98 lb) and is still found in fighters such as the MiG-17 'Fresco' and MiG-19 'Farmer', and is complemented by the **AM-23** improved version weighing 43 kg (94.8 lb), designed principally for **PV-23** turreted installations and capable of 1,300 rounds per minute. The weapon has also been made in China as the **Type 23-1** (NR-23) and **Type 23-2** (AM-23).

The NR-23 was introduced in 1949, and the gas-operated 30-mm **Nudel'man-Rikhter NR-30** followed in 1954: this is in essence a scaled-up NR-23 and, though for its calibre it is comparatively light at 66 kg (145.5 lb), the cannon still manages to fire a massive projectile, in this instance turning the scales at 410 g (14.46 oz), at a muzzle velocity of 780 m (2,559 ft) per second at a rate of 850 or 1,000 rounds per minute. The basic weapon has been developed in four subvariants, and still plays a significant part in Soviet tactical air power as the barrelled armament of aircraft such as late-model MiG-19 'Frescoes', early-model MiG-21 'Fishbeds' and in its **2NR-30** form, the Sukhoi Su-7/17/20/22 'Fitter' series.

Most prolific of the more modern weapons used by Soviet aircraft has been the medium-calibre 23-mm **Nudel'man-Kalashnikov GSh-23L**, a two-barrel weapon operating on the Gast principle with a mechanical linkage between the two breeches so that the firing of one barrel operates the breech mechanism of the other. This was introduced in 1959, and is in every respect an excellent weapon, offering the combination of a useful 200-g (7.05-oz) projectile fired with a muzzle velocity of 890 m (2,920 ft) per second at a rate of some 2,800 to 3,000 rounds per minute for an installed weight of just 72 kg (158.73 lb). The weapon is generally associated with the **GP-9** pack for installation in tactical fighters such as later models of the MiG-21 'Fishbed' and in the MiG-21 'Flogger' with ammunition capacity of 200, 320 or 400 rounds. The same GSh-23L is used in at least two different gun pods: that carried by the Su-21 'Flagon' is optimized for the high-speed air-to-air role, while that for the MiG-27 'Flogger' has barrels that can be angled down for fully effective use in the air-to-surface role.

In recent years a new generation of weapons has begun to enter service in the latest Soviet tactical aircraft, and for these no designations have yet been revealed. In the 23-mm calibre there is a new six-barrel 'Gatling' weapon whose primary applications are the MiG-23BN and MiG-27 'Flogger' variants: this weapon is located in the belly of each type as part of a semi-recessed installation, and a substantial ammunition tank is built into the fuselage above the weapon. At much the same time two new 30 mm weapons have made their appearance. The smaller of these is a twin-barrel weapon carried in the nose of the Su-25 'Frogfoot' close-support aircraft. This weapon may fire the powerful round already used in the NR-30, but may possibly have its own special ammunition optimized for the ground-attack role. The other new 30-mm weapon is a six-barrel 'Gatling' weapon, currently installed in the MiG-29 'Fulcrum', the Su-24 'Fencer' and (possibly) the Su-27 'Flanker'.

Royal Small Arms Factory Aden Mk 4 — UK

Type: aircraft cannon (air-to-air and air-to-surface)
Calibre: 30 mm

Operating method: gas operation with pyrotechnic re-cocking and electrical ignition.
Feed: left- or right-hand linked belt to a five-chamber revolver cylinder
Overall dimensions: length 62.6 in (1590 mm); width 9.45 in (240 mm); height 9.69 in (246 mm)
Barrel length: 42.5 in (1079.5 mm), or 47.8 calibres
Rifling: 16-groove progressive parabolic right-hand twist
Weight: complete weapon 192 lb (87.1 kg); barrel 27 lb (12.25 kg)
Recoil force: 7,060 lb (3202 kg)
Muzzle velocity: 2,430 ft (741 m) per second
Rate of fire: 1,200 to 1,400 rounds per minute
Effective range: 1,650 yards (1510 m)
Ammunition types: AP, API, HE and TP
Ammunition weights: complete round 17.46 oz (495 g) for AP and API, 17.5 oz (496 g) for HE, and 15.52 oz (440 g) for TP; projectile 7.76 oz (220 g)
Ammunition length: complete round 7.87 in (200 mm)

Variants and applications
Aden Mk 4: Like the DEFA series of 30-mm aircraft cannon, the Aden has its origins in the German Mauser MG213C revolver cannon of World War 2, and though developed in the late 1940s it is still a useful weapon in both the air-to-air and air-to-surface roles. The type takes its name from the Armament Development Establishment (where it was designed) and Enfield (where it was and is produced by the Royal Small Arms Factory). The oldest variant still in service is the Mk 4 described above. The type is used in a large number of fixed two and four-cannon installations, and can also be fitted in the French Matra SA 10 pod (see entry for the DEFA 553), as well as the electrically-controlled Swedish **FFV Aden** pod, which uses the cannon variant designation **M/55** in Swedish service. The FFV pod's overall dimensions include a length of 3850 mm (151.57 in) and a diameter of 500 mm (19.69 in), and its empty and loaded weights are 290 and 364 kg (639.3 and 802.5 lb) respectively; the standard ammunition capacity is 150 rounds, though there is an option for 200 rounds, and the cannon is cocked pneumatically on the ground, with a safety system to prevent firing until the aircraft has taken-off.
Aden Mk 5: This is an improved version designed for greater reliability and rate of fire increased to 1,500/1,700 rounds per minute. Development has not resulted in production Aden Mk 5 cannon, though converted Mk4s are designated **Aden Mk 5 Straden**, and several of the weapon's best features have been used in the more modern Aden 25 cannon.

UK

Royal Small Arms Factory Aden 25

Type: aircraft cannon (air-to-air and air-to-surface)
Calibre: 25 mm
Operating method: gas operation with pyrotechnic re-cocking and mechanical ignition
Feed: left- or right-hand linked belt to a five-chamber revolver cylinder
Overall dimensions: length 90.0 in (2286 mm); width 9.45 in (240 mm); height 9.69 in (246 mm)

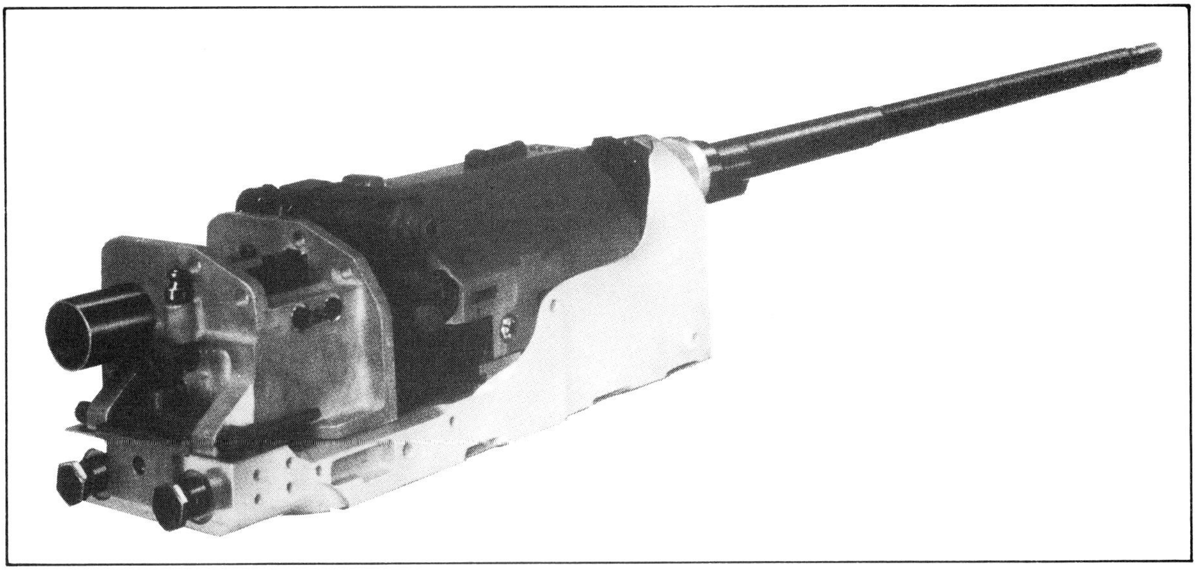

The Aden 25 is based on the earlier 30-mm Aden cannon, but for much the same weight and size offers three times the kinetic energy delivered on to the target.

Barrel length: 66.9 in (1700 mm), or 68 calibres
Rifling: data not available
Weight. complete weapon 203 lb (92.0 kg)
Recoil force: 5,845 lb (2651 kg) peak and 4,720 lb (2141 kg) average
Muzzle velocity: 3,445 ft (1050 m) per second
Rate of fire: 1,650 to 1,850 rounds per minute
Effective range: 1,650 yards (1510 m)
Ammunition types: PGU-20/U API, PGU-22/25 HEI and PGU-23 TP
Ammunition weights: complete round 18.8 oz (533 g) for API, and 17.71 oz (502 g) for HEI and TP; projectile 7.58 oz (215 g) for API, and 6.5 oz (184 g) for HEI and TP
Ammunition length: complete round 8.78 in (223 mm) for API, and 8.625 in (219.1 mm) for HEI and TP

Variant and application
Aden 25: The new Aden 25 is based on the Aden Mk 5 series, but fires the new range of NATO 25-mm ammunition at much higher rates of fire than its predecessor. The result is calculated as a threefold increase in killing power, yet the weapon can be installed in mountings designed for the earlier Aden cannon as it is dimensionally identical but has almost one-third less recoil force. The initial application of the Aden 25 is in the BAe Harrier GR.Mk 5, which has two such cannon in podded installations scabbed onto the underside of the fuselage: each pod contains one Aden 25 and 100 rounds of ammunition, and the two pods together weigh 948 lb (430 kg).

Ford Aerospace M39A3 — USA

Type: aircraft single-barrel cannon (air-to-air and air-to-surface)
Calibre: 20 mm
Operating method: gas operation with electrical ignition

Feed: left- or right-hand linked belt to a five-chamber revolver cylinder
Overall dimensions: length 72.09 in (1831 mm); width 8.0 in (203 mm); height 8.3 in (211 mm)
Barrel length: data not available
Rifling: right-hand twist
Weight: complete weapon 178.4 lb (80.92 kg)
Recoil force: data not available
Muzzle velocity: 2,850 ft (870 m) per second
Rate of fire: 1,700 rounds per minute
Effective range: data not available
Ammunition types: M50 series (including M53A1 API, M54 HPT, M55A1 Practice, M56 HE and M56A1 HEI) and PGU-17/B
Ammunition weights: complete round 9.07 oz (257 g); projectile 3.56 oz (101 g) for API, HE and HEI
Ammunition length: complete round 6.61 in (168 mm)

Variants and applications
Pontiac M39: The M39 was derived from the German Mauser MG213C of World War 2 in much the same fashion as the British Aden and French DEFA can trace their origins back to the same classic weapon. But the American designers at the Illinois Institute of Technology decided they could improve on the German concept, and thus produced a weapon of smaller calibre than the British, French and German weapons, and fitted it with a low-set barrel indexed in line with the chamber at the 6 o'clock position. The weapon began to mature just as the M61 Vulcan was making its first appearances, and thus featured on only a comparatively small number of US Navy and US Air Force tactical aircraft during the 1950s and early 1960s. However, with the advent of the Northrop F-5 Freedom Fighter/Tiger II series of light tactical aircraft the cannon found renewed favour, and was produced in substantial numbers by the Ford Aerospace & Communications Corporations as the **M39A2** and higher-rate **M39A3** variants for the twin-gun installation of the Northrop aircraft.
Ford Aerospace Tigerclaws: This is a developed version of the M39A2/A3 weapon with a number of important improvements: the Tigerclaws retains complete interchangeability with the M39 system, but has a 35% increase in rate of fire (from 1,700 to 2,300 rounds per minute), a 40 to 50% increase in spent case ejection velocity (from 40-50 ft/12.2.-15.2 m to 60-70/18.3-21.3 m per second), a 20% weight reduction (from 178.4 lb/80.9 kg to 128.4 lb/58.2 kg), a 100% endurance life increase (from 5,000 to 10,000 rounds) and a 30% increase in reliability (from 3,530 to 4,600 mean rounds between failures).

USA

General Electric GAU-2B/A Minigun

Type: aircraft six-barrel machine-gun (air-to-air and air-to-surface)
Calibre: 7.62 mm (0.3 in)
Operating method: General Electric 'Gatling' operation with external electric or ram-air power supply, and mechanical ignition
Feed: right-hand linkless or de-linked

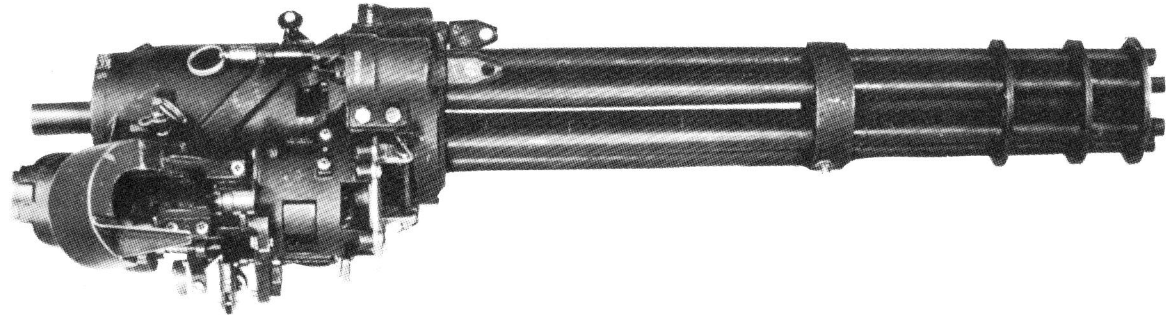

One of the earliest members of the General Electric 'Gatling' weapon family, the GAU-2B/A Minigun fires standard 7.62-mm (0.3-in) ammunition but has exceptional reliability and firepower in both fixed and flexible installations.

Overall dimensions: length 31.56 in (802 mm); with motor, or 29.5 in (749 mm) without motor; width 10.5 in (267 mm); height 7.87 in (200 mm)
Barrel length: 22 in (559 mm), or 73.33 calibres
Rifling: right-hand twist
Weight: complete weapon 56.5 lb (25.63 kg) with motor, or 35 lb (15.9 kg) without motor; barrels 2.4 lb (1.09 kg) each
Recoil force: 300 lb (136.1 kg) average at 6,000 rounds per minute, or 175 lb (79.4 kg) at 4,000 rounds per minute
Muzzle velocity: 2,750 ft (838 m) per second
Rate of fire: variable up to 4,000 or 6,000 rounds per minute (see below)
Effective range: 1,500 yards (1370 m)
Ammunition types: NATO standard types (M59 and M80 ball, M61 armour-piercing and M62 tracer)
Ammunition weights: complete round 1.32 oz (37.5 g); projectile 0.51 oz (14.4 g)
Ammunition length: complete round 2.8 in (71.2 mm)

Variants and applications

GAU-2B/A Minigun: This most useful lightweight machine-gun was developed on the conceptual basis of the M61 Vulcan aircraft cannon, and uses the General Electric 'Gatling' principle to secure very high rates of fire combined with great reliability (a figure of 500,000+ being quoted for mean rounds between failures). The system is based on a barrel assembly externally-powered to rotate round a common axis, each of the six barrels having its own bolt assembly operated by a cam follower which, riding in a cam path machined into the rotor assembly, converts the rotor's circular movement into the bolt's fore-and-aft movement for the cycle of chambering, firing and extracting one round per barrel per complete rotation. Thus even at the Minigun's maximum rate of 6,000 rounds per minute each barrel fires only 1,000 rounds, giving a barrel life of some 100,000 gun rounds.

The rotary system reduces internal loads on the gun, thereby increasing reliability, and the use of an external power system obviates the need for a cocking/re-cocking system as duds are automatically ejected which also promotes the utility of the weapon in remotely-controlled installations. The weapon is offered with a choice between two motors, one offering a maximum rate of 6,000 rounds and the other of 4,000 rounds per minute, though rates as low as 300 rounds per minute are also possible. The designation GAU-2B/A is used for the basic US Air Force weapon (developed from the single-rate **GAU-2/A**), the equivalent designation for the US Army version, which generally

Opposite *Manufacturer's artwork of the M21/MAMEE system reveals the primary features and fields of fire possible with this capable weapon system for medium helicopters.*

has the 4,000-round per minute rate, being **M134 Minigun**. There is also a longer-barreled **M139** version. General Electric has developed a pintle mounting to allow the installation of a Minigun in the Sikorsky UH-60 Black Hawk; complete with 4,000 rounds of ammunition this installation weighs 535 lb (242.7 kg).

Emerson Electric Dual Flexible Turret System: This is an advanced development of the M21/MAMEE system with two side-mounted flexible Miniguns with MAMEE ammunition modules, and the choice of seven-tube M158A1 or six-tube LAU-68 fixed launchers for 2.75-in (69.85-mm) rockets. The system is controlled by the co-pilot with the aid of an M60 reflex sight, though the pilot has the option of firing rockets and/or guns when the latter are fixed to fire directly forward. The guns can be traversed from 10° inboard to 70° outboard, and elevated from +3.5° to −70° at the rate of 80° per second. The standard rate of fire is 750 rounds per minute, though this is raised to 1,500 rounds per minute on the unhindered gun when either weapon is prevented from firing by reaching its inboard traverse limit. The system carries 6,000 rounds of ammunition, and the loaded weight is 1,360.6 lb (617.2 kg).

Emerson Electric Flexible Turret System: This trim remotely-controlled Minigun installation was developed as the **Mini-TAT**, one of the Emerson Tactical Armament Turret (TAT) series. The system is intended as an add-on system to provide light helicopters with effective supplementary armament, and as such is designed for easy installation in the field. Its basic form comprises a retractable electrically-powered mounting under the fuselage (occasionally on the side of the fuselage) to provide the Minigun with 360° traverse and elevation between +10° and −70° at rates of 80° per second. The weight of the complete installation with 1,000 rounds of ammunition in a rectangular box is 247.0 lb (112.0 kg), increasing somewhat dramatically when the 3,000-round MAMEE magazine system is used.

The Minigun in the FTS system has two rates of fire (750 and 1,500 rounds per minute) selected by varying pressure on the trigger by the co-pilot, who fires the system with the aid of a reflex sight; the pilot can operate the system when it is fixed to fire directly ahead. The FTS can also be configured for the installation of other weapons firing the NATO standard 7.62-mm (0.3-in) round, most notably the Belgian MAG 60, the French FN-1 and the West German MG3. In the event of power failure the Mini-TAT can be jettisoned mechanically.

Emerson Electric M21: This is a light armament system suitable for use by a variety of helicopters, most notably the Bell UH-1 series, with the co-pilot serving as the gunner. The system comprises an internal control system using the M60 reflex sight and two (occasionally one) pylon-mounted exterior armament installations each combining one Minigun and one M158A1 launcher for seven 2.75-in (69.85-mm) unguided rockets: the rocket launcher is fixed, but the Minigun is fitted on a hydraulically-powered mounting allowing traverse between 12° inboard and 70° outboard, and elevation between +5° and −90°. The standard ammunition capacity of the system is 6,000 rounds in 12 boxes carried within the helicopter cabin and supplied to the guns via flexible chutes. Each gun has a standard rate of 2,400 rounds per minute, and when either gun is automatically stopped at the inboard limit of its traverse, the other is automatically increased to a maximum rate of 4,000 rounds per minute.

Emerson Electric M21/MAMEE: This is an improved version of the M21 system with two MAMEE (Meyer Ammunition Module Emerson Electric) containers: these are 3,000-round cylindrical magazines attached to the weapon pylons to leave the cabin free. Each cylinder has a diameter of 11.0 in (279 mm) and a basic length of 34.0 in (864 mm) increased to 56.0 in (1422 mm) by the addition of aerodynamic nose and tail fairings, and weighs 60.0 lb (27.2 kg) empty rising to 250.0 lb (113.4 kg) fully loaded. The complete system weighs 566.4 lb (256.9 kg) empty and 1,231.2 lb (558.5 kg) loaded.

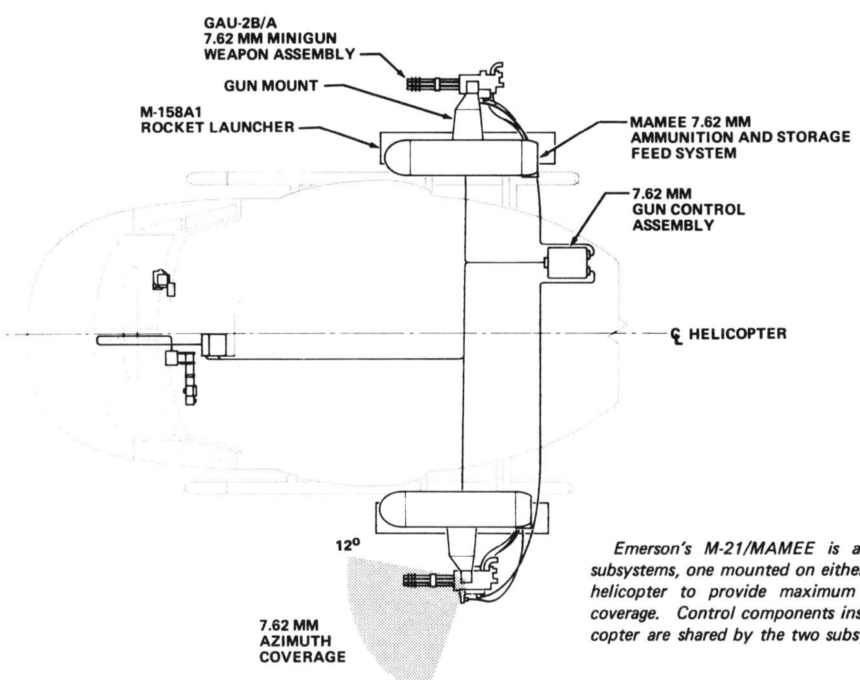

Emerson's M-21/MAMEE is actually two subsystems, one mounted on either side of the helicopter to provide maximum fire power coverage. Control components inside the helicopter are shared by the two subsystems.

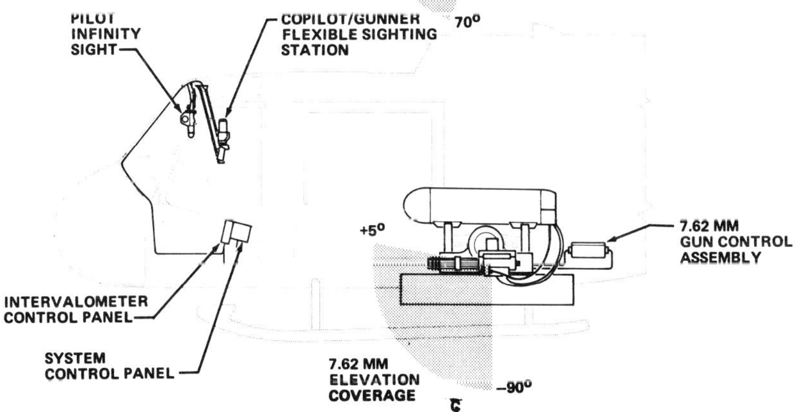

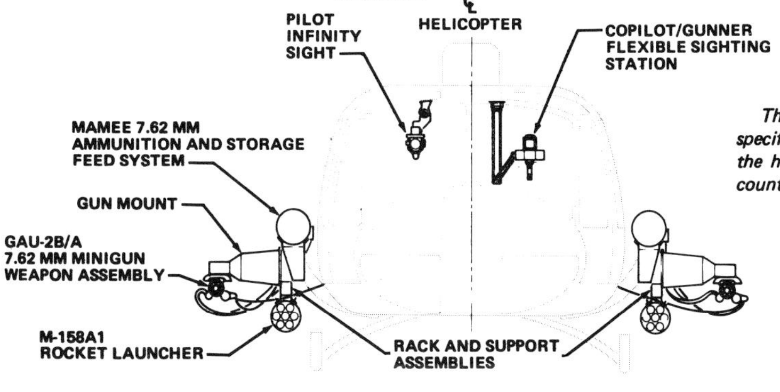

The M-21/MAMEE armament system was specifically designed for rapid installation on the hardpoints of utility helicopters in use by countries throughout the world.

Emerson Electric M28A1: Designed for the Bell AH-1G HueyCobra close-support helicopter, this is a chin-mounted installation able to accommodate one Minigun (on the right) and one M129 40-mm grenade-launcher (on the left), or alternatively (but very rarely) two of either weapon. The hydraulically-powered turret is controlled electrically by the gunner (or occasionally by the pilot in the rear seat when the weapons are in the stowed position) using an M73 series reflex sight. The Minigun has selectable rates of fire (2,000 or 4,000 rounds per minute) and 4,000 rounds of ammunition are accommodated in a large fuselage tank, being fed to the weapon through fixed and flexible chuting: the effective range of the Minigun in this application is 1,650 yards (1510 m). The turret can be traversed 114° left and right of the centreline, and the gun can be elevated between +12° and −50°. The weight of the installation is 450.0 lb (204.1 kg) empty and 900.0 lb (408.2 kg) loaded. The **M28E1** is a version slightly modified with a different drive motor.

Emerson Electric M53: Though developed for the abortive Lockheed AH-56A Cheyenne attack helicopter, this nose turret is suitable for retrofit on light helicopters such as early examples of the Bell UH-1 (Model 204) series. The powered turret accommodates a single Minigun, with ammunition fed from a magazine in the fuselage of the helicopter.

Emerson Electric TAT-102: This is a variant of the TAT-101 (see separate entry on the M60 machine-gun) designed to accommodate one Minigun (and in very rare circumstances, one M60 machine-gun) as the nose armament of light helicopters. The turret is hydraulically powered, and can be traversed through a maximum of 180° at a rate of more than 80° per second; the gun(s) can be elevated between −90° and +25° at a rate of 60° per second.

General Electric A/A49E3: This is a simple external mount for the Minigun, and is generally fitted in the UH-1N twin-engined version of the Huey family. The pintle mounting can accept the 7.62-mm (0.3-in) M60 or 0.5-in (12.7-mm) M2 machine-guns as alternatives to the Minigun; complete with 5,200 rounds per gun weighs 1,070 lb (485.4 kg).

General Electric M93: This is the semi-internal installation for one Minigun developed initially for the UH-1 series of light helicopters but since adapted for the Sikorsky SH-3 Sea King and Sikorsky HH-53 Super Jolly medium helicopters. When not required, the Minigun and its mounting are stowed out of the way in the cabin, but can swiftly be swivelled out for use: the mounting can also accept the M129 40-mm grenade launcher (in the **General Electric M94** system) or a 0.5-in (12.7-mm) M2 series heavy machine-gun. The installation is normally a double one, with one gun on each side of the cabin. When the guns are in the stowed position (aligned along the helicopter's flightpath) the pilot can fire them, though the more common operating method is for the two gunners to control the weapons on their flexible pintle mounts. A total of 10,600 or 12,000 rounds of ammunition is supplied in four cases under a common cover that incorporates an ammunition booster to lift the belts. The weight of the basic M93 installation with 5,200 rounds per gun is 1,176 lb (533.4 kg), and that of the improved **M93E1** installation with 6,000 rounds per gun 1,376 lb (624.2 kg).

General Electric SUU-11B/A: This is the standard podded installation of the GAU-2B/A Minigun intended for use on all manner of tactical aircraft with NATO standard 14-in (356-mm) suspension lugs. The pod can be carried at speeds up to Mach 1.2. The SUU-11B/A was developed to provide two firing rates (3,000 or 6,000 rounds per minute) rather than the single rate of the original **SUU-11A/A** model with the GAU-2/A Minigun. The forward third of the aluminium pod contains the Minigun and stowed MAU-69A/A hand-operated delinking loader, the muzzles of the barrels appearing through a stainless steel nosecap; the central section of the pod contains the drum magazine, which holds 1,500 rounds and the linkless feed system, while the rear section of the pod accommodates the battery and control system. The feed system is single-ended, the spent cases being ejected downward through a port in the bottom of the pod. The SUU-11B/A weighs 245 lb (111.1 kg) empty and 325 lb (147.4 kg) loaded, and the overall dimensions include a length of 85 in (2159 mm) and a diameter of 12 in (305 mm). A similar pod is used by US Army aircraft with the designation **M18E1**: this is basically identical to the SUU-11B/A but contains an M134 Minigun whose rates of fire are 2,000 and 4,000 rounds per minute.

McDonnell Douglas Helicopters (Hughes) M27E1: This system was developed at US Army behest to provide effective armament for the Hughes OH-6 Cayuse and Bell OH-58 Kiowa scout helicopters. The conceptual origin of the system was the experimental XM7 system, which carried two 7.62-mm (0.3-in) M60 machine-guns on the outboard end of the weapon beam attached to the cabin floor to project out through the door to the left of the helicopter. In its M27E1 form the system is revised to accommodate a specially modified Minigun with 2,000 rounds of belted ammunition. The gun is controlled by the co-pilot using an M70 reflex sight, and a two-step trigger allows selectable 2,000- or 4,000-round per minute fire rates. The gun is fixed in azimuth to fire along the line of flight, but can be controlled in elevation from +10° to −24° at the rate of 30° per second. The armament subsystem weighs 107 lb (48.54 kg) unloaded and 235 lb (106.6 kg) loaded, and the Minigun in this application has an effective range of 1,000 yards (1095 m). The system is also produced for the export market as the **HGS-5**, and is intended as an optional fitting for the McDonnell Douglas Helicopters (Hughes) Model 500 Defender series of light military helicopters.

General Electric GAU-8/A Avenger

USA

Type: aircraft seven-barrel cannon (air-to-surface)
Calibre: 30 mm
Operating method: General Electric 'Gatling' operation with two external hydraulic motors for a maximum of 77 hp (57.4 kW), and mechanical ignition
Feed: left-hand linkless
Overall dimensions: length 114 in (2896 mm); diameter 14 in (356 mm)
Barrel length: 90.5 in (2299 mm), or 76.63 calibres
Rifling: right-hand twist
Weight: complete weapon 620 lb (281.2 kg)
Recoil force: 19,000 lb (8618 kg) peak maximum, 14,000 lb (6350 kg) peak recurrent and 9,000 lb (4082 kg) average at 4,200 rounds per minute
Muzzle velocity: 3,240 ft (988 m) per second with API, and between 3,400 and 3,450 ft (1036 and 1052 m) per second with HEI and TP
Rate of fire: selectable 2,100 rounds per minute with one motor, or 4,200 rounds per minute with two motors
Effective range: 1,350 yards (1235 m)
Ammunition types: PGU-14/B API, PGU-13/B HEI and PGU-15/B TP
Ammunition weights: complete round 25.64 oz (727 g) for API, 23.35 oz (662 g) for HEI and 23.53 oz (667 g) for TP; projectile 15.0 oz (425 g) for API, 12.7 oz (360 g) for HEI and 12.87 oz (365 g) for TP
Ammunition lengths: complete round 11.42 in (290 mm); projectile 5.5 in (140 mm) for HEI and TP

A nose-on view of the Fairchild Republic A-10A Thunderbolt II reveals the impact of the General Electric GAU-8/A cannon on the design: to prevent asymmetric recoil effects when the 30-mm cannon is fired, the weapon has to be located as close to the centreline as possible, requiring the nose unit of the landing gear to be offset to starboard.

Variant and application

GAU-8/A Avenger: This is currently the world's most powerful aircraft gun by a very considerable margin, and was designed specifically to arm the winner of the US Air Force's A-X programme that resulted in selection of the Fairchild Republic A-10A Thunderbolt II over the Northrop YA-9A. The A-10A is a dedicated tank-killing aircraft, and its gun plus associated ammunition reflect the fact, the API round being built round a heavy Staballoy (depleted uranium) penetrator that also possesses excellent pyrophoric qualities. The 30-mm projectile has high velocity (allowing it to cover a typical slant range in some 30% less time that a standard 20-mm round). This translates into greater projectile accuracy and, given the additional mass of the 30-mm projectile, greater kinetic energy at the target.

The gun inevitably generates an enormous amount of heat when firing, and with the standard quantity of ammunition the GAU-8/A is designed to fire ten two-second bursts each separated by a one-minute cooling period. A simple calculation based on the concept of 'projectile horsepower' (the product of projectile mass and the square of velocity) indicates that at a slant range of 4,000 ft (1219 m), the GAU-8/A's projectile has fourteen times the kinetic energy of a 20-mm projectile from an M61 Vulcan cannon. One of the primary logistical advantages of the aluminium-cased ammunition specially developed for the Avenger is its salutory effect on barrel life, which is three times greater with this plastic-banded ammunition than it would have been with conventional ammunition using copper driving bands. The cannon is provided with two external motors, a single motor being used for the lower firing rate and both for the maximum rate.

Left *The massive quantity of heavy ammunition that must be loaded into modern attack aircraft as rapidly as possible calls for special equipment: here a mobile tanker feeds 30-mm ammunition into the externally-powered loader moving the linkless rounds into the ammunition drum of a Fairchild Republic A-10A Thunderbolt II's GAU-8/A cannon.*

In operating method the Avenger adheres to the basic General Electric 'Gatling' principle, though a new reverse-clearing technique is used to extract live but unfired rounds when the trigger is released; these rounds are returned within one second to the feed side of the cannon under control of an electro-hydraulic drive system which also ensures the correct positioning of the next live round on the feed side of the cannon. Weight is obviously very considerable with so massive a cannon and associated equipment, but a saving of 600 lb (272.2 kg) was made by choosing aluminium rather than steel for the ammunition cases. The magazine is a cylindrical unit located immediately behind the cannon and can contain 1,350 rounds, though the standard loading is 1,174 rounds. Ammunition is stored in a helical arrangement within the magazine, which has a length of 71.5 in (1816 mm) and a diameter of 34.5 in (876 mm).

Below *Currently the world's most powerful cannon for aircraft use, the General Electric GAU-8/A Avenger is seen here in the layout used in the weapon's sole operational application in the Fairchild Republic A-10A Thunderbolt II, with the seven-barrel cannon at the front, the closed-loop feed system in the centre, and the massive ammunition drum at the rear.*

The feed system is synchronized mechanically to the cannon, and comprises three major subassemblies: the exit unit at the front of the magazine to extract rounds and feed them into the closed-loop conveyor system to the cannon; the transfer unit to remove live rounds from the conveyor and feed them into the cannon, and to remove spent cases from the cannon and re-insert them into the conveyor; and the entrance unit at the rear of the magazine to feed the spent cases back into the magazine. The feed system weighs 1,150 lb (521.6 kg) including the magazine, and other weights in the system are 73.0 lb (33.1 kg) for the mount, 120.0 lb (54.4 kg) for the drive and controls, and 2,066 lb (937.1 kg) for the ammunition. The total weight of the system with maximum ammunition is thus 4,029 lb (1827.6 kg), declining to 3,759 lb (1705.1 kg) with 1,174 rounds of ammunition. The complete length of the system, from muzzle to rear of the entrance unit, is 238.5 in (6058 mm) and its height is 39.75 in (1010 mm).

USA

General Electric GAU-12/U Equalizer

Type: aircraft five-barrel cannon (air-to-air and air-to-surface)
Calibre: 25 mm
Operating method: General Electric 'Gatling' operation with 15-hp (11.1-kW) bleed air-drive external pneumatic power supply (or option of electric, hydraulic and mechanical drives), and mechanical ignition
Feed: right-hand linked or linkless

Working on the same 'Gatling' principle as most other General Electric cannon, the GAU-12/U uses the new and exceptionally powerful range of NATO standard 25-mm cannon ammunition.

Overall dimensions: length 83.2 in (2113 mm); diameter 11 in (279 mm)
Barrel length: data not available
Rifling: right-hand twist
Weight: complete weapon 270 lb (122.5 kg)
Recoil force: 9,000 lb (4082 kg) peak maximum, 8,000 lb (3629 kg) peak average and 4,000 lb (1814 kg) average at 3,600 rounds per minute
Muzzle velocity: 3,280 ft (1000 m) per second for API, and 3,400 ft (1036 m) per second for HEI and TP
Rate of fire: normally 3,600 rounds per minute, with 4,200 rounds per minute possible
Effective range: data not available
Ammunition types: M790 series, PGU-20/U API, PGU-22/25 HEI and PGU-23 TP
Ammunition weights: complete round 18.8 oz (533 g) for API, and 17.71 oz (502 g) for HEI and TP; projectile 7.58 oz (215 g) for API, and 6.5 oz (184 g) for HEI and TP
Ammunition length: complete round 8.78 in (223 mm) for API, and 8.625 in (219.1 mm) for HEI and TP

Variant and application

GAU-12/U Equalizer: This is the first of the General Electric 'Gatling' weapons to make use of the new NATO standard 25-mm ammunition also used by the British Aden 25. The weapon is derived technically from the GAU-8/A, and its reliability is attested by the quoted figure of 100,000 mean rounds between failures. The first application of the weapon is as the armament of the McDonnell Douglas AV-8B Harrier II STOVL close-support aircraft, which has the system incorporated into the two pods attached to the underside of the fuselage and thus forming part of its lift-improvement system. The port pod contains the GAU-12/U together with the pneumatic drive, whilst the starboard pod contains 300 rounds of ammunition fed to the cannon through a bridge structure connecting the two pods toward their after ends. The bridge also accommodates the drive

The finned pods under this McDonnell Douglas AV-8B Harrier II contain the GAU-12/U 25-mm cannon system, with the weapon in the port pod and its ammunition in the starboard pod. Also visible is the glass cover of the Hughes ASB-19 Angle-Rate Bombing Set in the nose of the aircraft.

shaft for the ammunition feed system. A double-ended (closed-loop) ammunition system is used so that spent cases are returned to the magazine, and the gun is cleared after each burst by a reversal of the feed system to extract any live rounds that might otherwise 'cook off'. The weight of the complete system is 900 lb (408.2 kg) empty and 1,230 lb (557.9 kg) loaded.

General Electric is also developing the Equalizer cannon as the weapon for use in the undernose turret of the Bell AH-1T Improved SeaCobra: the whole system is based conceptually on the Universal Turret System/M97 used in other Cobra helicopters with strengthening for the Equalizer's greater weights and recoil, and updated electronic features to match the systems of the Improved SeaCobra and AH-1W SuperCobra. The turret has a maximum height of 31.25 in (794 mm), and length from the muzzles to the turret centreline is 72.55 in (1843 mm). The cannon can be elevated from +20° to −50°.

USA

General Electric GAU-13/A

Type: aircraft four-barrel cannon (air-to-surface and air-to-air)
Calibre: 30 mm
Operating method: General Electric 'Gatling' operation with 22-hp (16.4-kW) external pneumatic power supply, and mechanical ignition
Feed: right-hand linkless
Overall dimensions: length 110 in (2794 m); diameter 12 in (305 mm)
Barrel length: 84.25 in (2140 mm), or 71.33 calibres
Rifling: right-hand twist
Weight: complete weapon 339 lb (153.8 kg)
Recoil force: (without muzzle brake) 12,000 lb (5443 kg) peak maximum, 10,000 lb (4536 kg) peak recurrent and 5,400 lb (2449 kg) average at 2,400 rounds per minute;

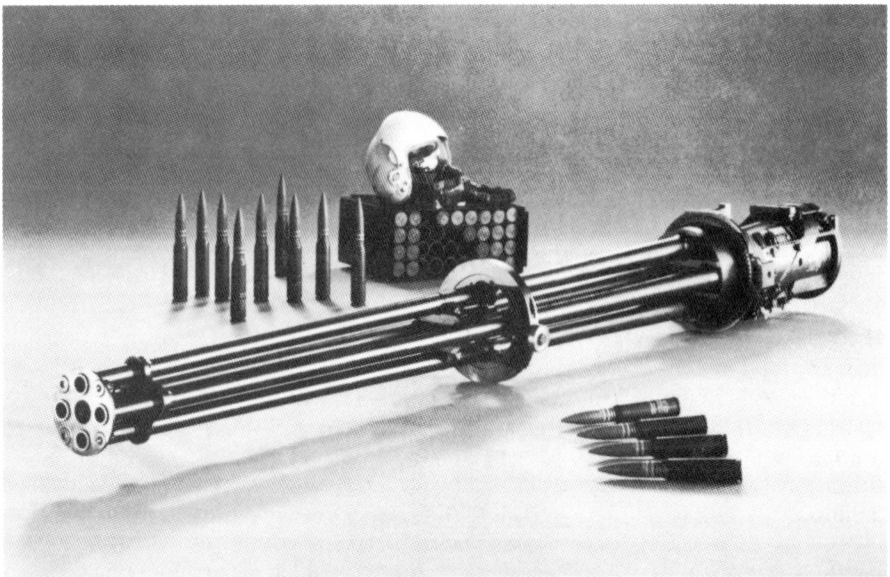

The General Electric GAU-13/A is essentially a four-barrel derivative of the seven-barrel GAU-8/A, and with its reduced weight and recoil loads is admirably suited to the GPU-5/A podded installation that provides the whole range of US tactical aircraft with a capable anti-tank weapon.

The GPU-5/A cannon pod uses the GAU-13/A 30-mm cannon, and can be carried by any tactical aircraft with NATO standard 30-in (762-mm) suspension lugs, or can alternatively be installed with the aid of its own dedicated adapter. The pod's ammunition capacity is a 'mere' 353 rounds, sufficient for just under 9 seconds of sustained fire at a cyclic rate of 2,400 rounds per minute, but such is the power of the projectiles that a few short bursts can achieve devastating results.

(with muzzle brake) 10,750 lb (4876 kg) peak maximum, 8,500 lb (3856 kg) peak recurrent and 5,400 lb (2449 kg) average at 2,400 rounds per minute
Muzzle velocity: 3,240 ft (988 m) per second with API, and 3,400 ft (1036 m) per second with HEI and TP
Rate of fire: normally 2,400 rounds per minute, with 3,000 rounds per minute possible
Effective range: data not available
Ammunition types: PGU-14/B API, PGU-13/B HEI and PGU-15/B TP
Ammunition weights: complete round 25.64 oz (727 g) for API, 23.35 oz (662 g) for HEI and 23.53 oz (667 g) for TP; projectile 15.0 oz (425 g) for API, 12.7 oz (360 g) for HEI and 12.87 oz (365 g) for TP
Ammunition length: complete round 11.42 in (290 mm)

Variants and applications
GAU-13/A: This is a lightweight version of the GAU-8/A Avenger, using four rather than seven barrels, and was developed in the late 1970s as the **GE 430** primarily for use in hard-hitting pods. The weapon uses the same ammunition and reverse-clearing system as the GAU-8/A, the latter ensuring that live rounds are extracted from the hot breeches when the trigger is released, thereby removing the possibility of 'cook-off' firing. Reliability is again very good, the figure quoted being 20,000 mean rounds between failures.
GPU-5/A: Developed in concert with the GE 430 as the **GEPOD 30**, this is the podded installation of the GAU-13/A complete with the drive motor, ammunition and micro-

A special powered ordnance trolley moves a GPU-5/A cannon pod up towards the McDonnell Douglas F-4E Phantom II flightline. The weapon carried in the pod is the GAU-13/A, a four-barrel derivative of the GAU-8/A Avenger tank-busting cannon.

processor-based control system. The pod is designed for carriage on NATO standard 30-in (762-mm) suspension lugs, and its primary data include a length of 170 in (4318 mm), a diameter of 24 in (610 mm), and empty and loaded weights of 1,318 lb (597.8 kg) and 1,850 lb (839.2 kg) respectively. The cannon is not limited to a maximum burst length (other than by ammunition capacity), and occupies much of the pod's length, with the control unit and high-pressure air bottle for the drive motor at the very rear. The 353 rounds of ammunition are carried in two concentric helical layers round the cannon, which is fed by a double-ended (closed-loop) system that carries away spent cases and makes possible the reverse-clearing system for the cannon.

USA

General Electric GE 225

Type: aircraft two-barrel cannon (air-to-surface and air-to-air)
Calibre: 25 mm
Operating method: 'Gast' principle operation with 25-hp (18.6-kW) external power supply (or tapped gun gas), and mechanical ignition
Feed: left- or right-hand linked belt
Barrel length: data not available
Rifling: right-hand twist

Overall dimensions: length 87 in (2210 mm); width 8.5 in (216 mm); height 8 in (203 mm)
Weight: complete weapon 180 lb (81.65 kg) with feeder and drive motor
Recoil force: 2,400 lb (1089 kg) peak maximum, 1,800 lb (816 kg) peak recurrent, 2,300 lb (1043 kg) average at 2,000 rounds per minute, and 860 lb (390 kg) average at 750 rounds per minute
Muzzle velocity: 3,600 ft (1097 m) per second with HEI, and 4,400 ft (1341 m) per second with APDS
Rate of fire: selectable up to 2,000 rounds per minute
Effective range: data not available
Ammunition types: PGU-20/U API, PGU-22/U HEI and PGU-23 U
Ammunition weights: complete round 18.8 oz (533 g) for API, and 17.71 oz (502 g) for HEI and TP; projectile 7.58 oz (215 g) for API, and 6.5 oz (184 g) for HEI and TP
Ammunition length: complete round 8.78 in (223 mm) for API, and 8.625 in (219.1 mm) for HEI and TP

Variant and application

GE 225: Like the Soviet 23-mm GSh-23L, this twin-barrel weapon works on the long-established but little-used Gast principle, with a reciprocating lever to extract the spent case from one breech while loading a fresh round in the other. A cam at the rear of the action provides the accelerative forces required for adequate action. The barrels fire alternately, and remain essentially fixed. Spent cases are ejected downwards, and live rounds are extracted from the breeches at the end of each burst. The considerable recoil force of the weapon dictates that any helicopter installation must use a 'soft mount' recoil-attentuation system.

An exception to the general use of the externally-powered 'Gatling' principle for General Electric barrelled weapons is the 25-mm GE-225, which still relies on external power but uses the 'Gast' operating principle also employed by the Soviet GSh-23L weapon.

General Electric GECAL 50

Type: aircraft three-barrel machine-gun (air-to-surface)
Calibre: 0.5 in (12.7 mm)
Operating method: General Electric 'Gatling' operation with external electric, hydraulic or ram-air power supply (or the option of tapped gun gas), and mechanical ignition
Feed: left- or right-hand de-linked or linkless
Overall dimensions: length 46.5 in (1181 mm) with 36-in (914-mm) barrels; diameter 8 in (203 mm)
Barrel lengths: options between 36 and 51 in (914 and 1295 mm) are envisaged
Rifling: right-hand twist
Weight: complete weapon 68 lb (30.85 kg) with 36-in (914-mm) barrels
Recoil force: 850 lb (385.6 kg) peak and 425 lb (192.8 kg) average at 2,000 rounds per minute
Muzzle velocity: 2,900 ft (884 m) per second
Rate of fire: up to 4,000 rounds per minute with linkless feed mechanism
Effective range: 1,600 yards (1465 m)
Ammunition types: APHC, M8 API, M17 Tracer, NM 140 and TP
Ammunition weights: complete round 4.23 oz (120 g); projectile between 1.4 and 1.65 oz (39.7 and 46.8 g) according to type
Ammunition length: complete round 5.45 in (138.4 mm)

The General Electric GECAL 50 is an important new aircraft gun, clearly destined for an important role in providing helicopters and tilt-wing aircraft with an exceptional air-to-surface fire capability. This is the three-barrel version, provision for the other three barrels being clearly visible.

Variants and applications

GECAL 50: This is another of General Electric's family of 'Gatling' weapons, in this instance an extremely powerful machine-gun designed to use the whole range of ammunition types developed for the classic M2/M3 series of heavy machine-guns. The basic three-barrel version is fitted with a controlled-burst device for salvoes of 10 rounds which, fired at the rate of 4,000 rounds per minute, arrive on the target almost simultaneously for maximum destructive effect. A six-barrel version has also been developed, and this has a firing rate of 8,000 rounds per minute at a weight of 96 lb (43.55 kg): dimensionally this version is identical to the three-barrel model. Both the three- and six-barrel versions have the option of a self-starting gas drive for use in situations where no external power is available. In common with General Electric's other 'Gatling' weapons, the GECAL 50 has provision for clearing unfired rounds, using a declutching feeder with linked ammunition and feed reversal with linkless ammunition. General Electric has developed a pintle mounting to allow the installation of a three-barrel GECAL 50 in the Sikorsky UH-60 Black Hawk: complete with 1,000 rounds of ammunition this weighs 724.0 lb (328.4 kg). A likely application for the weapon is the Bell/Boeing V-22 Osprey tilt-rotor aircraft, which is currently envisaged with a three-barrel GECAL 50 firing at 2,000 rounds per minute and provided with 1,200 rounds of ammunition in a closed-loop feed system.

General Electric Composite Turret Weapon System: This has been developed to allow the installation of the GECAL 50 (either three- or six-barreled) in a new generation of aircraft. The turret is made of composite materials for maximum strength combined with minimum weight, and can be installed for flexible forward and lateral arcs of fire. Ammunition capacity is variable with specific aircraft installation, but is maximized by the use of a linkless feed system, which can be topped up in flight by use of a delinking feeder for M9 belted ammunition. The turret can be traversed 100° left and right, and the gun elevated from +15° to −45°, the traverse and elevation rates each being 60° per second in this fully powered installation. The turret has a diameter of 20.0 in (508 mm), and the distance from the turret centreline to the muzzles is 41.0 in (1041 mm). Complete with 1,500 rounds of ammunition and a six-barrel GECAL 50, the installation weighs 313 lb (142.0 kg).

General Electric M61A1 Vulcan USA

Type: aircraft six-barrel cannon (air-to-air and air-to-surface)
Calibre: 20 mm
Operating method: General Electric 'Gatling' operation with 35-hp (26.1-kW) external electric or hydraulic power supply, and electrical ignition
Feed: right-hand de-linked or linkless
Overall dimensions: length 73.8 in (1875 mm); diameter 13.5 in (343 mm)
Barrel length: 60 in (1524 mm), or 76.2 calibres
Rifling: right-hand twist
Weight: complete weapon 265 lb (120.2 kg) with 40-lb (18.1-kg) motor; barrel 18 lb (8.165 kg) each
Recoil force: 3,818 lb (1732 kg) average at 6,000 rounds per minute, or 2,661 lb (1207 kg) average at 4,000 rounds per minute
Muzzle velocity: 3,400 ft (1036 m) per second
Rate of fire: variable up to 6,600 rounds per minute, with 4,000 or 6,000 rounds per minute generally preferred

Effective range: data not available
Ammunition types: M50 series (including M53A1 API, M54 HPT, M55A1 Practice, M56 HE and M56A1 HEI) and PGU-17/B
Ammunition weights: complete round 9.07 oz (257 g); projectile 3.56 oz (101 g) for API, HE and HEI
Ammunition length: complete round 6.61 in (168 mm)

Variants and applications

M61A1 Vulcan: This is the most widely used aircraft cannon of the Western world, and was the first member of the General Electric 'Gatling' family of externally-powered rotary-action weapons. Design of the weapon began in 1950 under the general heading Project 'Vulcan' after the start of conceptual analysis in 1946 when the US Air Force began to appreciate that conventional single-barrel weapons of the day were reaching the limits of their practical firing rates. It was clear from analysis of World War 2 combat that an adequate number of hits in the fleeting engagement times anticipated with turbojet-powered fighters could be provided only by the much-increased firing rates possible from batteries of multiple cannon, which were heavy and difficult to install, or from potentially more effective multi-barrel weapons. The latter are based on a group of rigidly-linked barrels turned (by a special rotor assembly) around a common longitudinal axis in front of a stationary gun housing. The bolt associated with each barrel is fitted with a cam follower which rides in a track machined inside the gun housing, so that while one barrel is being fired the others are each at a different part of the clearing/reloading cycle leading to the firing of the next round from that barrel.

It was also evident from the beginning of the programme that such an operating method would possess the advantages of reduced heating and wear of individual barrels. The General Electric system usually employs external power to ensure that the weapon continues to function effectively despite varying chamber pressures and the occasional misfire. The M61A1 current variant dates from 1956, when General Electric realized that the problems it was encountering with linked feed systems were basically insuperable at the firing rates being attained, and therefore turned to a linkless feed arrangement: this is sometimes of the single-ended variety, designed to eject spent cases and unfired rounds, but more commonly of the double-ended (closed-loop) variety in which the powered ammunition conveyor belt carries the spent cases and unfired rounds back to the magazine (where they deposited at the entrance end of the drum) before the conveyor moves to the exit end and collects fresh rounds. The ammunition is stored between radial partitions and slowly moved to the exit end of the magazine by a helical (Archimedean) screw arrangement driven by the gun motor.

The first installation of the Vulcan was on the Lockheed F-104 Starfighter interceptor, and trials soon confirmed that a single weapon of this type offered many advantages over the alternative of a multiple battery of guns. The Vulcan has thus become the primary fixed armament of US fighters designed from the mid-1950s onwards, the size of the weapon and its feed system/ammunition magazine dictating the design of individual installations for each type (see below).

GAU-4 Vulcan: This is the self-powered version of the M61A1, using gun gas tapped from four of the six barrels. This variant weighs some 10 lb (4.54 kg) more than the externally-powered Vulcan, and has the US Army designation **M130**.

M61A1 (A-7 Corsair): On the A-7D and A-7E variants of the Vought Corsair II tactical fighter and attack aircraft, the Vulcan installation is located on the port side of the forward fuselage with the gun port in the outside edge of the engine inlet lip and the forward edge of the horizontal magazine drum above and inside the gun's breech mechanism. The whole assembly is 42.25 in (1073 mm) wide and 81.5 in (2070 mm) high; the magazine assembly, with its entrance unit at the rear and exit unit at the front, is 55.75 in (1416 mm) long and contains 1,019 rounds. The magazine drum is fitted with a de-linking loader that strips the links from belted ammunition before the rounds are introduced into the magazine. The loaders of the A-7D and A-7E models differ slightly, that of the A-7D discarding the stripped links, while that of the A-7E stores the stripped links to belt the spent cases and cleared rounds as they are downloaded when the magazine is reloaded.

The characteristics of the whole hydraulically-powered installation include a gun weight of 252 lb (114.3 kg), a closed-loop feed system weight of 413 lb (187.3 kg) and an ammunition weight of 513 lb (232.7 kg) for an overall weight of 1,178 lb (534.3 kg). The reliability of the installation is 10,000 mean rounds between failures, and the time required for the gun to accelerate to and decelerate from its maximum firing rate of 6,000 rounds per minute is between 0.4 and 0.5 second. An alternative rate of 4,000 rounds per minute can be selected by the pilot.

M61A1 (F-4 Phantom): On the F-4E variant of the McDonnell Douglas Phantom tactical fighter, the Vulcan installation is located on the centreline of the forward fuselage with the gun port in a blister fairing under the nose just behind the antenna of the radar and the horizontal magazine drum above the gun's breech mechanism. The whole assembly is extremely compact, being a mere 24.5 in (622 mm) wide and 45 in (1143 mm) high. The magazine assembly, with its entrance unit at the rear and exit unit at the front, is just 28.5 in (724 mm) high from the pallet that mounts the gun below it to the magazine which contains 638 rounds. The magazine drum is fitted with a de-linking loader that strips the links from the belted ammunition before the rounds are introduced into the magazine.

The characteristics of the whole hydraulically-powered installation include a gun weight of 252.0 lb (114.3 kg), a closed-loop feed system weight of 345 lb (156.5 kg), an ammunition weight of 322 lb (146.0 kg) and a pallet weight of 60 lb (27.2 kg) for an overall weight of 979 lb (444.1 kg). The reliability of the installation is 10,000 mean rounds between failures, and the time required for the gun to accelerate to and decelerate from its maximum firing rate of 6,000 rounds per minute is 0.5 second. An alternative rate of 4,000 rounds per minute can be selected by the pilot.

M61A1 (F-14 Tomcat): On the F-14A variant of the Grumman Tomcat fleet air-defence fighter, the Vulcan installation is located on the port side of the forward fuselage with the gun port on the lower side of the aircraft just aft of the radar antenna, and the forward edge of the horizontal magazine drum behind and slightly inboard of the gun's breech mechanism. The whole assembly is 132 in (3353 mm) long, 24.25 in (616 mm) wide and 35.5 in (902 mm) high. The magazine assembly, with its entrance unit at the front and exit unit at the rear, is 40 in (1016 mm) long and contains 767 rounds. The magazine drum is used with a loading adapter interfaced with the US Navy's Linkless Ammunition Loading System to reduce onboard weight by obviating the need for an inbuilt system, and at the same time to allow ammunition replenishment at the rate of 400 rounds per minute. The characteristics of the whole hydraulically-powered installation include a gun weight of 252 lb (114.3 kg), a double-ended (closed-loop) feed system weight of 386 lb (175.1 kg) and an ammunition weight of 380 lb (172.4 kg) for an overall weight of 1,028 lb (466.3 kg). The reliability of the installation is 10,000 mean rounds between failures, and the time required for the gun to accelerate to and decelerate from its maximum firing rate of 6,000 rounds per minute is 0.5 second. An alternative rate of 4,000 rounds per minute can be selected by the pilot.

M61A1 (F-15 Eagle): On the F-15 variants of the McDonnell Douglas Eagle air-superiority and multi-role fighter, the Vulcan installation is located in the starboard leading-edge root extension between the starboard inlet and the starboard wing proper, with the gun port in the leading edge and the vertical magazine drum across the aircraft from the gun's breech mechanism just inboard of the port engine inlet trunking. The whole assembly is 81.75 in (2076 mm) long, 89.9 in (2283 mm) wide and 50 in (1270 mm) high. The magazine assembly, with its entrance unit at the top and exit unit at the bottom, is 50 in (1270 mm) high and contains 940 rounds. The magazine drum is fitted with a de-linking mechanism that connects with an external AGE loader to strip the links from the belted ammunition before the rounds are introduced into the magazine, while at the same time spent cases and unfired rounds are deposited into special containers in the loader.

The Vulcan installation's physical configuration dictates a somewhat complex system of three double-ended (closed-loop) conveyors between the magazine and the gun, with the first consisting of fixed chuting to move rounds from the magazine exit unit to the interface/hand-off assembly which connects the ammunition-handling elements and also synchronizes the three conveyors, the second consisting of flexible chuting to move the rounds from the interface/hand-off assembly to the transfer/feed unit on the gun, and the third consisting again of flexible chuting to return spent cases and unfired rounds to the magazine's entrance unit.

Considerable efforts were made to reduce the system's weight, the most important contribution being being the construction of the ammunition drum from plastics rather than steel, and this effort is reflected in the characteristics of the whole hydraulically-powered installation, which includes a gun weight of 252 lb (114.3 kg), a closed-loop feed system weight of 380 lb (172.4 kg) and an ammunition weight of 528 lb (239.5 kg) for an overall weight of 1,160 lb (526.2 kg). The reliability of the installation is 15,000 mean rounds between failures, and the time required for the gun to accelerate to and decelerate from its maximum firing rate of 6,000 rounds per minute is less than 0.5 second. An alternative rate of 4,000 rounds per minute can be selected by the pilot.

M61A1 (F-16 Fighting Falcon): On the F-16 variants of the General Dynamics Fighting Falcon tactical fighter, the Vulcan installation is located in the port leading-edge root extension with the gun port on the upper surface of the extension and the transversely-aligned horizontal magazine drum inboard of the gun's breech mechanism. The whole assembly is 90.5 in (2299 mm) long while the magazine assembly is 22.5 in (571.5 mm) in diameter and contains 512 rounds. The magazine drum is fitted with a de-linking mechanism that interfaces with two AGE external loaders and strips the links from the belted ammunition before the rounds are introduced into the magazine. The system can also be interfaced with the Linkless Ammunition Loading System (see F-14 section of this entry) for reduced turn-round time.

The same type of weight reduction exercise was undertaken on the F-16's Vulcan installation as on the F-15, so the characteristics of the hydraulically-powered system include a gun weight of 252 lb (114.3 kg), a double-ended (closed-loop) feed system weight of 380 lb (172.5 kg) and an ammunition weight of 528 lb (239.5 kg) for an overall weight of 1,160 lb (526.2 kg). The reliability of the installation is 15,000 mean rounds between failures, and the time required for the gun to accelerate to or decelerate from the maximum firing rate of 6,000 rounds per minute is between 0.4 and 0.45 second. An alternative rate of 4,000 rounds per minute can be selected by the pilot.

M16A1 (F-106 Delta Dart): On the F-106 variants of the Convair Delta air-defence fighter, the Vulcan installation is a palletized type designed for location in the fighter's belly bay (otherwise used for air-to-air missiles and/or rockets) with the gun muzzle under the fuselage and the rear edge of the horizontal magazine drum above the gun's barrel rotor mechanism. The whole assembly is 79.6 in (2022 mm) long, 26.8 in (681 mm) wide and 34 in (864 mm) high. The magazine assembly, with its entrance unit at the rear and exist unit at the front, is 29.2 in (742 mm) long and contains 650 rounds. The characteristics of the whole hydraulically-powered installation include a gun weight of 252 lb (114.3 kg), a double-ended (closed-loop) feed system and pallet weight of 436 lb (197.8 kg) and an ammunition weight of 371 lb (168.3 kg) for an overall weight of 1,059 lb (480.4 kg). The reliability of the installation is 10,000 mean rounds between failures, and the time required for the gun to accelerate to or decelerate from the sole firing rate of 4,500 rounds per minute is between 0.2 and 0.4 second.

M61A1 (F-111): On variants of the General Dynamics F-111 long-range strike and attack aircraft, a boxed Vulcan installation can be attached in the lower-fuselage weapons bay with the gun port on the outside of the bay and the horizontal magazine drum behind the breech mechanism. The magazine assembly, with its exit unit at the front, contains 2,084 rounds, the highest capacity for any Vulcan installation. The whole assembly is 145.2 in (3688 mm) long and 34.8 in (884 mm) high. Empty cases and cleared rounds are deposited in a container above the gun. The characteristics of the whole installation include a gun weight of 252 lb (114.3 kg), an ammunition system weight of 456 lb (206.8 kg) and an ammunition weight of 708 lb (321.1 kg) for an overall weight of 1,747 lb (792.4 kg). The reliability of the installation is 10,000 mean rounds between failures, and the time required for the gun to accelerate to and decelerate from its sole firing rate of 5,000 rounds per minute is 0.4 second.

M61A1 (F/A-18 Hornet): On the F/A-18 variants of the McDonnell Douglas Hornet tactical fighter and attack aircraft, the Vulcan installation is an extremely neat package located centrally in the nose just aft of the radar with the gun port on the upper side of the nose and the horizontal magazine drum under the gun. The whole assembly is 71.93 in (1827 mm) long, 25.8 in (655 mm) wide and 25.03 in (636 mm) high from the bottom of the magazine to the gun centreline at the rear of the installation. The magazine assembly is notably compact, being 20.7 in (526 mm) long and 22.36 in (568 mm) in diameter, and makes use of a new high-density stowage arrangement to contain 570 rounds. The magazine drum is fitted with a loading adapter for use with the Linkless Ammunition Loading System (see F-14 section above), while the double-ended (closed-loop) feed system brings spent cases and cleared rounds back to the drum for downloading during inter-mission reloading.

The characteristics of the whole installation include a gun weight of 252 lb (114.3 kg), a feed system weight of 270 lb (122.5 kg) and an ammunition weight 319 lb (144.7 kg) for an overall weight of 841 lb (381.5 kg). The reliability of the installation is 15,000 mean rounds between failures, and the time required for the gun to accelerate to and decelerate from its maximum firing rate of 6,000 rounds per minute is 0.5 second. An alternative rate of 4,000 rounds per minute can be selected by the pilot.

M61A1 (AMX): This installation has been designed for the Italian version of the joint Italo-Brazilian Aermacchi/Aeritalia/EMBRAER AMX attack fighter, and is unusual in using linked M50 or Raufoss Ammunisjons ammunition. The whole installation locates the cannon in the port side of the lower nose, with the ammunition drum stowed transversely above and inboard of the cannon with its exit unit forward of the cannon breeches. The complete assembly weighs 613.35 lb (278.2 kg), and the ammunition capacity is 403 rounds, whose links are stripped as the rounds are loaded into the cannon. A hydraulic drive system is used for a maximum rate of 4,000 rounds per minute, and the cannon acceleration and deceleration times are 0.35 and 0.4 seconds respectively.

M195 Vulcan: This is a version of the M61A1 with 40.4-in (1026-mm) barrels fitted with blast deflectors to suit the type for helicopter use. The only application of the weapon is in the **General Electric M35** fixed armament system for the Bell AH-1G HueyCobra close-support helicopter. The modified gun is attached under the left weapon sponson of the helicopter, with 950 rounds of linked ammunition fed to the right-hand side of the weapon from containers attached to the sides of the helicopter. The M195's rate of fire in the order of 750 rounds per minute, and the weights of the system are 540 lb (244.9 kg) empty and 1,170 lb (530.7 kg) loaded.

General Electric SUU-16/A: This is one of two standard podded installations of the Vulcan cannon, and is designed for use by tactical aircraft fitted with NATO standard 30-in (756-mm) suspension lugs. The SUU-16/A pod uses an M61A1 powered by a ram-air turbine: this mounted on a hinged panel in the skin of the pod, and is opened into the airstream just before the gun can be fired. The turbine spins up to speed in one second, but at air speeds below 400 mph (644 km/h) reduced turbine speed cuts the firing rate to a figure below the standard 6,000 rounds per minute. As the turbine also generates a high degree of drag, the SUU-16/A is better suited to the air-to-surface role at moderate speeds.

The cannon occupies the forward portion of the pod, and a cylindrical ammunition drum fills the rear, a single-ended linkless feed system being used to ferry rounds to the gun. Spent cases are ejected overboard, as are cleared rounds. The pod is 197.8 in (5004 mm) long and 22 in (559 mm) in diameter, and its weights are 965 lb (437.7 kg) empty and 1,615 lb (732.6 kg) loaded. The US Army designation of the pod is **M12**.

General Electric SUU-23/A: This is a variant of the SUU-16/A using the self-powered GAU-4 version of the Vulcan, producing a pod better suited to the high-speed air-to-air role as there is no projecting ram-air turbine. The cannon has an electric inertia starter to spin it up to speed, and the only other difference from the SUU-16/A is slightly increased weight as the GAU-4 is heavier than the M61A1. Thus the empty and loaded weights of the SUU-23/A are 1,045 lb (474.0 kg) and 1,720 lb (780.2 kg) respectively. Like the SUU-16/A, the SUU-23/A is attached to fire forward along the aircraft's line of flight, and the pilot uses his standard sight system to aim the weapon. The US Army designation of the pod is **M25**.

General Electric T171: This is the barbette installation of the M61A1 cannon to provide the Boeing B-52H Stratofortress strategic bomber with tail defence. The system was initially developed as a radar-supported manually controlled system for the out-of-service Convair B-58 Hustler supersonic bomber, and is a trim installation looking for all the world like a wasp's sting. The T171 in the B-52H is fully automatic, using the ASG-21 radar-directed fire-control system.

USA

General Electric M188E1

Type: aircraft three-barrel cannon (air-to-surface and air-to-air)
Calibre: 30 mm
Operating method: General Electric 'Gatling' operation with external hydraulic or electrical power supply, and mechanical ignition
Feed: right-hand linked belt or linkless
Overall dimensions: length 61.31 in (1557 mm); width 13.6 in (345 mm); height 10 in (254 mm)
Barrel length: 42 in (1067 mm), or 35.57 calibres
Rifling: right-hand twist
Weight: complete weapon 116 lb (52.6 kg); barrels 14.5 lb (6.6 kg) each
Recoil force: 4,000 lb (1814 kg) maximum at 2,000 rounds per minute, and 1,535 lb (696 kg) maximum at 730 rounds per minute
Muzzle velocity: 2,525 ft (770 m) per second with Aden/DEFA types
Rate of fire: variable between 200 and 2,000 rounds per minute
Effective range: data not available

Ammunition types: standard Aden and DEFA types (APHEI-SD, HEI, HEI-SD, TP and TP-T)
Ammunition weights: (DEFA range) complete round 449 g (15.84 oz) for HEI-SD, 450 g (15.87 oz) for HEI, TP and TP-T, and 472 g (16.65 oz) for APHEI-SD; projectile 243 g (8.57 oz) for HEI-SD, 244 g (8.61 oz) for HEI, TP and TP-T, and 270 g (9.52 oz) for APHEI-SD
Ammunition weights: (Aden range) complete round 17.46 oz (495 g) for AP and API, 17.5 oz (496 g) for HE, and 15.52 oz (440 g) for TP; projectile 7.76 oz (220 g)
Ammunition length: (DEFA range) complete round 200 mm (7.87 in)
Ammunition length: (Aden range) complete round 7.87 in (200 mm)

Variants and applications
M188: This is essentially an 'up-calibred' three-barrel version of the M61 Vulcan firing percussion-primed ammunition and designed to be completely interchangeable with the M197 (same dimensions and comparable weight) but offering considerably heavier fire-power. The weapon was developed as the Area Weapon Subsystem turreted gun for the Bell AH-63, which finally lost to the Hughes AH-64 Apache in the competition for the US Army's Advanced Attack Helicopter programme. The weapon is 58 in (1473 mm) long and weighs 120 lb (54.4 kg) complete with its feeder. The M188 was designed for the M552 HEDP round, with which a muzzle velocity of 2,200 ft (671 m) per second was attained.
M188E1: This was developed from the M188 to use the M788, M789 and M799 ammunition types as well as the standard Aden and DEFA ammunition ranges. The weapon has short barrels and is interchangeable with the M197, making it operationally attractive in installations requiring 30-mm rather than 20-mm weapons. The type can be fitted in helicopters as it features a recoil-attenuation system and the variable fire rate allows the firer to select a rate that does not result in resonance with the airframe. A typical installation (based on that for the AH-63) has 1,200 rounds contained in an ammunition tank measuring 80 in (2032 mm) in length, 40 in (1016 mm) in width and 11.5 in (453 mm) in height, a single-ended feed system taking rounds to the cannon located in an electrically-powered turret weighing 172 lb (78 kg) and having a diameter of 28 in (711 mm). The turret can traverse 110° left and right of the aircraft centreline at a rate of more than 80° per second, and the cannon can be elevated from −22° to −62° at a rate of more than 60° per second.

General Electric M197 — USA

Type: aircraft three-barrel cannon (air-to-surface and air-to-air)
Calibre: 20 mm
Operating method: General Electric 'Gatling' operation with external 3-hp (2.2-kW) electric power supply, and electrical ignition
Feed: right-hand linked belt or linkless
Overall dimensions: length 71.93 in (1827 mm); diameter 10.8 in (267 mm)
Barrel length: options up to 60 in (1524 mm)
Rifling: right-hand twist
Weight: complete weapon 132 lb (59.88 kg) excluding 13-lb (5.9-kg) drive motor
Recoil force: 1,500 lb (680 kg) average at 1,500 rounds per minute
Muzzle velocity: 3,380 ft (1030 m) per second with 60-in (1524-mm) barrels

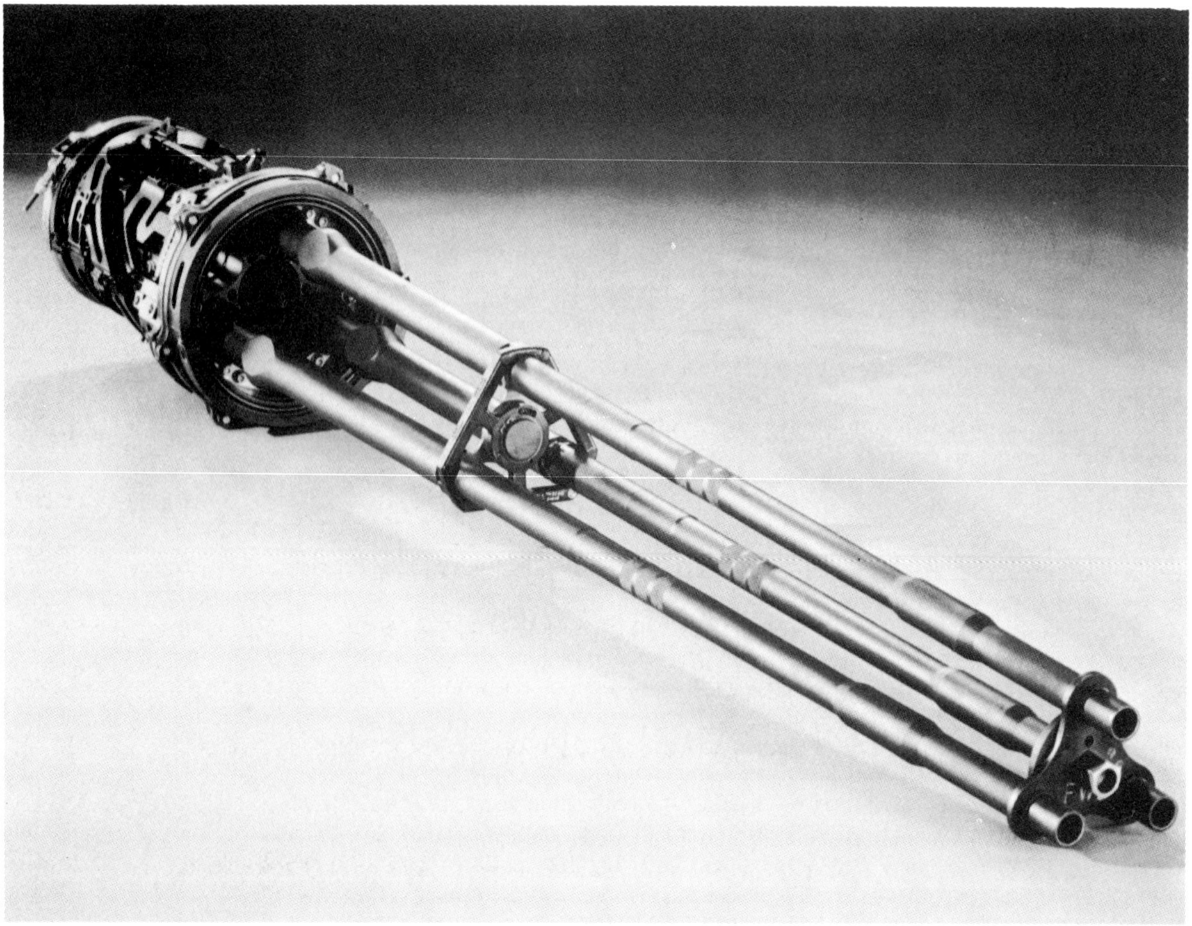

The 20-mm General Electric M197 is a three-barrel derivative of the classic M61A1 Vulcan six-barrel aircraft cannon, and is most effectively used as helicopter armament, either in a chin-mounted turret or on a pintle mounting in a doorway.

Rate of fire: selectable between 400 and 3,000 rounds per minute
Effective range: data not available
Ammunition types: M50 series (including M53A1 API, M54 HPT, M55A1 Practice, M56 HE and M56A1 HEI) and PGU-17/B
Ammunition weights: complete round 9.07 oz (257 g); projectile 3.56 oz (101 g) for API, HE and HEI
Ammunition length: complete round 6.61 in (168 mm)

Variants and applications
M197: This is essentially, a three-barrel version of the M61 Vulcan series, optimized for use in helicopter and light fixed-wing aircraft installations. Among its several attributes the gun offers a high degree of mechanical commonality with the M61 Vulcan series, use of the same ammunition, and extreme reliability. General Electric offers a pintle-mounted version of the M197 for helicopter and light aircraft installation, the latter epitomized by the use of the weapon in the door-mounted system of the Fairchild AU-23A Peacemaker counter-insurgency aircraft where the installation comprises a 500-round magazine feeding linked ammunition via flexible chuting to the M89A1 de-linking feeder on the pintle. Complete with ammunition the system weighs 765 lb (347 kg),

and provides the M197 with training from 15° forward to 60° aft, and elevation from 0° to −55°; the M197 cannon has 350- and 700-round per minute firing rates, and comes up to full rate in 0.2 seconds; the recoil force of the cannon in this installation peaks at 3,200 lb (1451.5 kg).

General Electric M97: This is the turreted installation of the M197 developed for the Bell AH-1J SeaCobra close-support helicopter operated by the US Marine Corps. The installation led to the more versatile Universal Turret System catering for a choice of several weapon types, but the M97 can use only the M197.

General Electric M97 (OV-10D): The M97 turret system as used on the Bell AH-1J SeaCobra helicopter has been developed for use in the Night Observation System nocturnal gunship version of the Rockwell OV-10 Bronco. This is installed in the underside of the OV-10D's fuselage, and when in the stowed position (with the gun trained to fire directly forwards) can be fired by either of the Bronco's two crew members. More usually, however, it is used flexibly, and in this capacity is slaved via the aircraft's fire-control system to the Texas Instruments AAS-37 forward-looking infra-red system carried by the OV-10D. The cannon is fixed to fire at 750 rounds per minute, and is provided with 1,500 rounds of linked ammunition. The turret can traverse 140° to each side of the aircraft centreline at the rate of 80° per second, and the cannon can be elevated between 0° and −50° at the rate of 60° per second. The magazine is located in the fuselage, and can be reloaded in 15 minutes.

General Electric Universal Turret System: This is an extremely versatile turret developed from the M97 (see above) for use mainly by the Bell AH-1 Cobra series of close-support and battlefield helicopters. The turret can be tied in with a number of cockpit and/or helmet-mounted sight systems, but is generally used with a stabilized reflex sight, the turret being slaved (via azimuth and elevation servo motors) to follow the movement of this pantograph unit. The UTS can be fitted with a variety of machine-gun or cannon types (including the 7.62-mm/0.3-in GAU-2B/A or M134 Minigun, the 0.5-in/12.7-mm GECAL 50, the 25-mm GE 225 and the 30-mm M188E1), but the standard installation is the M197. The turret has a diameter of 28 in (711 mm), and can be traversed 110° left and right of the aircraft centreline at a rate of more than 80° per second, while the gun can be elevated between +21° and −50° at a rate of 60° per second. The cannon comes up to its maximum rate of 750 rounds per minute in 0.3 second, and is fed with 750 rounds of linked ammunition from a magazine in the fuselage. this magazine is 42 in (1067 mm) long, 23.5 in (597 mm) wide and 15.5 in (394 mm) high, and can be reloaded in 15 minutes.

The complete length of the UTS with magazine and M197 cannon is 124.5 in (3162 mm), the weight of the installation is 1,032 lb (468.1 kg) and the peak recoil force is 2,000 lb (907 kg). The M197 can be replaced by the M188 (plus associated chuting and feed system) in just thirty minutes, this installation being characterized by 500 rounds of ammunition and a maximum rate of fire of 2,000 rounds per minute.

General Electric GPU-2/A: This is the standard podded installation of the M197, and is often called the **Lightweight Gun Pod**. The pod was designed for use by helicopters and a wide variety of tactical aircraft fitted with NATO standard 14-in (356-mm) suspension lugs. The weapon is fixed to fire along the flightpath and is aimed with the carrying aircraft's standard sighting system. The cannon has selectable 750- or 1,500-round per minute rates, and is located in the tapered nose of the pod, with its barrels protruding through a flash diffuser over which is the pod's projecting upper lip that further reduces the flash visible to the crew.

Behind the cannon is the electric drive assembly (which also has a reverse-clearing function), linkless single-ended feed system (spent cases being ejected downward out of the pod) and magazine drum, which holds 300 rounds inserted by means of a de-linking loader. At the rear of the pod is the control unit and rechargeable nickel/cadmium battery that can fire three full loads of ammunition before it must be recharged. The pod's characteristics include empty and loaded weights of 417.4 and 585.9 lb (189.3 and 266.0 kg) respectively, a length of 128 in (3251 mm) and a diameter of 19 in (483 mm). The recoil load of the system is 4,000 lb (1814 kg) peak maximum and 2,300 lb (1043 kg) peak recurrent.

USA

General Electric M214

Type: aircraft six-barrel machine-gun (air-to-surface)
Calibre: 5.56 mm (0.223 in)
Operating method: General Electric 'Gatling' operation with 3.2-hp (2.4-kW) external or hydraulic power supply, and mechanical ignition
Feed: linkless
Overall dimensions: length 27 in (686 mm)
Barrel length: data not available
Rifling: 4-groove right-hand twist
Weight: complete weapon 33 lb (14.97 kg) with drive motor
Recoil force: 100 lb (45.4 kg) at 4,000 rounds per minute and 220 lb (99.8 kg) at 10,000 rounds per minute
Muzzle velocity: 3,250 ft (991 m) per second
Rate of fire: variable up to 10,000 rounds per minute
Effective range: data not available
Ammunition types: M193 ball
Ammunition weights: complete round (no data available); projectile 0.14 oz (4 g)
Ammunition length: complete round 2.1 in (53.5 mm)

Variant and application
M214: This is the smallest of the General Electric 'Gatling' weapons, and is in essence a 7.62-mm (0.3-in) Minigun translated into 5.56-mm (0.223-in) calibre, offering approximately double the ammunition capacity of the Minigun within the same volume and weight limitations. The weapon has the highest firing rate of any weapon ever developed, but can also fire a burst of only three rounds. In so small a weapon reverse clearance is not feasible, and the system is designed in such a fashion that when release of the trigger halts the linkless feed system immediately, the few chambered rounds are fired off. The weapon has yet to find an application, but given the fact that the M214 can be used for economical firing rates of between 400 and 6,000 rounds per minute (and yet offer great reliability, maintainability and safety without special features) it will surely find a number of uses. General Electric has developed a podded installation weighing less than 100 lb (45.4 kg) complete with 1,500 rounds, making this a most attractive addition for even the lightest helicopter.

USA

Maremont M60D

Type: aircraft single-barrel machine gun (air-to-surface)
Calibre: 7.62 mm (0.3 in)

Operating method: gas operation with mechanical ignition
Feed: left-hand linked belt
Overall dimensions: length 43.5 in (1105 mm) with butt
Barrel length: 22 in (559 mm), or 73.33 calibres
Rifling: 4-groove right-hand twist
Weight: complete weapon 23.17 lb (10.51 kg); barrel 8.245 lb (3.74 kg)
Recoil force: data not available
Muzzle velocity: 2,805 ft (855 m) per second
Rate of fire: 550 rounds per minute
Effective range: 1,500 yards (1370 m)
Ammunition types: NATO standard types
Ammunition weights: complete round 1.32 oz (37.5 g); projectile 0.51 oz (14.4 g)
Ammunition length: complete round 2.8 in (71.2 mm)

Variants and applications

M60C: This is the version of the M60 infantry machine-gun developed for fixed use in light aircraft and helicopters. The type is basically similar to the ground machine-gun in all essential respects, but lacks the butt and is fitted with a firing solenoid. An enormous number of approved and field installations were effected during the Vietnam War with the M60 series to provide helicopters with a light armament capability. Typical of the approved installations is the **Emerson Electric M6** used on the UH-1B and UH-1C versions of the Bell 'Huey' helicopter, and comprising four M60C machine-guns (two on each of two powered mountings projecting from the sides of the helicopter) with a total of 6,000 rounds of standard belted ammunition fed from magazines in the lower fuselage: the guns can be traversed from 12° inboard to 70° outboard, and elevated from +11° to −63°, under the control of the pilot or co-pilot using reflex sights; there is an electro-mechanical cut-out device to prevent any of the guns from firing into the aircraft at the inboard limit of their traverse. A similar system is the **Emerson Electric M16** which adds a pair of M158 rocket launchers for a total of 14 2.75-in (69.85-in) rockets: these latter are fixed to fire along the helicopter flightpath, and are fired by the pilot; the machine-guns are cut out of the firing circuit when the rockets are being fired. Another use of the M60C was the pioneering **Emerson Electric TAT-101** chin turret for the UH-1E: this system introduced the Tactical Armament Turret (TAT) series, and is a self-contained hydraulically-powered system with the two M60C guns and ammunition contained within the electrically-controlled turret, which can be traversed a total of 115° (later modified to 180°) and provide the guns with elevation from +25° to −45°.
M60D: This is the version of the M60 designed for pintle mounting in helicopter doorways, and is fitted with with a special feed adapter. A typical installation is the **Emerson Electric M23** for use with the UH-1 helicopter: the whole system weighs 210 lb (95.3 kg) with ammunition and two M60D machine-guns (on in each doorway), and allows traverse between 2° inboard and 178° outboard, and elevation between +6.5° and −82°. The **Emerson Electric M59** variant of the system adds a pair of 0.5-in (12.7-mm) M2 heavy machine-guns to the armament fit. Other variants of the M23 system are the **Emerson Electric M24** and **M41** for use with the Boeing Vertol CH-47 Chinook (using 200-round ammunition boxes on each gun): the M24 system provides M60D machine-guns in a door on each side of the helicopter's fuselage, and the M41 provides for a single gun firing through the rear ramp of the helicopter. The M41 system weighs 90 lb (40,8 kg) without its 200-round ammunition box, and the machine-gun can be traversed 47° left and right, and elevated from +12.5° to −69°.

McDonnell Douglas Helicopters (Hughes) EX-34

Type: aircraft single-barrel machine-gun (air-to-surface)
Calibre: 7.62 mm (0.3 in)
Operating method: McDonnell Douglas Helicopters (Hughes) 'Chain Gun' operation with 0.3-hp (0.22-kW) external electrical power supply and mechanical ignition
Feed: left-hand de-linked belt
Overall dimensions: length 37 in (940 mm) with flash suppressor; width 6 in (152 mm); height 6.1 in (240 mm)
Barrel length: 22 in (559 mm), or 73.33 calibres
Rifling: data not available
Weight: complete weapon 30 lb (13.6 kg)
Recoil force: data not available
Muzzle velocity: 2,810 ft (856 m) per second
Rate of fire: variable up to 600 rounds per minute
Effective range: 1,000 yards (915 m)
Ammunition types: NATO standard types
Ammunition weights: complete round 1.32 oz (37.5 g); projectile 0.51 oz (14.4 g)
Ammunition length: complete round 2.8 in (71.2 mm)

Variants and applications

EX-34: This is the smallest of the Hughes-developed 'Chain Gun' weapons, and is characterized by very considerable reliability under all operating conditions. The need to secure an extremely high rate of fire was not considered as important as the reliability factor and an infinitely variable rate of fire so that the firer can satisfy tactical requirements fully yet economize on ammunition consumption. The basic 'Chain Gun' principle removes gun operation from the realm of tapped gun gas (which is subject to inconsistency factors derived from fouling, varying pressures etc) to an external power supply. But unlike the General Electric 'Gatling' weapons, which operate on a rotary principle with multiple barrels for very high rates of fire, the Hughes weapons use a chain running round a rectangular track for movement and positive control of the feed system, bolt carrier and bolt throughout the entire cycle of stripping a fresh round out of the belt, ramming the round, locking the bolt, firing the round, unlocking the bolt, extracting the spent case, ejecting the spent case and then starting the cycle over again. A distinct advantage of the 'Chain Gun' action is that it provides considerable 'dwell' periods at the firing and ejecting/stripping stages of the cycle, which contributes to safety in the former stage and to minimum build up of noxious gases. The machine-gun's reliability is attested by the quoted figure of 50,000 mean rounds between failures, and the 'Chain Gun' operating system also provides for a time-to-rate figure of just 0.15 second.
McDonnell Douglas Helicopters (Hughes) HGS-55: This is the primary helicopter application of the EX-34, and is based on the M27 system (exported as the HGS-5) to provide the McDonnell Douglas Helicopters (Hughes) Model 500 and Model 530 Defender series of light military helicopters with a powerful area-suppression capability. The system is fixed to the port side of the fuselage and fixed in azimuth to fire directly forward. The gun is controlled by the co-pilot using the M70 reflex sight, and can be elevated from +10° to −24° for precise control of trajectory with the aid of an optical beam-splitting rangefinder mechanically linked into the fire-control system. The machine-

gun is fed from a 2,000-round magazine in the fuselage, and the complete system weighs 90 lb (40.8 kg) empty and 218 lb (98.8 kg) loaded.
Royal Ordnance 'Chain Gun' pod: In the UK, Royal Ordnance is developing a podded installation of one EX-34 machine-gun (or optionally two such weapons) for tactical aircraft fitted with NATO standard 14-in (356-mm) suspension lugs. Some 2,000 rounds of ammunition are carried in the pod, which is trimly streamlined for carriage by higher-speed aircraft as well as helicopters.

McDonnell Douglas Helicopters (Hughes) M129 USA

Type: aircraft single-barrel grenade launcher (air-to-surface)
Calibre: 40 mm
Operating method: externally-powered electric motor
Feed: linked belt
Overall dimensions: data not available.
Barrel length: data not available
Rifling: data not available
Weight: complete weapon 45.0 lb (20.4 kg)
Recoil force: data not available
Muzzle velocity: 790 ft (240 m) per second
Rate of fire: 230 to 450 rounds per minute
Effective range: 2,400 yards (2195 m)
Ammunition types: M384 HE and M430 HEDP
Ammunition weights: data not available
Ammunition length: data not available

Variants and applications
M129: This is the US Army's high-velocity grenade launcher, an unusual but highly effective 'gun' able to fire its various grenade payloads with considerable accuracy to a useful range. The weapon was derived closely from the original **M75** grenade launcher (able to fire grenades to a range of 1,875 yards/1715 m at the rate of 230 rounds per minute), and is also used in the Emerson Electric M28 undernose turret system in partnership with the M134 Minigun: the launcher is located to the left of the Minigun with elevation limits of +17.5° to −50°, and is supplied with 300 rounds of ammunition.
M94: This is a pintle-mounted installation for two M75 or M129 grenade launchers in the left- and right-hand doorways of battlefield helicopters. The launchers can be fixed to fire directly forwards under control of the pilot, or can be individually operated as flexible weapons with traverse from 0° forward to 135° aft, and elevation between +3.5° and −60°. Each launcher is provided with a 400-round ammunition box, and the weight of the complete installation is 1,190 lb (539.8 kg) without ammunition.
Emerson Electric M5: This was the first grenade-launcher turret developed for US Army battlefield helicopters, and was initially used by the UH-1B, UH-1C and UH-1M variants of the Bell 'Huey' light helicopter. The installation is a ball turret cantilevered in front of the helicopter's nose, and fitted with an M75 grenade launcher. This has a rate of fire of 230 rounds per minute out to a range of 1,876 yards (1715 m). The co-pilot/gunner uses his reflex sight to aim the installation, the turret being capable of traverse

60° left and right of the centreline and the launcher being capable of elevation from +15° to −35°. The ammunition supply for the system is of a double nature using an ammunition booster, and comprises a rotary magazine with 156 rounds and a box magazine with 300 rounds.

Emerson Electric M120: This is an armament turret designed for the Bell AH-1 HueyCobra series of close-support/attack helicopters. The turret is intended to accommodate one M129 grenade launcher plus 300 rounds of ammuntion, and one gun (7.62-mm/0.3-in M60C machine-gun, or 7.62-mm/0.3-in M134 Minigun, or 20-mm M197 or 30-mm M188).

McDonnell Douglas Helicopters (Hughes) M8: This is the helicopter installation associated with the M129 grenade launcher, and is closely related to the M27, HGS-5 and HGS-55 machine-gun installations in being an M6 quick-release unit for attachment to the port side of the helicopter. The launcher is attached via an effective hydro-spring system to reduce recoil loads, and though it is fixed in azimuth it can be elevated between +10° and −24° at the rate of 25° per second for optimum shaping of the trajectory in conjunction with the M70 reflex sight and associated optical beam-splitting rangefinder. The launcher is fed (without any need for boosting) from a quickly-changed magazine holding 156 spiral-wound linked rounds. The system weighs 120 lb (54.4 kg) empty and 238 lb (107.95 kg) loaded.

USA

McDonnel Douglas Helicopters (Hughes) M230 E1

Type: aircraft single-barrel cannon (air-to-surface and air-to-air)
Calibre: 30 mm
Operating method: McDonnell Douglas Helicopters (Hughes) 'Chain Gun' operation with 6.5-hp (4.8-kW) external electrical power supply and electrical ignition
Feed: left-hand linkless
Overall dimensions: length 64.5 in (1638 mm) without the muzzle brake, and 74.4 in (1889 mm) with the muzzle brake; width 10 in (254 mm); height 11.5 in (292 mm)
Barrel length: data not available
Rifling: right-hand twist
Weight: complete weapon 123 lb (55.8 kg); barrel 35 lb (15.9 kg)
Recoil force: data not available
Muzzle velocity: 2,600 ft (792 m) per second
Rate of fire: variable up to 650 rounds per minute
Effective range: data not available
Ammunition types: M788 TP, M789 HEDP, M799 HEI and all standard Aden and DEFA types (APHEI-SD, HEI, HEI-SD, TP and TP-T)
Ammunition weights: (DEFA range) complete round 449 g (15.84 oz) for HEI-SD, 450 g (15.87 oz) for HEI, TP and TP-T, and 472 g (16.65 oz) for APHEI-SD; projectile 243 g (8.57 oz) for HEI-SD, 244 g (8.61 oz) for HEI, TP and TP-T, and 270 g (9.52 oz) for APHEI-SD
Ammunition weights (Aden range) complete round 17.46 oz (495 g) for AP and API, 17.5 oz (496 g) for HE, and 15.52 oz (440 g) for TP; projectile 7.76 oz (220 g)
Ammunition length: (DEFA range) complete round 200 mm (7.87 in)
Ammunition length: (Aden range) complete round 7.87 in (200 mm)

Variants and applications

M230E1: This powerful single-barrel cannon was developed in parallel with the General Electric M188 in a competitive programme designed to provide the barrelled armament for the US Army's Advanced Attack Helicopter competition (eventually won by the Hughes AH-64 from the Bell AH-63). The weapon entered development as the **M230** 20-mm weapon, it being thought in the late 1960s and early 1970s that 30 mm was an uneconomic calibre for a helicopter-carried cannon intended primarily for the ground-attack role. The M230 was the first cannon to make use of the excellent Hughes-developed 'Chain Gun' operating principle (see entry on EX-34 above) for positive control of all phases of the loading, firing and clearing cycle. Among the many advantages of the system are a fully variable rate of fire, a simplified operating cycle using an open bolt without any need for chargers, declutching feeders and other such devices, and a high feed system pull eliminating all need for a booster. From 1976 the M230 was recast as the M230E1 to use standard 30-mm Aden and DEFA ammunition (and their M780 series US counterparts). The result is a weapon of ruthless efficiency and mechanical reliability, the latter being attested by the quoted figures of 50,000 mean rounds between failures. The weapon accelerates to full firing rate in 0.2 second, and decelerates to a halt in 0.1 seconds.

McDonnell Douglas Helicopters (Hughes) Area Weapon Subsystem (AH-64A): This is the most important of the current applications of the M230E1 cannon, and is designed specifically to match the AH-64A Apache attack helicopter. The cannon is installed in a wholly exposed underfuselage 148-lb (67.1-kg) Lightweight Turret. The mounting is powered by the helicopter's hydraulic system to provide traverse between 100° left and right of the helicopter's centreline, and the cannon can be elevated between the stowed position of +11° and −60°. The ammunition tank is a semi-permanent fixture in the lower fuselage of the helicopter (directly under the main rotor to present no problems of centre-of-gravity movement as the ammunition is used) and is charged by means of a de-linking loader which can install a full complement of 1,200 rounds in a mere 10 minutes. The linkless feed system is of the double-ended (closed-loop) variety, taking fresh rounds from the ammunition tank along the port side of the helicopters and returning spent cases along the starboard side.

The system was designed with the possibility of crash landings in mind, and in the event of a heavy descent the whole cannon installation is designed to be shunted straight up between the two cockpits. The weights of the complete Area Weapon Subsystem are 585 lb (265.4 kg) empty and 1,509 lb (684.5 kg) loaded. The primary role of the Area Weapon Subsystem is fire-suppression and the destruction of light armoured vehicles, but a significant secondary role is the destruction of enemy helicopters. In both the air-to-surface and air-to-air roles the Apache's sensors and fire-control system are all-important for acquisition and accurate designation of the correct targets. The primary instrument in this highly complex system is the Target-Acquistion Designation Sight (TADS). This is a gyro-stabilized sight unit for day/night target acquisition, and is located in a chin turret. The whole unit can be traversed 120° left and right of the centreline as well as elevated from +30° to −60°, both at the rate of 60° per second. For daylight use the TADS includes direct-view optics, a daylight TV camera, a laser rangefinder and designator, a laser spot tracker and an automatic tracking function, while for night use it has a forward-looking infra-red sensor. This provides data to the co-pilot/gunner (through his cockpit displays or the Integrated Helmet And Display Sight System) which slaves the weapon to the operator's line of sight), allowing him to designate targets which are then engaged by the cannon via the fire-control system, the latter providing the necessary range, bearing and other information.

The operation of the gun and feed arrangements are controlled by a Lear Siegler system that re-aims the gun ten times per second to compensate for recoil effects. The result is great accuracy to the maximum effective range of the cannon. The pilot can also aim and control the cannon by means of video information from the TADS, but generally by using his IHADSS and the Pilot's Night-Vision Sensor (PNVS): this is a small gyro-stabilized forward-looking infra-red sensor mounted above the TADS unit in the nose, trainable 90° left and right of the helicopter centreline and capable of elevation between +20° and −45° at the rate of 120° per second.

McDonnell Douglas Helicopters (Hughes) Area Weapon Subsystem (AH-1S): A derivative of the Apache's AWS has been developed for the AH-1S version of the Bell HueyCobra: this has 426 rounds of ammunition, and the weights of the complete outfit are 430 lb (195.0 kg) empty and 749 lb (339.75 kg) loaded.

McDonnell Douglas Helicopters (Hughes) HGS-30: McDonnell Douglas Helicopters has also developed a light underfuselage fitting for its Model 500 and Model 530 Defender series of light military helicopters. This installation is similar to the HGS-55 in that the cannon is fixed in azimuth but controllable in elevation. The M230 E1 cannon is reduced in firing rate to 300 rounds per minute to reduce recoil forces and permit ammunition economies. It is provided with a tank containing 200 rounds which are supplied to the cannon by a single-ended feed system that allows spent cases to be ejected overboard. The weights of the system are 240 lb (108.9 kg) empty and 395 lb (179.2 kg) loaded.

USA

McDonnell Douglas Helicopters (Hughes) Mk 11 Mod 5

Type: aircraft twin-barrel cannon (air-to-air and air-to-surface)
Calibre: 20 mm
Operating method: combination of gas and recoil operation, and electrical ignition
Feed: left- or right-hand linked belt to an eight-chamber revolver cylinder
Overall dimensions: length 70.5 in (2776 mm); without the loader, and 78.5 in (1994 mm) with the loader; width 10.62 in (418 mm); height 14 in (356 mm)
Barrel length: 56.5 in (1435 mm) each, or 71.76 calibres
Rifling: none
Weight: complete weapon 240 lb (108.9 kg) with 45-lb (20.4-kg) loader; barrels 51 lb (23.1 kg) each
Recoil force: 8,000 lb (3629 kg) peak and 2,500 lb (1134 kg) average
Muzzle velocity: data not available
Rate of fire: 700 or 4,200 rounds per minute
Effective range: 1,050 yards (960 m)
Ammunition types: M50 series (including M53A1 API, M54 HPT, M55A1 Practice, M56 HE and M56A1 HEI) or M100 series
Ammunition weights: (M50 series) complete round 9.07 oz (257 g); projectile 3.56 oz (101 g) for API, HE and HEI
Ammunition length: (M50 series) complete round 6.61 in (168 mm)

Variant and application

Mk 11 Mod 5: This unusual weapon was developed in the mid-1950s, and comprises two vertically-arranged smooth-bore barrels (one Mk 19 and one Mk 20), a breech assembly, a revolving cylinder assembly, a receiver assembly and a Mk 2 loader. The loader draws rounds from linked belts entering from each side, and the operation simultaneously rams two rounds, fires the rounds at the 12 and 6 o'clock positions, and ejects two spent cases.

McDonnell Douglas Helicopters (Hughes) Mk 4: This is the sole application of the twin-barrel Mk 11 cannon, a pod designed specifically for use by US Navy tactical aircraft fitted with NATO standard 14- or 30-in (356- or 762-mm) suspension lugs. The pod contains the appropriate control unit, one Mk 11 cannon and 750 rounds of ammunition, the last in a rotary magazine at the rear of the pod and fed forward through dual feed chutes. The pod weighs 787 lb (357 kg) empty and 1,389 lb (630.1 kg) loaded with a full complement of ammunition, which weighs 593 lb (269.0 kg).

Mauser BK 27

West Germany

Type: aircraft cannon (air-to-air and air-to-surface)
Calibre: 27 mm
Operating method: gas operation with pyrotechnic re-cocking and electrical ignition
Feed: left- or right-hand linked belt to a five-chamber revolver cylinder
Overall dimensions: length 2310 mm (7.58 ft); width 2960 mm (11.65 in); height 246 mm (9.69 in)
Barrel length: 1700 mm (66.93 in) or 62.96 calibres
Rifling: right hand twist
Weight: complete weapon 100 kg (220.5 lb)
Recoil force: 2855 kg (6,294 lb) with Tornado blast deflector, or 920 kg (2,028 lb) with Alpha Jet damping system
Muzzle velocity: 1025 m (3,363 ft) per second
Rate of fire: selectable 1,700 or 1,000 rounds per minute; other rates are also possible
Effective range: data not available
Ammunition types: AP, APHE, APHE-SD, HE, HE-SD, TP and TP-FP
Ammunition weights: complete round 516 g (18.2 oz); projectile 260 g (9.17 oz)
Ammunition length: complete round 243 mm (9.57 in)

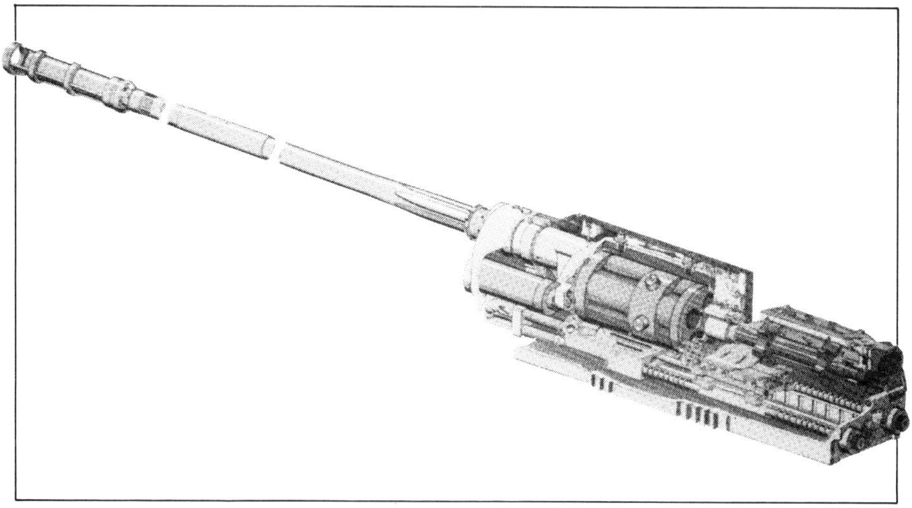

Using belted ammunition and the latest developments in revolver action, the Mauser BK 27 is a remarkably effective weapon. Its primary disadvantage is its odd calibre, adding further to the logistic headache already faced by NATO forces.

Variant and application

BK 27: This is a powerful dual-role weapon currently installed in the Dassault-Breguet/Dornier Alpha Jet A (podded installation under the fuselage), the Panavia Tornado (internal installation in the lower fuselage) and the Saab-Scania JAS 39 Gripen (internal installation in the lower fuselage). Notable features of the weapon are its high rate of fire and high muzzle velocity. The ballistically-matched projectiles of various types are fitted with a fuse system that operates effectively at high angles of impact. Less fortunate, however, is the choice of calibre, for the round used by this important cannon is 3 mm smaller than the standard 30-mm Aden/DEFA round, and 2 mm larger than the new high-powered 25-mm round selected for the GAU-12/U and Aden 25. This fact adds yet more complexity to NATO's already chaotic logistic infrastructure.

West Germany

Mauser MK 30F

Type: aircraft cannon (air-to-air and air-to-surface)
Calibre: 30 mm
Operating method: gas operation with electric re-cocking and electrical ignition
Feed: individual left- or right-hand, or dual linked belt feed
Overall dimensions: length 3350 mm (131.89 in) with the muzzle brake, or 3220 mm (130.71 in) without the muzzle brake; width 205 mm (8.07 in) with feed mechanism; height 274 mm (10.79 in) with feed mechanism
Barrel length: 2458 mm (96.77 in) without the muzzle brake, or 81.93 calibres
Rifling: right-hand twist
Weight: complete weapon 143 kg (315.26 lb); barrel 57.5 kg (126.76 lb)
Recoil force: 1835 kg (4,045 lb) peak and 1630 kg (3,595 lb) recurrent
Muzzle velocity: 1100 m (3,609 ft) per second
Rate of fire: 800 rounds per minute
Effective range: 2000 m (2,185 yards)
Ammunition types: API, HEI and TP (Mauser types), or APDS-T, HE-SD-T and TP-T (Aerojet General types)
Ammunition weights: (Mauser types) complete round 1422 g (50.16 oz) for HEI and TP; projectile 370 g (3.05 oz) for HEI and TP, and 390 g (13.75 oz) for API
Ammunition weights: (Aerojet General types) complete round 707 g (24.94 oz) for APDS-T, and 811 g (28.61 oz) for HEI-SD-T and TP-T; projectile 252 g (8.89 oz) for APDS-T, and 363 g (12.8 oz) for HEI-SD-T and TP-T
Ammunition length: complete round 290 mm (11.416 in)

Variant and application

MK 30F: This is an extremely powerful, though heavy, weapon designed originally as the armament of self-propelled anti-aircraft vehicles, but aerial applications are also possible and planned. The weapon can also fire potent 30 × 173 ammunition developed from that used in the General Electric GAU-8/A Avenger cannon: this latter gives a flight time of 1.87 seconds to 2000 m (2,185 yards) with the APDS-T projectile fired at a muzzle velocity of 1250 m (4,100 ft) per second, and 2.7 seconds to 2000 m (2,185 yards) with the HEI-SD-T projectile fired at a muzzle velocity of 1100 m (3,609 ft) per second. The cannon has the option of left- or right-hand single feeds, or a twin-feed system to allow the selection of two ammunition types for maximum tactical effect.

Rheinmetall MG3

West Germany

Type: aircraft machine-gun (air-to-surface)
Calibre: 7.62 mm (0.3 in)
Operating method: short recoil
Feed: left-hand linked belt
Overall dimensions: length 1097 mm (43.19 in) without butt
Barrel length: 565 mm (22.24 in) with extension, or 74.15 calibres
Rifling: polygonal right-hand twist
Weight: complete weapon 10.5 kg (23.15 lb); barrel 1.8 kg (3.97 lb)
Recoil force: data not available
Muzzle velocity: 820 m (2,690 ft) per second
Rate of fire: 1,300 rounds per minute
Effective range: 800 m (875 yards)
Ammunition types: NATO standard types
Ammunition weights: complete round 37.5 g (1.32 oz); projectile 14.4 oz (0.51 oz)
Ammunition length: complete round 71.2 mm (2.8 in)

Variant and application
MG3: This is West Germany's standard ground-use medium machine-gun, essentially the MG42 of World War 2 remanufactured to take NATO standard 7.62-mm (0.3-in) ammunition in place of the original 7.92-mm (0.315-in) type. The weapon is widely used by other countries, and is also found on a number of pintle mountings to provide light helicopters with a fire-suppression capability. The weapon's high cyclic rate of fire is offset by the need to change barrels fairly frequently, so a more practical rate is about 250 rounds per minute.

Rheinmetall MK 20 Rh202

West Germany

Type: aircraft cannon (air-to-surface and air-to-air)
Calibre: 20 mm
Operating method: gas operation with electro-hydraulic re-cocking and electrical ignition
Feed: overhead feed from single or double linked belts
Overall dimensions: length 2612 mm (102.83 in); width 110 mm (4.33 in), or 155 mm (6.1 in) with Type 3 single feed, or 241 mm (9.5 in) with Type 2 double feed; height 259 mm (10.2 in) with Type 3 feed or 261 mm (10.28 in) with Type 2 feed
Barrel length: 1840 mm (72.44 in) without the muzzle brake, or 92 calibres
Rifling: data not available
Weight: complete weapon 76.7 kg (169.1 lb) with Type 3 feed, or 83.2 kg (183.4 lb) with Type 2 feed, and barrel 28 kg (61.75 lb)
Recoil force: 750 kg (1,653.5 lb) peak and 550 kg (1,212.5 lb) average
Muzzle velocity: 1050 to 1150 m (3,445 to 3,773 ft) per second depending on ammunition type
Rate of fire: 800 to 1,000 rounds per minute
Effective range: 2000 m (2,185 yards)
Ammunition types: APDS-T, API-T, HEI, HEI-T, TP and TP-T

Ammunition weights: complete round 310 g (10.93 oz) for APDS-T and API-T, or 317 g (11.18 oz) for the other types; projectile 108 g (3.81 oz) for APDS-T, 111 g (3.92 oz) for API-T and 120 g (4.23 oz) for the other types
Ammunition lengths: complete round 212 mm (8.35 in); projectile 92 mm (3.62 in)

Variant and application
MK 20 Rh202: This capable cannon was developed for a number of ground-based applications, and has also been developed for naval AA use, but its sole air application is the **HBS 202** fixed installation under variants of the MBB BO105 helicopter, in which it is aimed with the aid of a reflex sight. The mounting itself weighs 96 kg (211.6 lb), and with the gun and 525 rounds the complete system thus turns the scales at 354.2 kg (780.9 lb). The API-T projectile can penetrate 32 mm (1.26 in) of armour at a range of 1000 m (1,095 yards) at an incidence of 0°, decreasing to 24 mm (0.95 in) at an incidence of 30° and 8 mm (0.315 in) at an incidence of 60°.

BOMBS

Avibras bombs

Brazil

Avibras produces under licence in Brazil a number of general-purpose and incendiary bombs modelled on US originals. These are intended for export as well as home forces' use. The smallest of these is the 127-kg (250-lb) **BAFG-120** general-purpose bomb based on the Mk 81 and measuring 230 mm (9.06 in) in diameter and 1800 mm (70.87 in) in length. Next in size comes the 177-kg (390-lb) **BI-200** incendiary bomb based on the US 400-lb (181-kg) weapon and measuring 420 mm (16.5 in) in diameter and 1810 mm (71.25 in) in length. After this comes the 250-kg (551-lb) **BAFG-250** general-purpose bomb based on the Mk 82 and measuring 270 mm (10.6 in) in diameter and 2200 mm (86.6 in) in length. One step up in size is the 270-kg (595-lb) **BI-250** incendiary bomb based on the American BLU-32 and measuring 400 mm (15.75 in) in diameter and 3460 mm (136.2 in) in length. Finally, there is the 362-kg (798-lb) **BI-375** incendiary bomb modelled on the US BLU-27 and measuring 480 mm (18.9 in) in diameter and 3630 mm (142.9 in) in length.

Cardoen CB 60-K

Chile

Type: light-weight cluster bomb
Dimensions: diameter 253 mm (9.96 in); length 2050 mm (80.71 in) without fuse; fin span 400 mm (15.75 in)

The Avibras range of general-purpose bombs is modelled on the US Mk 81 to Mk 84 series based on the Aero 1A low-drag profile. This is an AV-BAFG-250 227-kg (500-lb) weapon.

Weight: total round 60 kg (132.29 lb)
Warhead: 50 0.74-kg (1.63-lb) bomblets

Variant and application
CB 60-K: This is a comparatively simple yet highly effective cluster bomb, being designed for carriage on NATO standard 356-mm (14-in) or Warsaw Pact standard 250-mm (9.84-in) suspension lugs, and intended for dropping over a wide combination of speeds and altitudes. The weapon's fusing allows delays of between 3 and 18 seconds before the cluster is broken up to generate different elliptical coverage patterns over a maximum area of 30000 m^2 (35,880 sq yards). Each of the bomblets carried by the CB 60-K has a diameter of 48 mm (1.89 in) and length of 360 mm (14.17 in). An armour penetration capability of 150 or 230 mm (5.91 or 9.06 in) is claimed by differing sources. Each bomblet has 150 g (5.29 oz) of explosive containing zirconium for enhanced incendiary effect, and the body also shatters into 500 fragments for secondary capability against personnel and soft-skinned vehicles.

Chile

Cardoen CB 250-K

Type: medium-weight cluster bomb
Dimensions: diameter 446 mm (17.56 in); length 2640 mm (103.94 in) without fuse; fin span 1140 mm (44.88 in)
Weight: total round 245 kg (540.1 lb)
Warhead: 240 0.74-kg (1.63-lb) bomblets

Variant and application
CB 250-K: This is essentially the larger brother of the CB 60-K and, like its companion, is a comparatively simple yet highly effective cluster bomb designed for carriage on NATO standard 356-mm (14-in) or Warsaw pact standard 250-mm (9.84-in) suspension lugs. The weapon is intended for dropping over a wide combination of speeds and altitudes. The bomb's fusing allows delays of between 3 and 18 seconds before the cluster is broken up to generate different elliptical coverage patterns up to a maximum 50000 m^2 (59,800 sq yards). Each bomblet is 48 mm (1.89 in) in diameter and 360 mm (14.17 in) in length. An armour penetration capability of 150 or 230 mm (5.91 or 9.06 in) is claimed by differing sources. Each bomblet has 150 g (5.29 oz) of explosive containing zirconium for enhanced incendiary effect, and the body also shatters into 500 fragments for secondary capability against personnel and soft-skinned vehicles.

Chile

Cardoen CB 500-K

Type: heavy-weight cluster bomb
Dimensions: diameter 497 mm (19.57 in); length 2265 mm (89.17 in); fin span 703 mm (27.68 in)
Weight: total round 454 kg (1,000 lb)
Warhead: 450 0.74-kg (1.63-lb) bomblets

Variant and application

CB 500-K: This is the largest of the cluster bombs produced by Cardoen and, like its smaller companions, a comparatively simple yet highly effective cluster bomb. The weapon is designed for carriage on NATO standard 356-mm (14-in) or Warsaw Pact standard 250-mm (9.84-in) suspension lugs, and intended for dropping over a wide combination of speeds and altitudes. The fusing allows delays of between 3 and 18 seconds before the cluster is broken up to generate different elliptical coverage patterns up to a maximum 80000 m^2 (95,680 sq yards). Each bomblet is 48 mm (1.89 in) in diameter and 360 mm (14.17 in) in length, and armour penetration capability of 150 or 230 mm (5.91 or 9.06 in) is claimed by differing sources. Each bomblet has 150 g (5.29 oz) of explosive containing zirconium for enhanced incendiary effect, and the body also shatters into 500 fragments for secondary capability against personnel and soft-skinned vehicles.

The CB 500-K differs from the CB 250-K is having fixed rather than opening fins. The casing of the weapon is made of glassfibre reinforced with steel frames, and is broken open by a controlled explosion along the external structure. The bomb can be dropped at speeds up to 1050 km/h (652 mph), and Cardoen is offering the type with a new programmable electronic fuse that can open the dispenser after a delay variable from 2 to 99 seconds, so offering the possibility of very low- to very high-altitude delivery. This electronic fuse is offered as a retrofit for the mechanical fuse of the earlier CB 60-K and CB 250-K weapons.

Cardoen general-purpose bombs

Chile

Type	Diameter	Length	Weights Nominal	Weights Actual
Mk 81	229 mm (9 in)	1814 mm (71.4 in)	113 kg (250 lb)	117.9 kg (260 lb)
Mk 82	273 mm (10.75 in)	2207 mm (86.9 in)	227 kg (500 lb)	240.4 kg (530 lb)
Mk 83	356 mm (14 in)	3008 mm (118.42 in)	454 kg (1,000 lb)	446.8 kg (985 lb)
Mk 84			907 kg (2,000 lb)	

Variants and application

Mks 81, 82, 83 and 84: Based in the Chilean capital Santiago, Cardoen produces a series of bombs identical in all essential respects with the US 250-lb (113-kg) Mk 81, 500-lb (227-kg) Mk 82, 1,000-lb (454-kg) Mk 83 and 2,000-lb (907-kg) Mk 84 'slick' bombs. These bombs can be hung on NATO standard 14-in (356-mm) or Warsaw Pact standard 250-mm (9.84-in) suspension lugs, and their tactical flexibility is considerably enhanced by their facility for addition of a retarding device (of the airbrake or ballute variety) or a laser-homing kit of the Texas Instruments Paveway or Matra BLG, or an Elbit Opher infra-red homing kit.

The Mk 81 clone has a fin diameter of 320 mm (12.6 in) and contains 45.4 kg (100 lb) of explosive, while the comparable figures for the Mk 82 are 384 mm (15.1 in) and 87.1 kg (192 lb) respectively, and for the Mk 83 498 mm (19.62 in) and 201.9 kg (445 lb) respectively.

Other free-fall bombs from Cardoen are multi-purpose penetration bombs whose weight has not yet been disclosed, but featuring a 0.25-second delay fuse in the tail, and the **CFB 27-300** high-fragmentation bomb. This latter is a 300-kg (661.4-lb) weapon with proximity and impact fuses, which ensure optimized detonation of the explosive filling

A display rack shows off Cardoen's three copies of US general-purpose bombs: from top to bottom these are the 1,000-lb (454-kg) Mk 83, the 500-lb (227-kg) Mk 82 and the 250-lb (113-kg) Mk 81.

to shatter and disperse the multiple layers of pre-fragmented steel inside a tempered steel casing.

Chile

Cardoen PJ-1

Type: light-weight anti-personnel bomb
Dimensions: diameter 60 mm (2.36 in); length 500 mm (19.69 in); fin span 85 mm (3.35 in)
Weight: total round 2.4 kg (5.29 lb)
Warhead: 0.8-kg (1.76-lb) hollow-charge fragmentation

Variant and application

PJ-1: This is a neat hand-launched anti-personnel weapon designed to be dropped from its container mainly from light aircraft, helicopters and even transport aircraft. The warhead is contained in a pre-fragmented casing that bursts into 110 fragments, lethal within a radius of 25 m (82 ft) and highly damaging within a radius of 50 m (165 ft). The nose has a copper cup over the explosive charge, giving the PJ-1 useful capability against light armoured vehicles, and the basic variant can be supplied with zirconium fragments for an enhanced incendiary effect.

FAMAE 3.5-kg fragmentation bomb

Chile

Type: light-weight anti-personnel bomb
Dimensions: diameter not available; length 428 mm (16.89 in)
Weight: 3.5 kg (7.72 lb)
Warhead: 0.4-kg (0.88-lb) HE fragmentation

Variant and application
3.5-kg bomb: This is a small anti-personnel bomb, and is intended primarily for light aircraft use. The warhead casing shatters into lethal fragments on detonation, and the efficiency of the weapon is improved by the incorporation of a package of steel balls in the core of the charge.

FAMAE general-purpose bombs

Chile

Little detailed information has been vouchsafed about the FAMAE series of general-purpose bombs, which are available in 250-, 500-, 750 and 1,500-lb (113, 227, 340- and 680-kg) nominal sizes. These weapons have nose and tail fuses, the former being plugged when the bombs are used in the anti-armour role.

Ferrimar WB-500-F Avispa

Chile

Type: medium-weight cluster bomb
Dimensions: diameter 442 mm (17.4 in); length 2490 mm (98.03 in) without fuse; fin span 630 mm (24.8 in)
Weight: total round 227 kg (500 lb)
Warhead: 248 0.65-kg (1.43-lb) bomblets

The 227-kg (500-lb) Cardoen cluster bomb is a simple but effective and cheap solution to the problem of major armoured attacks.

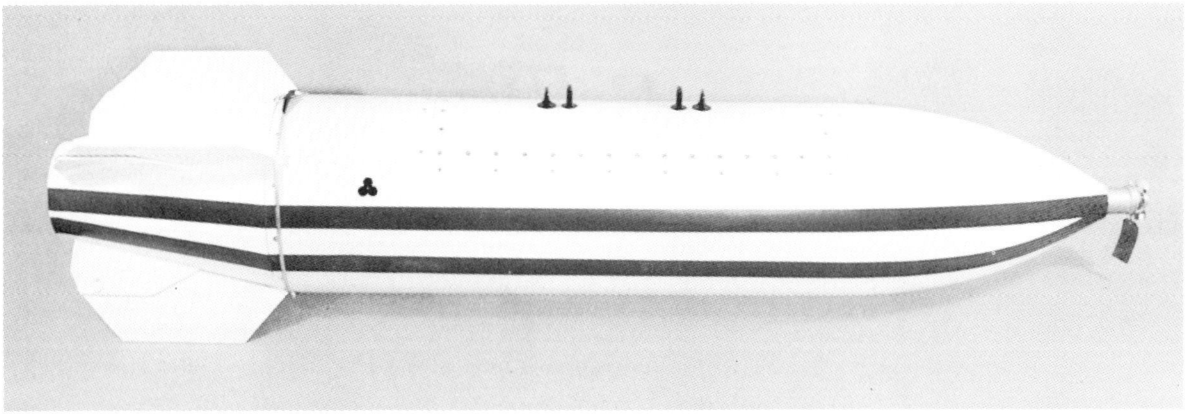

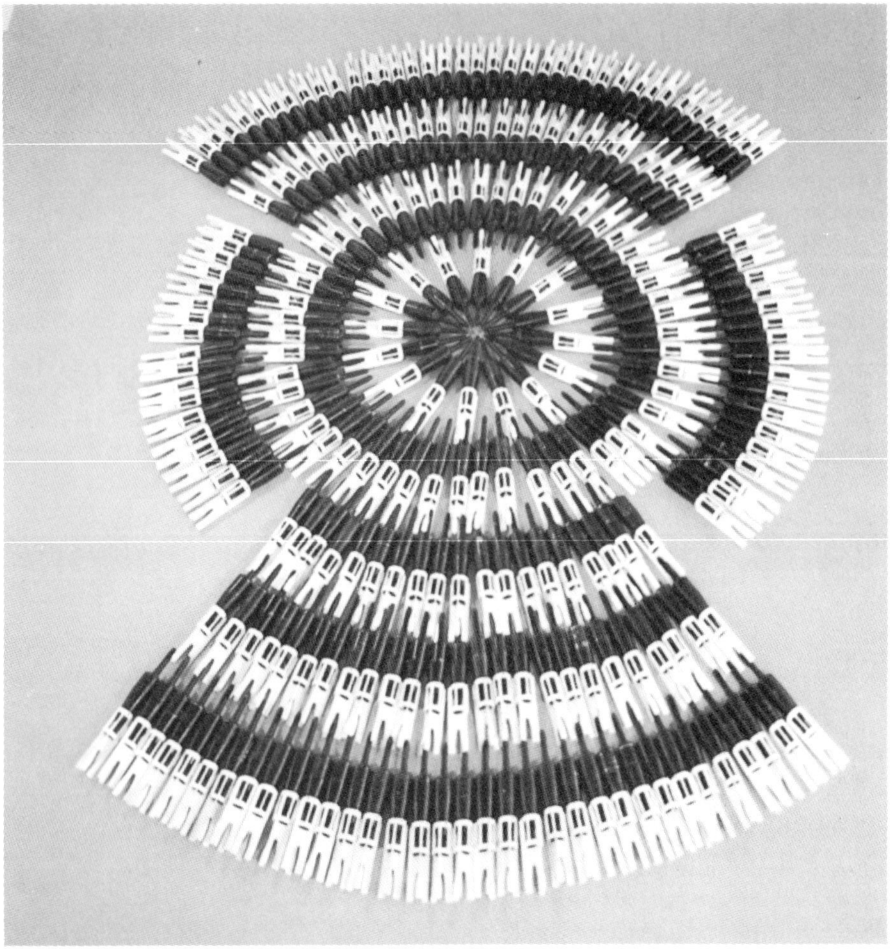

Laid out in a neat geometrical pattern, the bomblets carried by the Cardoen 227-kg (550-lb) cluster bomb give a good indication as to the area that can be saturated by such a weapon.

Variants and application

WB-500-F Avispa: Ferrimar is based in Santiago, and makes a series of cluster bombs in 500-lb (227-kg), 1,500-lb (680-kg) and 2,250-lb (1021-kg) nominal loaded sizes as the WB-500-F, **WB-1500-F** and **WB-2250-F** respectively; the large cluster bombs are made of aluminium alloy, carbon fibre, glassfibre and moulded plastic, and according to size carry between 100 and 248 Avispa (wasp) bomblets. These latter have been produced in diameters ranging from 48 to 53 mm (1.89 to 2.09 in), a length of 380 mm (14.96 in) and a weight of 0.65 kg (1.43 lb), and the bomblets are available in eight types (HEAT, HEAT-AP30, HEAT-AP50, HE Area-Denial, HEAD-AP30, HEAD-AP50, AP-Prox30 and AP-Prox50).

The Ferrimar bombs can be carried on NATO standard 356-mm (14-in) suspension lugs or, in a useful attempt to increase sales potential, Warsaw Pact standard 250-mm (9.84-in) suspension lugs. When dropped, the bombs are spin-rolled by their fins (those of the 227-kg type spanning 1140 mm/44.88 in when sprung open to retard the bomb) to improve the dispersion of the bomblets over an area of 60000 m^2 (71,760 sq yards).

The bomblet of the original series has impact fusing and carries a 165-g (5.82-oz) charge of Pentolite explosive, capable of penetrating between 150 and 210 mm (5.91 and 8.27 in) of armour, while that of the later series carries a 180-g (6.35-oz) charge and can be programmed for detonation delayed by an interval of anything between 30 seconds and 72 hours, so offering an area-denial capability. Improved capability is to be added by the development of an anti-submarine type of larger size and weight. The electronic fuse of the cluster bomb must be programmed before take-off with a delay between 1.2 and 18 seconds, though a fuse programmable in flight is under development.

NORINCO 100-kg Type 2 — China

Type: light-weight general-purpose bomb
Dimensions: diameter 280 mm (11.02 in); length 1080 mm (42.52 in) without nose fuse; fin span 311 mm (12.24 in)
Weight: total round 140 kg (308.6 lb) without fuse
Warhead: 39.8 kg (87.74 lb) HE

Variant and application
100-kg Type 2: This light general-purpose bomb is used by the whole spectrum of Chinese bombers, attack aircraft and fighter-bombers, and can most charitably be described as an obsolescent weapon. The 100-kg Type 2 is a blast fragmentation bomb, has single-point suspension, and takes 21.44 seconds to reach the ground from 8000 m (26,245 ft).

NORINCO 250-kg Type 2 — China

Type: light-weight cluster bomb
Dimensions: diameter 325 mm (12.8 in); length 1520 mm (59.84 in); fin span 410 mm (16.14 in)
Weight: casing 74 kg (163.14 lb)
Warhead: see below

Variant and application
250-kg Type 2: This is a simple dispenser (cluster) weapon designed to open at a preset altitude and strew its bomblets over a large area. The weapon is used by all of China's tactical aircraft, and has three basic loadings: for an overall weight of 200 kg (440.92 lb) the weapon can hold 28 5-kg (11.02-lb) anti-personnel and anti-vehicle fragmentation bomblets, or for a weight of 197 kg (434.3 lb) a total of 42 2.5-kg (5.51-lb) anti-personnel

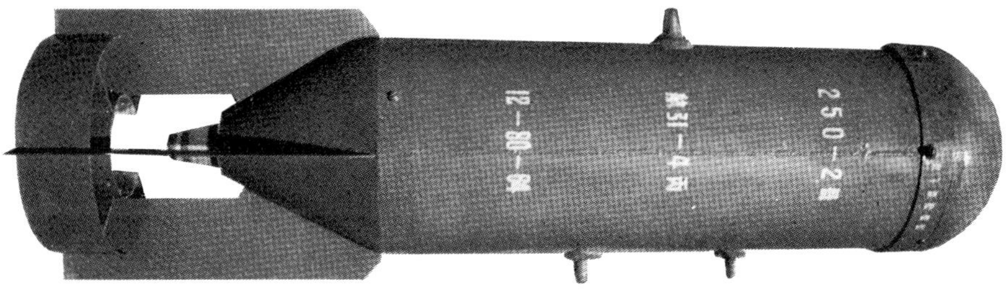

Simple yet effective, the Chinese Sub-Munition Dispenser Type 2 carries between 26 and 42 bomblets, and can be carried by most of China's indigenous tactical aircraft types.

bomblets, or for a weight of 125 kg (275.57 lb) a total of 26 2-kg (4.41-lb) anti-tank bomblets. The three varieties are quoted in Chinese promotional literature as having fall times of 20.78, 20.81 and 21.14 seconds respectively, but no information is given about release altitude. No penetration figure is quoted for the anti-tank bomblet.

France

CEA AN-22

Very little is known with certainty about this free-fall nuclear weapon which is obsolescent and used only by the dwindling number of Dassault-Breguet Mirage IVA strategic bombers left to the French air force. The weapon is a highly streamlined design with a cruciform of rear fins, and is believed to have a yield in the order of 60 kilotons.

France

CEA AN-52

Very little is known of this comparatively small nuclear weapon, a free-fall tactical device carried by Dassault-Breguet Mirage IIIEs and SEPECAT Jaguar As of the French air force, and by Dassault-Breguet Super Etendards of the French naval air arm. The weapon is of conventional 'bomb' layout with a cruciform of rear fins, weighs about 600 kg (1,322.75 lb), and has a selectable yield believed to be in the order 14 to 18 kilotons.

France

Matra Durandal

Type: medium-weight anti-runway bomb
Dimensions: diameter 223 mm (8.78 in); length 2700 mm (106.3 in); fin span 430 mm (16.93 in)
Weight: total round 219 kg (482.8 lb)
Warhead: 100 kg (220.5 lb) including 15 kg (33.1 lb) of explosive

Variant and application
Durandal: This important runway-cratering weapon was developed by Matra in conjunction with SAMP and Thomson Brandt, and began to enter service in 1977. The weapon is carried on NATO standard 356-mm (14-in) suspension lugs, and requires no special control equipment. On release from the parent aircraft at any altitude down to 60 m (195 ft) at speeds between 650 and 1020 km/h (404 and 634 mph), the Durandal is retarded by a double parachute and its nose begins to pitch down. When the optimum angle for penetration without the chance of ricochet has been reached (−20° to −30°), the parachute retarder is released and the 9175-kg (20,277-lb) thrust internal rocket motor is ignited. The weapon then accelerates to impact the runway at a velocity of some 260 m (853 ft) per second, sufficient for the forged steel warhead to pierce 400 mm (15.75 in) of concrete. A one-second delay is built into the fusing system, whch detonates the warhead under the runway's surface to form a crater 2 m (6.56 ft) deep and 5 m (16.4 ft) in diameter, with heave extending the diameter of the damage to 7 m (23 ft); fractures further extend the damage area to a total of 250 m^2 (300 sq yards).

The weapon is designed for use in multiple launches for maximum destructive effect, double diagonal attacks each delivering six, nine or ten Durandals being considered the best way of rendering a standard 2400- by 45-m (2,625- by 50-yard) runway inoperative for one day. The Durandal is used by the US Air Force with the designation **CBU-15**, and by another ten export customers.

The Matra Durandal is an effective anti-runway bomb, and has secured useful export sales, including sales to the USA where the weapon is classified as the CBU-15.

Matra/Thomson Brandt BLG 66 Belouga

France

Type: medium-weight cluster bomb
Dimensions: diameter 360 mm (14.17 in); length 3300 mm (129.92 in); fin span 580 mm (22.83 in)
Weight: total round 290 kg (639.33 lb)
Warhead: 181.2 kg (399.5 lb) comprising 152 1.2-kg (2.65-lb) bomblets

Variant and application
BLG 66 Belouga: This is the standard cluster bomb used by the French air force and seven export customers. The type was developed in the 1970s as successor to the Giboulée

cluster bomb, the design objectives being lower drag, a larger ground pattern for the dispensed bomblets, and the ability to be launched at higher speeds (anything up to 1020 km/h; 634 mph). The weapon is delivered as a certified round of ammunition and the cylindrical mid-section of the weapon is filled with 19 rows of eight rearward-facing ejector tubes for 152 66-mm (2.6-in) bomblets (called '*grenades*' by the French). These bomblets are of three types, namely general-purpose fragmentation for attacks on motor transport, parked aircraft and dumps, HE anti-tank for attacks on armoured concentrations, and area interdiction for attacks on fixed sites such as airfields, harbours, marshalling yards and road junctions.

Before releasing the weapon the pilot chooses between two operating patterns, namely a length of 120 or 240 m (395 or 785 ft) by a width of 40 or 60 m (130 or 195 ft). On release down to a minimum altitude of 60 m (195 ft), the sequencer in the weapon's nose deploys a drag chute to retard the bomb and so allow the aircraft to clear the area before the release of the bomblets, which are spewed out at regular intervals to each side of the dispenser. The bomblets are themselves retarded, and reach the ground vertically: the general-purpose bomblet type then detonates into a disc of horizontal fragments, and the HEAT bomblet type explodes downwards on hitting the top of an armoured vehicle, the penetration capability being 250 mm (9.84 in).

France

Thomson Brandt 125-kg retarded bomb

Type: light-weight retarded fragmentation bomb
Dimensions: diameter 195 mm (7.68 in); length not available; fin span 310 mm (12.2 in)
Weight: total round 125 kg (275.6 lb)
Warhead: approx 25 kg (55.1 lb) of HE

Variants and application
125-kg retarded bomb: This is a useful weapon based on the thick-wall Type 18A anti-personnel fragmentation bomb. The special sections can be added to the standard bomb in the field using standard armourer's tools, and comprise a 350-mm (13.8-in) nose probe with associated FP78 fuse, and the 770 mm (30.31 in) retarder tail; this latter is of the parachute type, and spans 640 mm (25.2 in) when fully deployed after the preset interval of between 13 and 17 seconds from bomb release. The bomb then falls vertically, being detonated at the optimum height above the ground by the probe/fuse combination.
250-kg retarded bomb: This is a medium-weight version of the 125-kg (275.6-kg) weapon, and based on the 250-kg (551.1-lb) fragmentation bomb using the same combination of add-on nose probe and retarder tail.

France

Thomson Brandt BAP 100

Type: light-weight anti-runway bomb
Dimensions: diameter 100 mm (3.94 in); length 1780 mm (70.08 in)
Weight: total round 32.5 kg (71.65 lb)

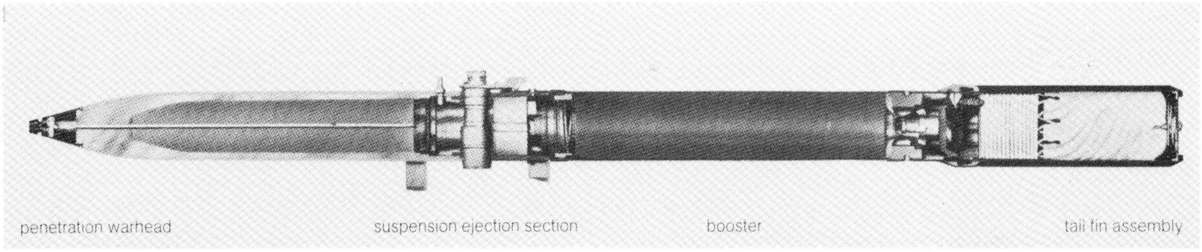

Above *This cutaway illustration of the BAP 100 runway-cratering bomb clearly reveals the retarding parachute, the booster rocket and the penetrating warhead.*

Left *A French air force SEPECAT Jaguar A releases a stick of BAP 100 anti-runway or BAT 120 anti-vehicle bombs, their retarder parachutes deploying almost immediately after each has left the rack.*

Warhead: 18 kg (39.7 lb) including 3.6 kg (7.94 lb) of explosive

Variant and applications

BAP 100: Development of the BAP 100 (**Bombe Accélérée de Pénétration 100 mm** or 100-mm accelerated penetration bomb, otherwise the **Bombe Anti-Piste 100 mm** or 100-mm anti-runway bomb) began in the mid-1970s to provide the French air force (plus a current five export customers) with a lightweight anti-runway weapon as counterpart to the heavier Durandal. The weapon is released as the launch aircraft approaches the target runway on a diagonal course, the launch altitude being between 50 and 80 m (165 and 260 ft), and the launch speed between 630 and 1040 km/h (391 and 646) respectively.

The weapon's onboard sequencer is activated by a condenser after 0.5 second and opens the tail to allow deployment of a braking parachute with the bomb some 3 m (10 ft) below the aircraft. Thereafter the warhead is armed, the motor ignition sequence started, the parachute released and the motor fired to boost the bomb obliquely downward at 65° through the runway's concrete surface at 230/260 m (755/853 ft) per second. The delay fuse allows warhead detonation after any interval up to six hours. The warhead can penetrate 1000 mm (39.4 in) of reinforced concrete, and is claimed to be capable of destroying a 50-m^2 (60-sq yard) area of concrete 300 mm (11.8 in) thick.

Thomson Brandt 14-3-M2: This is the special mutli-bomb carrier system for use by aircraft with NATO standard 356-mm (14-in) suspension lugs, and can carry three rows of three BAP 100 bombs, for a total of nine bombs. The 25-kg (55.1-lb) carrier is 911 mm (35.87 in) long, 380 mm (14.96 in) wide and 87 mm (3.425 in) high, and when fully loaded has a length of 1780 (70.08 in), a width and height of 465 mm (18.31 in), and a weight of 320 kg (705.5 lb). The BAP 100 bomb has its own ejection device, and a full cluster of nine weapons can be loaded in less than five minutes.

A French air force SEPECAT Jaguar A carries the maximum 18 BAP 100 anti-runway bombs on the tandem adapter rack under its fuselage.

Thomson Brandt 30-6-M2: This is the special multi-bomb carrier system for use by aircraft with NATO standard 762-mm (30-in) suspension lugs, and can carry two clusters each of three rows of three BAP 100 bombs, for a total of 18 bombs. The 125-kg (275.6-lb) carrier is 2300 mm (90.55 in) long, 380 mm (14.96 in) wide and 115 mm (4.53 in) high, and when fully loaded has a length of 3700 mm (145.67 in), a width of 465 mm (18.31 in), a height of 495 mm (19.49 in) and a weight of 710 kg (1.565.25 lb). The BAP 100 bomb has its own ejection device , and two full clusters (18 weapons) can be loaded in less than ten minutes. As with the 14-3-M2 adaptor unit, the rear of the adaptor has a special intervalometer, on which are entered (before take-off) the aiming release delay (between 0.16 and 1.6 seconds), the firing rate (between 0.08 and 1 second) and (when more than one adapter is carried) the release distribution of 0.5 or 1 second.

France

Thomson Brandt BAT 120

Type: light-weight anti-armour bomb
Dimensions: diameter 120 mm (4.72 in); length 1500 mm (59.06 in)
Weight: total round 34 kg (74.96 lb)
Warhead: 24 kg (52.9 lb) blast fragmentation including 6 kg (13.2 lb) of explosive

Variant and applications
BAT 120: Conceptual brother of the BAP 100 and used by the French air force plus

several export customers, the BAT 120 (**Bombe d'Appui Tactique 120 mm** or 120-mm tactical attack bomb) has a retarding parachute in the tail but no booster rocket. The warhead is based on that of a Thomson Brandt 120-mm (4.72-in) mortar bomb, and has two sections (one forward and one aft of the central suspension/ejector unit). After release at low altitude and any speed between 650 and 1020 km/h (404 and 634 mph), the bomb is retarded by the parachute and falls at 80° to reach the ground at a speed of 20 m (66 ft) per second after a six-second descent from a launch altitude of 80 m (260 ft).

The weapon is detonated just above the ground by a piezo-electric fuse, and the focussed burst of the front and rear charges produces two nearly horizontal discs of fragments moving at 1200 m (3,937 ft) per second to a radius of 30 m (100 ft).

The BAT 120 comes in two subvariants, namely the **BAT 120ABL** for use against light armour, and the **BAT 120AMV** for use against unarmoured targets. The former sprays 800 12.5-g ((0.44-oz) main and many smaller fragments capable of penetrating 7 mm (0.28 in) of steel over a radius of 20 m (66 ft), and the latter disburses 2,600 4-g (0.14-oz) fragments capable of penetrating 4 mm (0.16 in) of steel over the same radius.

Thomson Brandt 14-3-M2: This is the special multi-bomb carrier system for use by aircraft with NATO standard 356-mm (14-in) suspension lugs, and can carry three rows of three BAT 120 bombs, for a total of nine bombs. The 25-kg (55.1-lb) carrier is 911 mm (35.87 in) long, 380 mm (14.96 in) wide and 87 mm (3.425 in) high, and when fully loaded has a length of 1500 mm (59.06 in), a width and height of 465 mm (18.31 in), and a weight of 330 kg (727.5 lb). The BAT 120 bomb has its own ejection device, and a full cluster of nine weapons can be loaded in less than five minutes.

Thomson Brandt 30-6-M2: This is the special multi-bomb carrier system for use by aircraft with NATO standard 762-mm (30-in) suspension lugs, and can carry two cluster each of three rows of three BAT 120 bombs, for a total of 18 bombs. The 125-kg (275.6-lb)

Like the smaller BAP 100, the BAT 120 can be carried on a tandem adapter rack under the fuselage of the SEPECAT Jaguar A, the bomb type having been combat proven as highly effective against unarmoured motor transport such as the tanker illustrated here.

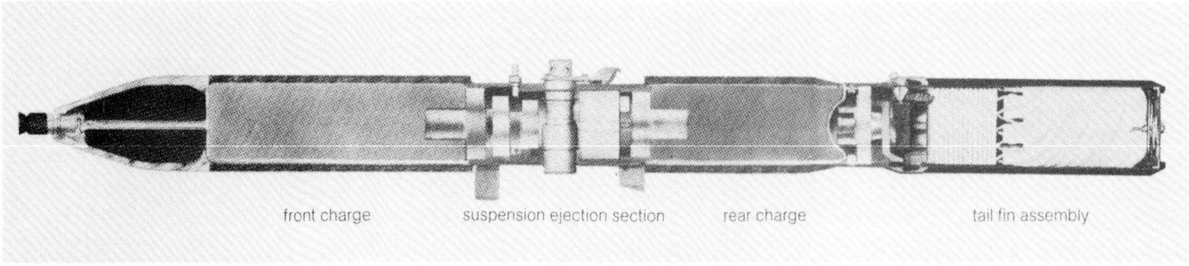

Above *This cutaway illustration of the BAT 120 anti-vehicle weapon clearly shows the two explosive sections, their detonations creating a pair of rapidly expanding fragmentation discs capable of piercing light armoured vehicles.*

Right *The tandem adapter rack for the BAT 120 can also be carried on underwing hardpoints; this BAe Hawk carries 36 such bombs in four nine-round groups.*

Right *The maximum under-fuselage load for the SEPECAT Jaguar A is 18 BAT 120 anti-vehicle bombs. These are individually retarded by a tail-mounted parachute to descend almost vertically, ground impact causing the two explosive sections to detonate and throw out two horizontal discs of evenly spaced fragments capable of disabling vehicles and light armour.*

carrier is 2300 mm (90.55 in) long, 380 mm (14.96 in) wide and 115 mm (4.53 in) high, and when fully loaded has a length of 3375 mm (132.87 in), a width of 465 mm (18.31 in), a height of 495 mm (19.49 in) and a weight of 735 kg (1,620.4 lb). The BAT 120 bomb has its own ejection device, and two full clusters (18 weapons) can be loaded in less than ten minutes. As with the 14-3-M2 adaptor unit, the rear of the adaptor has a special intervalometer, on which are entered (before take-off) the aiming release delay (between 0.16 and 1.6 seconds), the firing rate (between 0.08 and 1 second, and (when more than one adaptor is carried) the release distribution of 0.5 or 1 second. Four BAT 120 bombs can effectively carpet an area 150 m (492 ft) long and 40 m (131 ft) wide.

Thomson Brandt BM 400 — France

Type: heavy-weight modular bomb
Dimensions: diameter 320 mm (12.6 in); length 3300 mm (129.92 in)
Weight: total round 390 kg (859.8 lb)
Warhead: three 100-kg (220.5-lb) parachute-retarded modules

Variants and application
BM 250: This is the two-module lightweight variant of the BM 400, suited to lighter tactical aircraft engaged against smaller target areas.
BM 400: The BM 400 was designed in the late 1970s and early 1980s for the destruction of high-value hard and area targets after release at any speed between 650 and 1100 km/h (404 and 684 mph) at an altitude of 30 m (100 ft). The concept behind the weapon is greater efficiency than a single 'slick' or retarded bomb combined with lower cost than a cluster bomb. The weapon can be released horizontally, or in a loft manoeuvre for a stand-off range of 5000 m (5,470 yards) or 10000 m (10,935 yards) when the weapon is fitted with the optional rocket booster.

The BM 400 consists of a cylindrical 90-kg (198.4-lb) fin-stabilized sleeve containing three explosive modules. As the weapon is released, a windmill-powered sequencer starts the process whereby the modules are parachute-extracted from the rear of the casing at pre-set intervals. Each module has one of two types of pre-fragmented warhead, and drops vertically to detonate at a height of 700 mm (27.6 in) as the nose probe touches the ground.

Of the two warhead types, one is designed to deal with harder targets (concentrations of light armour, ammunition dumps and radar installations) and bursts into some 700 or 800 80-kg (2.82-oz) fragments that fly out in a horizontal 360° disc at an initial velocity of 2000 m (6,562 ft) per second: these fragments are able to penetrate 17 mm (0.67 in) of armour at a radius of 50 m (165 ft), or 12 mm (0.47 in) of armour at 100 m (330 ft).

The other warhead type is designed for softer targets (transport and troop concentrations) and bursts into 1,500 larger and many smaller fragments: the larger fragments can penetrate 12 mm (0.47 in) at a radius of 50 m (165 ft) or 7 mm (0.28 in) at 100 m (330 ft). A spacing of 200 m (220 yards) between the modules allows a single BM 400 to saturate an area 600 m (655 yards) long by 100 m (110 yards) wide with a devastating hail of heavy fragments. The manufacturer is currently working on two important new submunitions for the type, namely the SMABL and the SMAP: the SMABL is derived from the warhead of the BAT 120 bomb, and at a weight of 20 kg (44.1 lb) and length of 670 mm (26.38 in) shatters into 500 fragments; the SMAP is derived from the warhead of the BAP 100, and at a weight of 23 kg (50.7 lb) and length

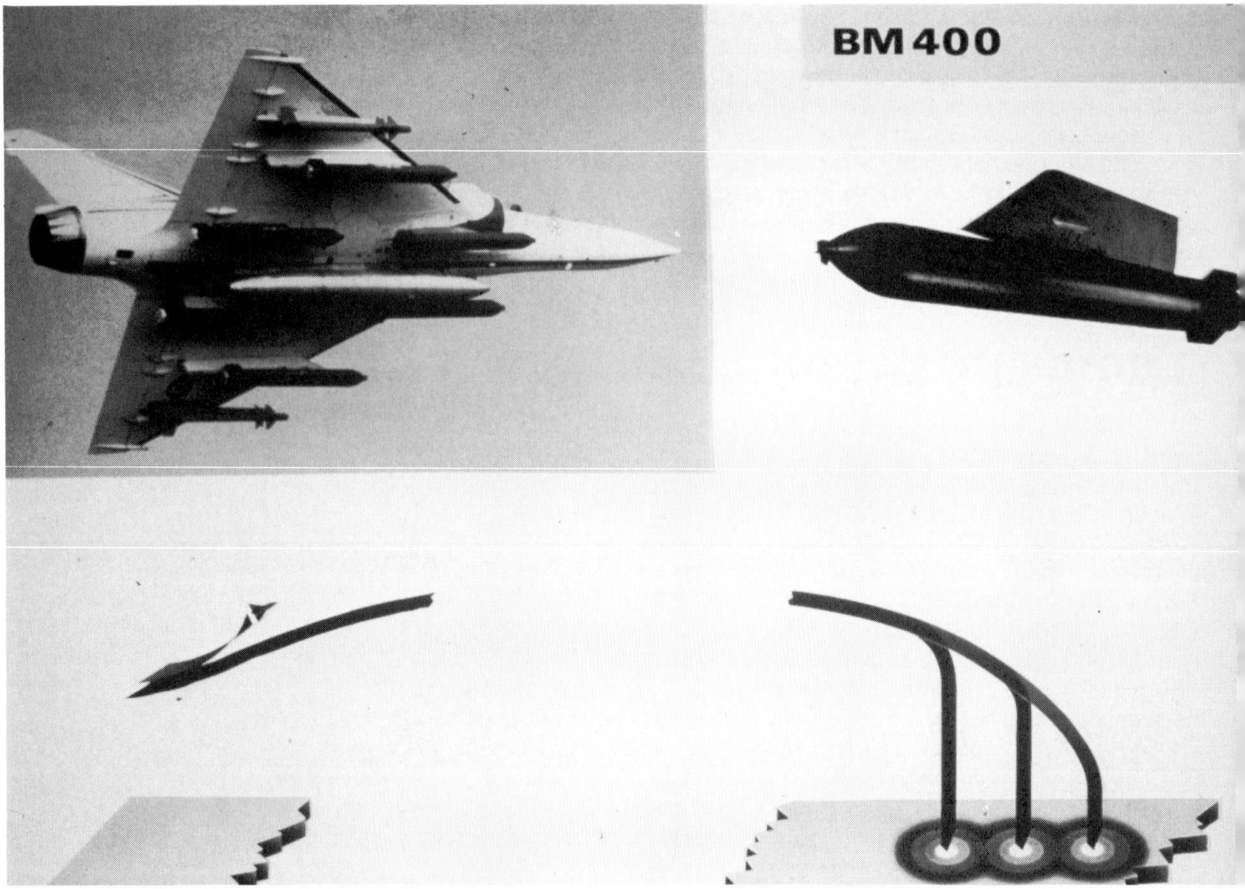

of 1000 mm (39.37 in) has the same performance as the BAP 100. The BM 400 is used by France and has also been exported to several countries.

France

Thomson Brandt TB 74

Type: light-weight anti-tank grenade
Dimensions: diameter (cartridge) 74 mm (2.91 in) and grenade 70 mm (2.76 in); length (cartridge) 214 mm (8.425 in) and (grenade) 190 mm (7.48 in)
Weight: whole cartridge 2.0 kg (4.41 lb)
Warhead: grenade 1.6 kg (3.53 lb)

Variant and applications
TB 74: This is a powerful anti-tank grenade designed to be carried in special grenade launchers. The complete cartridge is loaded into the launcher, and on command fires a gas generator to eject the grenade proper at an altitude of 50 m (165 ft) at any speed between 650 and 1110 km/h (404 and 690 mph). After ejection the grenade descends vertically under a retarding parachute, and can penetrate 230 mm (9.06 in) of armour

if it lands on the upper surfaces of a tank or other AFV. The grenade also possesses a significant secondary fragmentation effect to a radius of 20 m (65 ft).

Alkan 501: This is the lightest dispenser for 74-mm (2.91-in) cartridges, and is designed to be carried on any light tactical aircraft fitted with NATO standard 356-mm (14-in) suspension lugs. The dispenser has a 20-chamber centre section (which can be aligned vertically or angled at 5°, 10° or 15° to give greater lateral dispersion, with the cartridges disposed to fire to the left or right) in a streamlined pod. Loaded with a full complement of cartridges, this dispenser weighs 74 kg (163.14 lb) and is designed for use mainly with counter-insurgency aircraft.

Alkan 530: This is essentially a larger version of the 501 model with provision for forty cartridges. It weighs 135 kg (297.62 lb) fully loaded, and can be carried by most tactical aircraft.

Opposite *Modular bombs such as the Thomson Brandt 400 provide a cost effective way of saturating an area target, as indicated in the accompanying diagram. The coverage of the six BM 400s carried by the Dassault-Breguet Mirage 2000 is very considerable.*

Left *The Thomson Brandt 74-mm (2.91-in) grenade was designed expressly for the Alkan series of cartridge dispensers; It is a useful anti-armour weapon relying on a powerful hollow-charge warhead lowered vertically on to the target by its tail-mounted parachute.*

France

Thomson Brandt/Matra Giboulée

Type: heavy-weight cluster bomb
Dimensions: width 400 mm (15.75 in); height 700 mm (27.56 in); length 3850 mm (151.57 in)
Weight: complete 24-tube round 490 kg (1,080.25 lb)
Warhead: 12 or 24 tubes each accommodating five 0.7-kg (1.54-lb) bomblets with 10 g (3.53 oz) of explosive

Variant and application
Giboulée: This was a pioneering cluster weapon, used only by France and developed from 1966 to provide tactical aircraft with an area saturation weapon for use mainly against armour concentrations. The result is a useful but high-drag system in which the 12 or 24 50-mm (1.97-in) axial tubes are fitted in a Matra streamlined pod for carriage under the wing or fuselage of the attacking aircraft. Each tube contains five Thomson Brandt bomblets, which are ejected rearward by a La Technique Electronique timing system. The tubes are arranged in azimuth at slightly different angles to ensure the correct dispersion, and the speed of the launch aircraft should be 830 km/h (516 mph) to ensure maximum efficiency by cancelling the 230-m (755-ft) per second rearward velocity of the ejected bomblets to produce zero speed over the ground.

The result with the 24-tube launcher and the ejection sequence for optimum anti-tank effect is a pattern measuring 80 m (260 ft) in length and 20 m (65 ft) in width, with the 120 bomblets in a mesh measuring 4 m (13.1 ft) by 3.3 m (10.8 ft). Increasing the interval between bomblet ejections can increase the length of the pattern to 300 m (985 ft). Each bomblet is retarded by drag vanes and reaches the ground vertically, and the impact fusing system allows the anti-tank warhead to punch through 250 mm (9.84 in) of armour. The system can also be used with HE fragmentation warheads for use against motor transport and other 'soft' targets.

France

SAMP fragmentation bombs

Type	Diameter	Length	Weights	
			Nominal	Actual
T15	119 mm (4.69 in)	752 mm (29.6 in)	15 kg (33.1 lb)	15 kg (33.1 lb)
BL8	195 mm (7.68 in)	1584 mm (62.4 in)	120 kg (264.6 lb)	120 kg (264.6 lb)
BL9	195 mm (7.68 in)	1380 mm (54.33 in)	120 kg (264.6 lb)	120 kg (264.6 lb)
BL18	195 mm (7.68 in)	1805 mm (71.1 in)	125 kg (275.6 lb)	125 kg (275.5 lb)

Variants and application
T15: This light fragmentation bomb is designed for carriage (generally in clusters of four such bombs) by light tactical aircraft, and contains 3 kg (6.61 lb) of explosive.
BL8: This light fragmentation bomb is designed for carriage by light tactical aircraft, and contains 20.4 kg (45 lb) of explosive.
BL9: This is the version of the BL8 designed for carriage only by the SIAI-Marchetti SF.260 light attack and counter-insurgency aircraft, and contains 20.4 kg (50 lb) of explosive.

BL18: This light fragmentation bomb is designed for carriage by most tactical aircraft, and contains 25 kg (55.1 lb) of explosive. The type can be fitted with a proximity fuse for air-burst capability.

SAMP general-purpose bombs

France

Type	Diameter	Length	Weights	
			Nominal	Actual
BL5	176 mm (6.93 in)	1320 mm (51.97 in)	50 kg (110.2 lb)	50 kg (110.2 lb)
BL6	228 mm (8.98 in)	1778 mm (70.0 in)	120 kg (264.6 lb)	118 kg (260.1 lb)
BL7	228 mm (8.98 in)	1660 mm (65.35 in)	120 kg (264.6 lb)	115 kg (253.5 lb)
EU2	273 mm (10.75 in)	2253 mm (88.7 in)	250 kg (551.2 lb)	250 kg (551.2 lb)
T25	324 mm (12.75 in)	2121 mm (83.5 in)	250 kg (551.2 lb)	247 kg (544.5 lb)
T200	403 mm (15.87 in)	2195 mm (86.4 in)	400 kg (881.8 lb)	345 kg (760.6 lb)
EU3	356 mm (14.02 in)	2855 mm (112.4 in)	500 kg (1.102 lb)	452 kg (996.5 lb)
BL4	457 mm (18.0 in)	3500 mm (137.8 in)	1000 kg (2,205 lb)	1000 kg (2,205 lb)

Variants and applications

BL5: This light general-purpose bomb can be carried by any aircraft fitted with NATO standard 356-mm (14-in) suspension lugs, and on multiple ejector racks. The bomb has 20 kg (44.1 lb) of explosive.

BL6: This light general-purpose bomb can be carried by any aircraft fitted with NATO standard 356-mm (14-in) suspension lugs, and on multiple ejector racks. The bomb has 53 kg (116.8 lb) of explosive.

BL7: This is a version of the BL6 for carriage only by the SIAI-Marchetti SF.260 series of light attack and counter-insurgency aircraft. The bomb has 57.5 kg (126.8 lb) of explosive.

EU2: This light general-purpose bomb can be carried on any NATO standard suspension lugs, and has 100 kg (220.5 lb) of explosive. The type can be fitted with the **Matra Type 200** or **Type SFA** retarding tail systems. The Type 200 comprises a container-packed cruciform parachute and the fusing system (appropriate to the mission and parachute characteristics) built into the tail of the weapon. The retarding system allows the bomb to be released at an altitude of 30 m (100 ft) by an aircraft flying at between 740 and 1110 km/h (460 and 690 mph), the retarding system slowing the bomb sufficiently for the launch aircraft to be some 480 m (525 yards) beyond the bomb's detonation point; the Type SFA retarding unit is a more capable type, allowing bomb drop at slower speeds and at altitudes down to a mere 15 m (50 ft). The Type SFA retarder also permits air-burst detonation of the weapon.

T25: This light general-purpose bomb can be carried on any NATO standard suspension lugs, and has 123.5 kg (272.3 lb) of explosive. The type can be fitted with the Matra Type 200 or Type SFA retarding tail systems. The Type 200 comprises a container-packed cruciform parachute and the fusing system (appropriate to the mission and parachute characteristics) built into the tail of the weapon. The retarding system allows the bomb to be released at an altitude of 30 m (100 ft) by an aircraft flying at between 740 and 1110 km/h (460 and 690 mph), the retarding system slowing the bomb sufficiently for the launch aircraft to be some 480 m (525 yards) beyond the bomb's detonation point; the Type SFA retarding unit is a more capable type, allowing bomb drop at slower speeds and at altitudes down to a mere 15 m (50 ft). The Type SFA retarder also permits air-burst detonation of the weapon.

T200: This medium general-purpose bomb can be carried on any NATO standard suspension lugs, and has 172.5 kg (380.3 lb) of explosive. The type can be fitted with the Matra Type 200 or Type SFA retarding tail systems. The Type 200 comprises a container-packed cruciform parachute and the fusing system (appropriate to the mission and parachute characteristics) built into the tail of the weapon. The retarding system allows the bomb to be released at an altitude of 30 m (100 ft) by an aircraft flying at between 740 and 1110 km/h (460 and 690 mph), the retarding system slowing the bomb sufficiently for the launch aircraft to be some 480 m (525 yards) beyond the bomb's detonation point. The Type SFA retarding unit is a more capable type, allowing bomb drop at slower speeds and at altitudes down to a mere 15 m (50 ft). The Type SFA retarder also permits air-burst detonation of the weapon. The basic bomb is also used as the core of the Matra 400-kg (882-lb) laser-guided bomb (see Air-to-Surface Missile section entry).

EU3: This medium general-purpose bomb can be carried on any NATO standard suspension lugs, and has 203.4 kg (448.4 lb) of explosive. The type can be fitted with the Matra Type 200 or Type SFA retarding tail systems. The Type 200 comprises a container-packed cruciform parachute and the fusing system (appropriate to the mission and parachute characteristics) built into the tail of the weapon. The retarding system allows the bomb to be released at an altitude of 30 m (100 ft) by an aircraft flying at between 740 and 1110 km/h (460 and 690 mph), the retarding system slowing the bomb sufficiently for the launch aircraft to be some 480 m (525 yards) beyond the bomb's detonation point. The Type SFA retarding unit is a more capable type, allowing bomb drop at slower speeds and at altitudes down to a mere 15 m (50 ft). The type SFA retarder also permits air-burst detonation of the weapon.

BL4: This heavy general-purpose bomb can be carried on any NATO standard 762-mm (30-in) suspension lugs, and has 550 kg (1,212.5 lb) of explosive. The type can be fitted with the Matra Type 200 or Type SFA retarding tail systems. The Type 200 comprises a container-packed cruciform parachute and the fusing system (appropriate to the mission and parachute characteristics) built into the tail of the weapon. The retarding system allows the bomb to be released at an altitude of 30 m (100 ft) by an aircraft flying at between 740 and 1110 km/h (460 and 690 mph), the retarding system slowing the bomb sufficiently for the launch aircraft to be some 480 m (525 yards) beyond the bomb's detonation point. The Type SFA retarding unit is a more capable type, allowing bomb drop at slower speeds and at altitudes down to a mere 15 m (50 ft). The Type SFA retarder also permits air-burst detonation of the weapon. The basic bomb is also used as the core of the Matra 1000-kg (2,205-lb) laser-guided bomb (see Air-to-Surface Missile section entry).

France/USA

Luchaire/Goodyear LU250 EG-FT

Type: medium-weight general-purpose bomb (with retarding option)
Dimensions: diameter 10.75 in (273 mm); length 84.8 in (2154 mm) with conical tail or 85.1 in (2162 mm) with ballute retarder tail; fin span 15.0 in (381 mm) with conical tail or 15.16 in (385 mm) with ballute tail
Weight: total round 530 lb (240.4 kg) with conical tail or 560 lb (254.0 kg) with ballute retarder tail

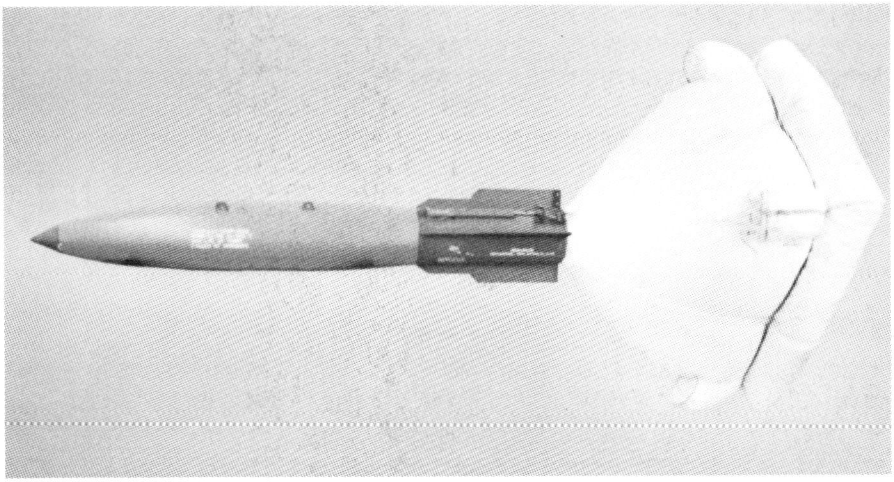

The Goodyear ballute (balloon parachute) retarder offers a simple alternative to the more usual finned retarder.

Warhead: mechanically- or electrically-fused HE

Variant and application
LU250 EG-FT: A joint venture between French and US companies (with an optional Italian outlet for a version with Borletti fuses), this is a general-purpose bomb (relying for its primary results on a combination of blast and fragmentation) based on the specification for the American Mk 82 low-drag weapon and designed for carriage on NATO standard 356-mm (14-in) suspension lugs. The weapon is offered with a choice of standard conical tail (and in general with mechanical fusing) or with a Goodyear BSU-49/B ballute (balloon parachute) retarder: the retarder tail weighs 56 lb (25.4 kg) and inflates under the pressure of ram air into a large-volume cruciform with a span of 15 in (381 mm).

Air-sown land mines

International

Mines designed for use on the land battlefield are considerably smaller than their maritime counterparts: this results from the close proximity of the mine to the target whose influence detonates it (requiring a comparatively small charge to inflict decisive damage), and also from the need to sow comparatively dense fields of mines that are difficult to detect and thus to neutralize. The two primary types of land mine are the anti-tank and anti-personnel varieties, generally laid in mixed fields so that the presence of the anti-personnel mines makes it more dangerous (and thus slower) for the enemy's mine warfare teams to clear the anti-tank mines which are generally the main reason for the field. There is no doubt of the efficacy of such minefields in denying the enemy terrain that is easily passable to armoured and other mechanized formations (so channelling his progress into areas better suited to the defenders' purposes), but such fields are of necessity large. These take time to lay and, though dummy or nuisance minefields can be useful, only a proper minefield (laid in the right density, depth and location) will usefully fulfil the defence's requirements.

Such fields are time-consuming to lay even with the aid of mechanized systems, and thus need to be laid in advance of the enemy's move, which may in the event bypass such pre-planned operational-level minefields. Here air-laying of minefields has an important part to play, for comparatively large minefields can be laid as the tactical situation develops if the right launch platforms (helicopters) and delivery systems are available. Initial efforts in this direction involved the wholesale dropping of mines from transiting helicopters, generally with adverse effects on mine density and fuse reliability even when the field was laid accurately.

Next came ramp launchers derived from those used on ground minelayer vehicles. This technique was pioneered by the Soviets, who still use it for the scattering of **PGMDM** anti-tank and **PFM-1 'Green Parrot'** anti-personnel mines. The PFM-1 has been extensively used in Afghanistan, where a less precise but nonetheless effective laying system has been used: helicopter-carried packs are used to dispense cardboard or comparable frangible containers which carry substantial numbers of mines that fall onto guerrilla paths etc. The main tactical disadvantage of the system is that any helicopter fitted with a ramp launcher has to fly slowly, very close to the ground, meaning that mines cannot be laid close to obstacles such as trees, rocky outcrops, buildings etc.

The West has progressed from this intermediate system in two differing forms epitomized by Italian and US laying techniques: the Italians have opted for the suspended magazine dispenser, and the Americans for the fixed tube dispenser.

The Italian magazine system resembles a crate slung under the carrying helicopter, and makes use of a sling/swivel-hook system incorporating the necessary electrical circuits to release the mines. The specially designed mines are dropped in pre-programmed or manually-controlled sequences. The whole magazine can be hand-loaded in a mere five minutes by two men, the plastic-cased mines are very strong and difficult to detect, and the mines are fitted with fuses that are not armed until after the mines have landed. The **Tecnovar DAT** system makes use of the modello A dispenser, which can accommodate 64 magazines with a total of 1,536 **TS/50** anti-personnel scatter mines, or 32 magazines with a total of 128 **MATS/2** anti-tank scatter mines, or a mix of 32 and 16 magazines with 768 anti-personnel and 64 anti-tank mines respectively. The doors at the base of each magazine section are controlled individually by hand or by a pre-set automatic programmer, the interval between sets determining the density of the resultant field. The opening of the doors two by two allows the drop of 96 anti-personnel

Mines are an important technical feature of the modern battlefield, carefully placed fields being designed to dent territory and thus to channel the enemy into 'killing grounds' of the defence's choosing. However, the fluidity of modern warfare demands that minefields be laid at a moment's notice, and here the air-laid field comes into its own. The most developed European system is the Tecnovar DAT, an Italian system for laying anti-tank and anti-personnel mines from a slowly moving helicopter. Here a crew man hitches a DAT modular dispenser (loaded with both types of mine) on to the lifting hook of a Westland Lynx helicopter.

or eight anti-tank mines, or a mixture of 48 anti-personnel and four anti-tank mines. The helicopter generally flies at 100 m (330 ft) at any speed up to 200 km/h (124 mph). The TS/50 mine is 45 mm (1.77 in) high and 90 mm (3.54 in) in diameter, weighs 190 g (6.7 oz) including 50 g (1.8 oz) of explosive, and is detonated by a pressure of 12.5 kg (27.6 lb). The MATS/2 mine is 90 mm (3.54 in) high and 260 mm (10.24 in) in diameter, weighs 3.5 or 4.5 kg (7.7 or 9.9 lb) including respectively 1.6 or 2.6 kg (3.5 or 5.7 lb) of explosive, and is detonated by a pressure of about 180 kg (396.8 lb).

Valsella Meccanotecnica produces a comparable system, in this instance designated the **Valsella VS-MD H.** This is designed for carriage under a light/medium helicopter, and its capacity amounts to 40 individual magazines carrying a maximum of 2,080 **VS-50** or **VS Mk 2** light anti-personnel mines, or 200 **VS-1.6** anti-tank mines, or a mix of the two types. The anti-personnel mines are dropped in groups of 52, and the anti-tank mines in clusters of five. The mine-release programmer automatically sequences the drop to obviate any alterations in the dispenser's centre of gravity, and a sequencing interval variable between 0.1 and 0.6 seconds allows modification of the minefield's density. The mines are armed in mid-air as the slipstream blows away a protective cap. The VS-50 mine is 45 mm (1.77 in) high and 90 mm (3.54 in) in diameter, and weighs 185 g (6.53 oz) including 43 g (1.52 oz) of explosive. The VS Mk 2 mine is 32 mm (1.26 in) high and 90 mm (3.54 in) in diameter, weighs 135 g (4.76 oz) including 33 g (1.16 oz) of explosive, and is detonated by a pressure of 10 kg (22.05 lb). The VS-1.6 mine is 92 mm (3.62 in) high and 222 mm (8.74 in) in diameter, weighs 3 kg (6.61 lb) including 1.85 kg (4.08 kg) of explosive, and is detonated by a pressure between 180 and 220 kg (396.8 and 485 lb).

MISAR has the comparable **MISAR SY-AT** dispenser system, which comprises a core section attached to the helicopter sling, plus two hook-on auxiliary sections. The core section contains 32 magazines for a maximum load of 2,496 **SB-33** anti-personnel or 160 **SB-81** anti-tank mines, while the two hook-on sections each contain eight magazines for 624 anti-personnel or 40 anti-tank mines. Overall capacity is therefore 3,744 anti-personnel mines, or 240 anti-tank mines, or a mixture of the two types. The SB-33 mine is 32 mm (1.26 in) high and 88 mm (3.46 in) in diameter, weighs 140 g (4.94 oz) including 35 g (1.23 oz) of explosive, and is detonated by a pressure of between 10 and 12 kg (22.05 and 26.5 lb). The SB-81 mine is 90 mm (3.54 in) high and 230 mm (9.06 in) in diameter, weighs 3.15 kg (6.94 lb) including 2 kg (4.41 lb) of explosive, and is detonated by a pressure of between 180 and 200 kg (396.8 and 440.9 lb).

The US approach is epitomized by the first-generation **M56** and second-generation **Honeywell Volcano** systems. The first was designed in the 1960s and entered service in the early 1970s as part of the armament kit available for the Bell UH-1 'Huey' helicopter: the M56 system allowed the carriage and scattering of 150 mines from two skid-mounted dispensers, the pressure-activated mines being scattered in a pattern typically 875 yards (800 m) long. The Volcano is being developed for service on the Sikorsky UH-60 Black Hawk battlefield helicopter from 1992. Originally known as the UMIDS (Universal MIne Dispensing System), the Volcano comprises four 40-tube launcher units fitted in the helicopter's cabin doorways (two launchers on each side) and firing members of the 'Gator' family of mines. The helicopter flies fast and low over the target area, and the mines are ejected laterally under the control of a microprocessor-based control unit in the cabin: this latter is fed with factors such as aircraft speed and height to ensure a pattern of uniform density. Each mine is armed by an electrical pulse passed through the flexible strap separating the mines in the dispenser, the unwinding strap ensuring that the mines are randomly but evenly scattered within the designated target area.

The West Germans have developed (but not yet adopted) a similar system under the designation **Dornier/Dynamit Nobel MinenWerfer System**. This MiWS and comparable weapons constitute very much a developing sector of the air armament field, and several other countries are working on air-launched mine systems for the modern battlefield. The UH-1 can carry 100 AT2 DM1274 anti-tank mines using the MiWS, these mines being laid in a 500-m (545-yard) long strip during a twenty-second pass by the helicopter as it flies at 90 km/h (56 mph) at a height between 5 and 15 m (16 and 49 ft).

Other moves along similar lines include the French **Alsetex Mitral** for use by helicopters (with a triangular-section rather than the standard disc-shaped mine), and a similar system under consideration by GIAT using magnetically-activated anti-tank mines.

Israel

Israel Military Industries ATAP 1000

Type: medium-weight cluster bomb
Dimensions: data not available
Weight: total round 450 kg (992.1 lb)
Warhead: 320.6 kg (706.8 lb) of anti-tank/anti-personnel bomblets

Variant and application
ATAP 1000: Revealed in 1988, the ATAP (Anti-Tank and Anti-Personnel) 1000 cluster bomb was designed to provide Israeli tactical aircraft with a dual-role cluster weapon possessing capabilities against both armour and infantry. The spur for the development of the ATAP 1000 was the US refusal to allow Israeli purchases of US cluster weapons after the use of such bombs in the 1982 Operation 'Peace for Galilee' invasion of southern Lebanon. The ATAP 1000 contains 1,098 0.292 kg (0.64 lb) bomblets each containing a shaped charge of 44.5 g (1.57 oz) of pressed explosive inside a pre-fragmented casing. The bomb can be released at speeds up to 1020 km/h (634 mph), and after release flies a ballistic trajectory with spin stabilization imparted by fin-mounted tabs. The fuse is activated at the moment of release, and after a pre-set interval initiates the pyrotechnic release system, which cuts the casing into four sections that are flung off by centrifugal force to allow the dispersion of the bomblets in a pattern claimed to neutralize an area of 5000 m^2 (5,980 sq yards) with a bomblet density of 0.22 per 1 m^2 (1.2 sq yards). Each bomblet is detonated by an impact fuse or a self-destruct system.

Israel

Israel Military Industries Condib 120

Type: light-weight runway-cratering bomb
Dimensions: data not available
Weight: total round 120 kg (164.6 lb)
Warhead: data not available

Variants and application

Condib 70: Introduced in time for the 'Six-Day War' of 1967, in which it proved decisive in denying the decimate Arab air forces the use of their runways, the 70 kg (154.3 lb) Condib (concrete dibber) is typical of runway-cratering bombs on which a parachute is streamed immediately after weapon release to retard and stabilize the bomb while its nosed is pitched down, and then released as a solid-propellant rocket booster is ignited to accelerate the bomb through the runway surface before the warhead detonates to cause considerable crater and heavy damage.
Condib 120: This larger and updated version of the Condib 70 was introduced in 1987, and features a drogue parachute to ensure clean separation from the launch aircraft, a retarding parachute for deceleration and damping of any oscillations, a solid-propellant booster rocket for acceleration, a hardened steel nose to ensure runway penetration at impact angles as oblique as 45°, and a powerful warhead for extensive damage to runways up to 400 mm (15.75 in) thick.

Rafael Armament Authority TAL-1 and TAL-2

Israel

Type: light-weight cluster bomb
Dimensions: diameter 406 mm (16.0 in); length 2345 mm (92.32 in); fin span 560 mm (22.05 in)
Weight: total round (TAL-1) 250 kg (551.1 lb) or (TAL-2) 236 kg (520.3 lb)
Warhead: (TAL-1) 139.5 kg (307.5 lb) or (TAL-2) 126 kg (277.8 lb) of bomblets

Variants and application

TAL-1: Introduced in 1981, this is a simple but yet highly effective cluster bomb containing 279 0.5-kg (17.64-oz) bomblets. Each of the latter contains 160 g (5.64 oz) of HE, sufficient to shatter the casing into some 1,800 anti-personnel and anti-vehicle fragments which are lethal to a radius of about 8 m (26 ft). The weapon can be carried singly or in multiples on NATO standard 356-mm (14-in) suspension lugs and, before the casing is opened, it is spun by its fins as an aid to bomblet dispersal. The fuse system allows the casing to open only when the bomb is well clear of the launch aircraft, and the impact-fused bomblets are individually wire-safed to become armed only after they have fallen away from the dispenser casing. The coverage of a single bomb can be as great as 53000 m^2 (63,390 sq yards).
TAL-2: This improved variant of the basic weapon was introduced in 1983, and differs from its predecessor in having 315 0.4-kg (14.1-oz) bomblets (but with the same explosive charge and fragmentation quotient as the slightly larger bomblets of the TAL-1) and aerodynamic rather than mechanical fusing.

Cometna bombs

Portugal

Cometna is the Portuguese national steel company, and amongst its varied output is a wide range of bombs. These are in general basically similar to US bombs of various

The Portuguese national steel company Cometna produces a wide range of bombs modelled on the Mk 81 to Mk 84 series evolved round the Aero 1A low-drag shape in the USA.

series, and like their US counterparts can be fitted with different types of retarding tail (the American Goodyear BSU-49/B and BSU-50/B ballute types, the British Hunting Engineering Types 117 and 118, and the French Matra Type 200) or laser-guidance and control surface kits (the American Texas Instruments or French Matra systems) to turn the weapons into analogues of the American Paveway and French BLG laser-guided bombs. The Cometna bombs can also be fitted with hardened steel nose caps to turn them into semi-armour-piercing weapons, and a full range of electronic fuses is offered. Cometna also makes licensed copies of two American practice bombs, namely the **AA EX 2.5** 2.5-kg (5.5-lb) version of the Mk 106 and the AA EX 11.5 11.5-kg (25.4-lb) version of the BDU-33.

Demolition bombs: These are thin-cased weapons for maximum blast effect. There are two models in the series, namely the 290-kg (693.3-lb) **AA LC 300/1** with 215 kg (474.0 lb) of explosive and the 490-kg (1,080.2-lb) **AA LC 500/1** with 360 kg (793.7 lb) of explosive.

Fragmentation bombs: These are thick-walled bombs with a comparatively small detonating charge, gaining their maximum effect from the hail of steel fragments hurled outward as the forged casing is shattered. There are three basic examples in the series, ranging from the 15-kg (33.1-lb) **AA FR 15** with 3 kg (6.6 lb) of explosive, via the 50-kg (110.2-lb) **AA FR 50** with 7 kg (15.4 lb) of explosive to the 200-kg (440.9-lb) **AA FR 200** with 40 kg (88.2 lb) of explosive.

Low-drag general-purpose bombs: These are medium-walled bombs similar to the US Mk 80 series, gaining their effect from a combination of blast and fragmentation. The family has four members, namely the 120-kg (264.6-lb) **AA GP/LD 125/1** with 45 kg (99.2 lb) of explosive, the 235-kg (518.1-lb) **AA GP/LD 250/1** with 87 kg (191.8 lb) of explosive, the 447-kg (985.4-lb) **AA GP/LD 500/1** with 185 kg (407.8 lb) of explosive, and the 894-kg (1,971.0-lb) **AA GP/LD 1000/1** with 429 kg (945.8 lb) of explosive.

Standard general-purpose bombs: These are similar to the earlier American general-purpose bombs. The two members of the subclass are the 59-kg (130.0-lb) **AA GP 50/2** with 26 kg (57.3 lb) of explosive, and the 114-kg (251.3-lb) **AA GP 100/2** with 56 kg (123.5 lb) of explosive.
Supersonic general-purpose bombs: These are well-streamlined general-purpose bombs for supersonic carriage. There are four members of the subclass, namely the 254-kg (560.0-lb) **AA GP 250/2** with 112 kg (246.9 lb) of explosive, the 326-kg (718.7-lb) **AA GP 300/2** with 177 kg (390.2 lb) of explosive, the 356-kg (784.8-lb) **AA GP 400/1** with 164 kg (361.6 lb) of explosive, and the 481-kg (1,060.4-lb) **AA GP 500/1** with 243 kg (535.7 lb) of explosive.

Armscor CB 470

South Africa

Type: medium-weight cluster bomb
Dimensions: diameter 419 mm (16.5 in); length 2600 mm (102.36 in); fin span 560 mm (22.05 in)
Weight: total round 465 kg (1,025.1 lb)
Warhead: 248 kg (546.7 lb) of submunitions

Variant and application

CB 470: This is a particularly ingenious and effective cluster bomb whose origins lie with the peculiar circumstances of the Rhodesian air force in the period of bush war against black guerrillas after the white unilateral declaration of independence. The nature of this bush war clearly called for the use of air-burst cluster munitions to tackle elusive insurgent forces, yet the arms embargo placed on Rhodesia by the rest of the world made it impossible for such weapons to be imported. Rhodesia lacked the technological and manufacturing bases for indigenous production of such a weapon (especially its electronic fuses), and so had recourse to the locally-designed Alpha bouncing bomblet. This weighed 6.2 kg (13.7-lb) and was a steel sphere round a 1.4-kg (3.1-lb) bursting charge of RDX/TNT.

Dropped from trays in the bomb bays of elderly English Electric Canberras, the Alpha bomblets were designed to land on the ground (or mud or even water), spin and bounce upwards and forwards to detonate exactly 0.65 second later at a height of between 3 and 5 m (10 and 16.5 ft), hurling outwards a hail of steel fragments at a velocity of 1300 m (4,265 ft) per second. South Africa has now refined the idea, with 40 such bomblets packed into a 185-kg (407.8-lb) dispenser casing for use by the whole range of tactical aircraft available to the South African air force.

The bomb can be hung on NATO standard 356-mm (14-in) or Warsaw Pact standard 250-mm (9.84-in) suspension lugs, and can be released at any altitude between 25 and 300 m (80 and 985 ft) at speeds between 850 and 1100 km/h (528 and 684 mph). The first radial row of four bomblets is ejected with a velocity of 12 m (39.4 ft) per second just 0.8 second after the bomb has been released, but successive rows of bomblets are ejected at velocities declining to 2 m (6.6 ft) per second. Each bomb is armed only 1 second after it has been ejected, this staggering of the sequence ensuring that the bomblets are not detonated by collisions. The 40 bomblets in each container can saturate a 17500-m^2 (20,930-sq yard) area 70 m (230 ft) wide and 250 m (820 ft) long, and the ripple of four pairs of CB 470s at 0.4-second intervals saturates an area of 54000 m^2 (64,585 sq yards). The formal designation of this weapon is the **Mk II Fragmentation Weapon System.**

South Africa

Armscor general-purpose bombs

Increasingly severed from the rest of the world by politically-motivated embargoes imposed because of the country's policy of *apartheid*, South Africa has moved steadily towards a policy of military self-sufficiency on the grounds that the country may have to fight a war of survival against its black neighbours, and that in such a war only superior training and superior weapons will allow success for South Africa's considerably outnumbered forces. For this reason, therefore, Armscor has developed a series of operational bombs (plus the appropriate training types) for the South African air force.

12.5-kg practice bomb: This neat practice weapon is designed to provide pilots with experience of dropping the 450-kg (992-lb) general-purpose bomb, and though it has the same ballistic qualities as the larger weapon, it weighs a mere 12.5 kg (27.65 lb), and its dimensions include a diameter of 95 mm (3.74 in) and a length of 765 mm (30.12 in). The tail contains a 920-g (32.45-oz) pyrotechnic charge to mark the impact point with smoke by day or a flash by night.
120-kg general-purpose bomb: This is the standard light general-purpose bomb of the South African air force. The weapon comprises a 60-kg (132.3-lb) casing whose weight increases to 115 kg (253.5 lb) filled with a 60/40 RDX/TNT mixture, fitted with its impact- or delayed-actions fuses and carrying its 8-kg (17.6-lb) tail section.
120-kg shrapnel bomb: This is essentially the anti-personnel version of the 120-kg general-purpose bomb. The core of the 122-kg (269.0-lb) glassfibre-skinned weapon is 27 kg (59.5 lb) of an RDX/TNT mixture covered with cast resin containing 42,000 6.7-mm (0.26-in), 26,000 7.9-mm (0.31-in) or 19,500 8.7-mm (0.34-in) steel balls.
250-kg general-purpose bomb: This is in essence a scaled-up 120-kg general-purpose bomb, with the same basic proportions and fusing arrangements, but possessing an overall weight of 228 kg (502.6 lb).
250-kg practice bomb: This is a large practice bomb built of glassfibre and polyurethane, and filled with steel balls. The bomb breaks up on impact leaving no debris on which a subsequent weapon can ricochet.
480-kg general-purpose bomb: This 466-kg (1,027.3-lb) weapon is South Africa's largest standard bomb, a general-purpose typed filled with Torpex 4B and fitted with a hardened steel nose cap for semi-armour-piercing capability.

Spain

Explosivos Alaveses BR bombs

Type	Diameter	Length	Weights	
			Nominal	Actual
50	180 mm (7.09 in)	1395 mm (54.92 in)	50 kg (110.2 lb)	
125	240 mm (9.45 in)	1820 mm (71.65 in)	125 kg (275.6 lb)	
250	290 mm (11.42 in)	2150 mm (84.6 in)	250 kg (551.1 lb)	
375	330 mm (12.99 in)	2805 mm (110.4 in)	375 kg (826.7 lb)	
500	360 mm (14.17 in)	2940 mm (115.75 in)	500 kg (1,102.3 lb)	
1000	460 mm (18.1 in)	3820 mm (150.4 in)	1000 kg (2,204.6 lb)	

Variants and application

BR: This is the standard series of general-purpose low-drag bombs used by the Spanish

Comparison of the Explosivos Alaveses BR series of low-drag bombs reveals the origin of the type in the same Aero 1A shape as the US Mk 81 to Mk 84 series. From top to bottom these weapons are of the 1000-, 500-, 375-, 250-, 125- and 50-kg) sizes, most of which can be converted into BRP parachute-retarded bombs by the substitution of a different tail section with a wider rear containing the retarder.

air force, and the weapons are designed for carriage by high-performance aircraft on either a NATO standard 356-mm (14-in) or (for the BR-1000) NATO standard 762-mm (30-in) suspension lugs. Nose or tail fuses, of the impact- or delay-action varieties, can be fitted and the launch parameters of the series include an aircraft speed between 370 and 1110 km/h (230 and 690 mph). The explosive content and fin diameter of the **BR-50** are 20 kg (44.1 lb) and 275 mm (10.83 in), of the **BR-125** 55 kg (121.25 lb) and 360 mm (14.17 in), of the **BR-250** 115 kg (253.5 lb) and 440 mm (17.32 in), of the **BR-375** 170 kg (374.8 lb) and 500 mm (19.69 in), of the **BR-500** 210 kg (463.0 lb) and 500 mm (19.69 in), and of the **BR-1000** 475 kg (1,047.2 lb) and 538 mm (21.18 in). The BR series is produced in several types, notably general-purpose (with a filling of TNT, Tritonal or H6), anti-personnel (with a pre-fragmented casing over the explosive charge), smoke (with a filling of white phosphorus), and inert (a practice round filled with an inert substance possessing the same density as TNT).

Explosivos Alaveses BRFA-330 — Spain

Type: medium-weight runway-cratering bomb

Dimensions: diameter 300 mm (11.81 in); length 3044 mm (119.84 in); fin span 600 mm (23.62 in)
Weight: total round 330 kg (727.5 lb)
Warhead: 75 kg (165.3 lb) of HE

Variant and application
BRFA-330: This is a powerful runway-attack weapon for use by high-performance aircraft fitted with NATO standard 356-mm (14-in) suspension lugs, though the weapon can also be supplied to order with Warsaw Pact standard 250-mm (9.84-in) attachment capability. The weapon is designed for low-altitude attack with a minimum release height of 80 m (265 ft) at any speed between 520 and 1110 km/h (323 and 690 mph). Immediately after it has been released the weapon deploys its braking parachute to allow the launch aircraft to clear the area, and to pitch down the bomb's nose. At this point the parachute is released and a rocket motor in the tail ignited to accelerate the weapon through the runway and into the earth to a depth of 600 mm (23.6 in), where the warhead is detonated by the two Infa fuses to create crater, fracture and heave damage over an area of some 200 to 250 m^2 (240 to 300 sq yards).

Spain

Explosivos Alaveses BRP bombs

Type	Diameter	Length	Weights	
			Nominal	Actual
50	180 mm (7.09 in)	1425 mm (56.1 in)	50 kg (110.2 lb)	
125	240 mm (9.45 in)	1820 mm (71.65 in)	125 kg (275.6 lb)	
250	290 mm (11.4 in)	2285 mm (90.0 in)	250 kg (551.1 lb)	
375	330 mm (13.0 in)	2855 mm (112.4 in)	375 kg (826 lb)	
500	360 mm (14.17 in)	3000 mm (118.1 in)	500 kg (1,102.3 lb)	

Variants and applications
BRP: This is the parachute-retarded version of the BR bomb series, the tail of the standard bomb being detached and replaced by the para-tail, which is activated by the withdrawal of the arming wire as the bomb drops away from the launch aircraft. The bombs are fitted with Kappa fuses, and with comprehensive safety devices to prevent arming in the event that any part of the system fails, or if the bomb is dropped at too low an altitude for the tail retarder to function successfully. Launch parameters include a minimum release height of 50 m (165 ft) and an aircraft speed between 520 and 1110 km/h (323 and 690 mph). The explosive content and fin diameter of the **BRP-50** are 20 kg (44.1 lb) and 250 mm (9.84 in), of the **BRP-125** 57 kg (125.7 lb) and 337 mm (13.27 in), of the **BRP-250** 112 kg (264.9 lb) and 404 mm (15.91 in), of the **BRP-375** 170 kg (374.8 lb) and 470 mm (18.5 in), and of the **BRP-500** 206 kg (454.1 lb) and 510 mm (20.08 in).

The BRP series is produced in several types, notably general-purpose (with a filling of TNT, Tritonal or H6), anti-personnel (with a pre-fragmented casing over the explosive charge), smoke (with a filling of white phosphorus), and inert (a practice round filled with an inert substance possessing the same density as TNT).
BRP.S: This is the version of the BRP designed for very low-altitude release, and thus fitted with a fast-acting parachute retarder. The series comes in the **BRP.S 125**, **BRP.S 250** and **BRP.S 500** variants with nominal weights of 125, 250 and 500 kg (275.6, 551.1 and 1,102.3 lb) respectively.

The BRP series is the parachute-retarded equivalent to the BR series produced in Spain by Explosivos Alaveses (EXPAL). Like their free-fall brothers, these weapons come in a variety of tactical sizes, and have secured useful export sales by their low cost without any appreciable sacrifice of quality.

Explosivos Alaveses special-purpose bombs

Spain

Apart from its main range of general-purpose, general-purpose retarded and runway-cratering munitions, Explosivos Alaveses also produces a number of free-fall weapons for specialized roles.

BIN: This prefix covers the company's range of fire bombs. The series is based on an aerodynamically shaped aluminium casing filled with an incendiary mixture (kerosene and petrol thickened with naphthalic acid) and designed for delivery at altitudes as low as 30 m (100 ft). The weapon can be fitted with stabilizing fins, but is designed for use without such surfaces in order that the casing should tumble and so disperse the incendiary gel farther on impact and subsequent break-up. The individual members of this series are the **BIN-100, BIN-200** and **BIN-375**, containing 100, 200 and 375 litres (22, 44 and 82.5 Imp gal) of incendiary gel respectively.

BME-330: This is a cluster bomb designed for the standard area-denial role after delivery from high or low altitudes. The weapon has a nominal weight of 330 kg (727.5 lb) and, after release, is stabilized and retarded by a parachute attached to the frame carrying the submunition payload. This frame is decelerated by the parachute and pulled backwards out of the casing; once the frame is completely clear of the casing the 180 bomblets are individually ejected by gas pressure to descend under the parachutes. The three types of bomblet available are anti-personnel (a pre-fragmented body filled with Hexolite), incendiary (filled with white phosphorus) and anti-tank (based on a hollow-charge warhead of RDX). From a low-altitude delivery the BME-330 saturates an area of 40 m (130 ft) by 100 m (330 ft), the area increasing but the density decreasing with greater delivery altitude.
Practice bombs: The Spanish manufacturer also produces a range of practice bombs based on American types. The **Mk 76** and **BDU-33** are basically similar miniature practice bomblets with a weight of 11.4 kg (25 lb), a diameter of 102 mm (4 in) and a length of 570 mm (22.4 lb) for the BDU-33 or 585 mm (23 in) for the Mk 76; the signal cartridge to indicate impact point is loaded in the nose of the Mk 76 and the tail of the BDU-33. The **Mk 87** is tailored aerodynamically to simulate the Mk 80 series of general-purpose bombs; the type's dimensions include a length of 2311 mm (91 in) and diameter of 273 mm (10.75 in), and its weights are 46 kg (101.5 lb) empty and 157 kg (346 lb) filled with wet sand; provision is made in the tail for a 30-mm signal cartridge to indicate the bomb's impact point. The **Mk 106** is another type designed to provide training with the Mk 80 series of bombs when fitted with a tail retarder, but is a light and inexpensive type; length is 488 mm (19.2 in), diameter 97 mm (3.8 in) and weight 2.3 kg (5 lb) complete with the nose-mounted 21-mm signal cartridge.

Sweden

FFV bombs

As part of its national policy of producing weapons tailored to the specific requirements of its armed forces, Sweden has produced a range of free-fall bombs matched to its primary combat aircraft, most notably the Saab 105 light tactical aircraft, the Saab 35 Draken fighter-bomber and the Saab 37 Viggen multi-role aircraft, soon to be joined by the Saab JAS 39 Gripen multi-role fighter. The development and production of these bombs was entrusted to FFV, the country's primary manufacturer of ordnance items, and the current weapons (and appropriate training items) are the 600-kg (1,322.8-lb) **M50 HB** general-purpose bomb, the 250-kg (551.1-lb) **M50 SB** general-purpose bomb, the 120-kg (264.6-lb) **M51 SB** general-purpose bomb, the 500-kg (1,102.3-lb) **M56 MB** general-purpose bomb, the 80-kg (176.4-lb) **M60 LYSB** blast/fragmentation bomb, the 75-kg (165.3-lb) **M62** flash bomb, and a 100-kg (220.5-lb) flash bomb.

Sweden

FFV M71 Virgo

Type: light-weight fragmentation bomb
Dimensions: diameter 214 mm (8.425 in); length 1900 mm (74.8 in) with nose fuse and parachute retarder pack; fin span 368 mm (14.49 in)
Weight: total round 123 kg (271.2 lb)

Warhead: 30 kg (66.1 lb) of RDX/TNT inside a casing of steel fragments

Variant and application
M71 Virgo: Though small and comparatively light, this is a powerful and apparently effective fragmentation weapon with undoubted capability against targets as diverse as personnel, anti-aircraft guns, light armoured vehicles and landing craft (in short, the types of weapon and vehicles likely to be used by the first wave of any invading force). The Virgo has a single-point attachment (including a safety device) to the launch aircraft, and possesses three main sections connected by the weapon's central tube. The forward end is the sensitive electronic proximity fuse, 947 mm (37.3 in) long and weighing 1.7 kg (3.75 lb), to detonate the weapon a few metres above the ground or any thick vegetation. In the centre is the bomb proper with an explosive filling inside a pre-fragmented casing. And at the rear is the GRP-reinforced plastic tail section, 290 mm (11.4 in) long and weighing 1.2 kg (2.65 lb); the 530-mm (20.87-in) diameter parachute retarder emerges from this container 0.7 seconds after bomb release, though there is an option for free-fall delivery from a higher-altitude release.

SPICE CH-TABO

Switzerland

Type: medium-weight bomblet dispenser
Dimensions: data not available
Weight: data not available
Warhead: typically seven bomblets

Variant and application
CH-TABO: This is a joint development by the Swiss companies F + W (the Federal Aircraft Factory, which has design leadership) and the Federal Ammunition Factory at Thun with marketing undertaken by the French company Matra. The dispenser takes the form of a pylon-mounted container which typically holds seven special bomblets optimized for attacks on soft (personnel, motor transport etc) and semi-hard (materiel etc) targets. Each bomblet is a spherical casing over an aluminium matrix for some 8,000 cast steel balls over an explosive core.

After ejection from the container by dynamic air pressure, each bomblet falls towards the ground, stabilized and retarded in this process by an integral cast aluminium tail fan, which spins the bomblet as an aid to warhead dispersion. The bomblet is fitted with either a highly sensitive mechanical fuse (using a telescopic stand-off probe) or an electronic proximity fuse for an air-burst detonation at a height of some 70 to 100 cm (27.6 and 39.4 in) above the ground.

Using the latter type of fusing after bomblet release at low altitude, two CH-TABO containers can saturate an area 50 m (165 ft) in width and 200 m (655 ft) in length with four fragments per 1 m^2 (10.75 sq ft). Each fragment is calculated as being able to penetrate 5 mm (0.2 in) of steel. The dispenser releases each submunition directly to the rear, thus causing minimum disruption to the airflow over the launch aircraft, and the dispenser itself is a low-drag installation that need not be jettisoned. The system can be used at any speed between 720 and 1000 km/h (447 and 621 mph), and the standard attack is in a shallow dive followed by pull-out at 50 m (165 ft). The CH-TABO system can be used by lightweight subsonic aircraft with simple sights, but the installation of the system of more advanced aircraft with laser rangefinders and a sophisticated fire-control system adds much to the accuracy with which the submunitions can be delivered.

Soviet conventional and nuclear bombs

It is believed that more than 1,000 types of free-fall ordnance are carried in the inventory of the Warsaw Pact forces, and it is estimated that some 75% of these are of Soviet design and, in most cases, manufacture. However, very little is known (or at least released for public consumption) in the West about these weapons, which are clearly available in large numbers and in an apparently bewildering number of forms and varieties for the whole gamut of tactical and operational tasks envisaged for the Warsaw Pact air forces, complemented by the Soviet air arms which also fulfil strategic tasks.

BETAB series weapons: The BETAB designation encompasses a comparatively small but highly important number of concrete-piercing weapon types for the anti-runway use in the counter-air role. The two most significant bombs of this type are the 250-kg (551.1-lb) **BETAB-250** and 500-kg (1,102.3-lb) **BETAB-500** weapons, each designed for release at low altitude before being retarded by a parachute and then, after the nose has dropped and the parachute freed, boosted through the surface of the target by a short-impulse rocket motor. There is also the 500-kg (1,102.3-lb) **M62** penetration bomb, which is also available in a 250-kg (551.1-lb) version.

Chemical weapons: The USSR has long been a proponent and exponent of chemical weapons for a host of tactical applications, and this fact is attested by the number and variety of chemical weapons in the Soviet inventory. These chemical weapons fall into two categories, an old and a new series. Though the four members of the old series might be thought obsolescent by Western standards, the Soviets seem reluctant to discard any weapon that may have even a vestigial capability in tactical warfare, so these weapons remain in the stockpile for use as and when appropriate. The smallest of the old series is the 200-kg (440.9-lb) **VAP-200**, a container weapon filled with persistent Toxic-B agents; next up in size is the 250-kg (551.1-lb) **AK-2**, an inline dispenser weapon for 240 1-kg (2.2-lb) bomblets filled with a mixture of mustard gas and Lewisite; the 500-kg (1,102.3-lb) **KhAB-500** (sometimes rendered **ChAB-500**) is an M1942 bomb filled with 180 kg (396.8 lb) of phosgene; and the 1000-kg (2,204.6-lb) **VAP-1000 (BATT)** is a dispenser weapon filled with 700 litres (154 Imp gal) of stabilized hydrogen cyanide, and generally carried in four-dispenser groups for low-altitude spraying into the air over the battlefield.

The new series of chemical weapons is more numerous than the old series, and inevitably includes some of the Soviets' most advanced chemical agents. Even less is known of these weapons than of some other types, but information available in the West suggests that there are several types of **VAP** dispenser for persistent Toxic-B agents, various types of **NOV-AB** bombs filled with non-persistent toxic agents, and various types of **SOV-AB** bombs filled with persistent toxic agents.

Cluster weapons: The Soviets were amongst the first in the field with dispenser weapons, having pioneered such munitions in the 1930s and used them extensively in World War 2 (in which the Germans came to appreciate the military virtues of the type and thus launched an intensive programme of development and use that subsequently paved the way for many modern Western developments). Throughout the 1940s and 1950s the Soviets continued to develop and produce cluster munitions, and their current inventory includes large numbers of such weapons with bombs containing hollow-charge anti-tank, fragmentation, high explosive, high explosive incendiary, and incendiary bomblets as their payloads. The designations of these weapons remain unknown for the most part, but a few have been publicly named: the **PTAB-2, PTAB-5** and **PTAB-5/1** anti-tank cluster bombs, the first containing 2-kg (4.41-lb) hollow-charge submunitions, and the latter pair containing 5-kg (11-lb) hollow-charge submunitions.

These older weapons have in general been replaced on current aircraft by the 250-kg (551.1-lb) **PTK-250** low-drag cluster bombs filled with hollow-charge submunitions. There are also two known low-drag fragmentation cluster bombs, the 180-kg (396.8-lb) **RPK-180** and 250-kg (551.1-lb) **RPK-500** each filled with specialized anti-personnel and anti-material submunitions. The last known cluster bomb is the **RRAB**, which is filled with a large number of incendiary submunitions. Current cluster bombs contain 'dumb' submunitions with no terminal guidance and/or intelligent sensor mechanism. But given the fact the Western nations have been developing 'smart' submunitions for some time, it is inevitable that the Soviets will also have been working along the same lines, using 'gleaned' information as well as indigenous resources.

DAP series weapons: These are underwing dispensers enabling tactical aircraft to lay battlefield smoke screens.

Demolition weapons: These are generally thin-walled weapons with a high explosive-to-casing ratio to provide maximum blast at the expense of fragmentation. The designations of these weapons remains unrevealed, but such weapons are known to come in 100-, 250-, 750- and 1100-kg (220.5-, 551.1-, 1,653.4- and 2,425.0-lb) sizes.

DV-AB series weapons: These are smoke bombs for use on the tactical battlefield, and are produced in a number of weights.

FAB series weapons: These are general-purpose bombs and, like most other types of Soviet free-fall ordnance, are still current in old- and new-series forms, the former characterized by a wide casing and comparatively blunt nose, and the latter by their better-streamlined casings for lower drag when carried by the higher-performance tactical aircraft that began to enter Soviet service in the later 1950s. In this regard the more modern Soviet FAB bombs may be regarded as the counterparts of the US Mk 80 series based on the Aero 1 shape with a length/diameter ratio of about 15/1. The older weapons are the 8-kg (17.6-lb) **FAB-8**, 10-kg (22.05-lb) **FAB-10**, 15-kg (33.07-lb) **FAB-15**, 20-kg (44.1-lb) **FAB-20**, 25-kg (55.1-lb) **FAB-25**, 50-kg (110.2-lb) **FAB-50**, 110-kg (242.5-lb) **FAB-110**, 500-kg (1,102.3-lb) **FAB-500** and 1000-kg (2,204.6-lb) **FAB-1000**. The newer series includes the 100-kg (220.5-lb) **FAB-100**, 250-kg (551.1-lb) **FAB-250**, 500-kg (1,102.3-lb) **FAB-500**, 750-kg (1,653.4-lb) **FAB-750** and 1000-kg (2,204.6-lb) **FAB-1000**. It is likely also that the Soviets have followed the Western lead into retarded versions of their general-purpose bombs, the nature of modern battlefield air defences making it all but suicidal for tactical aircraft to make anything but the lowest-level attacks. It is also known that the Soviets have produced laser guided versions of the FAB-500, FAB-750 and FAB-1000 bombs, the associated laser designator being carried in a podded installation.

FOTAB series weapons: The FOTAB prefix covers photo-flash bombs for clear visual and photographic reconnaissance. The only currently known version is the 250-kg (551.1-lb) **FOTAB-250**.

Fuel/air explosive weapons: Nothing detailed has been revealed publicly in the West about the Soviet development of such weapons, though it is known that such weapons are operational with a number of Soviet tactical aircraft types. Despite the technical difficulty of creating the right fuel/air aerosol mixture and then igniting it at precisely the correct moment, it is certain that the Soviets have followed the American lead in these weapons, which have a number of extremely important tactical applications, especially the clearance of minefields and the destruction of personnel and/or armour concentrations.

Guided bombs: The Soviets have also copied the Western lead into the development of 'iron' bombs with a nose-mounted guidance package (either electro-optical or laser) to promote terminal guidance and so facilitate their effective use against point targets. It is also likely that these Soviet 'smart' bombs have been fitted with add-on aerodynamic surfaces to provide greater stand-off range, and in certain weapons, a degree of cross-range capability.

N series weapons: The designation N covers Soviet strategic thermonuclear weapons. It is not known how many types of thermonuclear free-fall bomb the Soviets number in their inventory, which is limited by the SALT II treaty to 1,080 weapons to the Americans' 4,080 (as the Soviets have a 6,420 to 2,100 superiority in warheads carried by intercontinental ballistic missiles). It is believed that bombs with 5-, 20- and 50-megaton yields are carried by long-range bombers such as the Myasishchyev M-4 'Bison' (four bombs per aircraft), Tupolev Tu-95 'Bear' (three bombs per aircraft) and Tupolev Tu-? 'Blackjack' (an unknown number of bombs, possibly four or six, per aircraft), and by medium-range bombers such as the Tupolev Tu-16 'Badger' (one bomb per aircraft), Tupolev Tu-22 'Blinder' (one bomb per aircraft) and Tupolev Tu-26 'Backfire' (two bombs per aircraft). The bombs are themselves guarded by KGB troops, and only allocated to aviation units after strict clearance from the national command authorities.

OFAB series weapons: This designation embraces two types of low-drag blast fragmentation bombs, namely the 250-kg (551.1-lb) **OFAB-250** and the 500-kg (1,102.3-lb) **OFAB-500**.

PKPE series weapons: Little is known of this series, which is believed to encompass the USSR's dispenser weapons designed for carriage by helicopters such as the Mil Mi-8 'Hip' and Mi-24 'Hind', and by tactical aircraft such as the Sukhoi Su-25 'Frogfoot'. In 1986 a Soviet armoured personnel carrier knocked out in Afghanistan by the Moslem resistance was revealed to have been carrying two aerodynamically-contoured **PKPE-1** units apparently field-modified for the role from aircraft dispensers. Each launcher has 29 tubes able to contain two or three submunitions for lateral ejection by a small charge. The diameter of the tubes is in the order of 70 mm (2.76 in), and fin-stabilized blast fragmentation grenades were recovered. It is likely that the system can also carry a hollow-charge anti-tank submunition based on the warhead of the RPG-7/SPG-9 ground-launched anti-tank rockets. Two PKPE-1 units are probably carried for submunition ejections to left and right of the aircraft flightpath, and the provision of a sequencing unit in each dispenser indicates that a controlled laying pattern is achieved.

PLAB series weapons: These are napalm bombs of the low-drag variety, the two known members of the family being the 250-kg (551.1-lb) **PLAB-250** and the 500-kg (1,102.3-lb) **PLAB-500**.

PROSAB series weapons: The prefix PROSAB is used for parachute flares, which are available in several weights of which only the 250-kg (551.1-lb) **PROSAB-250** is definitely known.

TN series weapons: The prefix TN is used for Soviet tactical nuclear and thermonuclear weapons, of which there are known to be at least fourteen types. The weapons are subject to the same stringent security measures as the N series of strategic weapons, and little public information is available other than the fact that there is a 1000-kg (2,204.6-lb) weapon with a yield of 350 kilotons and that this is being slowly complemented by a newer 700-kg (1,543.2-lb) weapon with a yield of 250 kilotons. The medium-range bombers listed in the N series entry (see above) can carry these weapons, and purely tactical aircraft qualified for these weapons are the Mikoyan-Gurevich MiG-21 'Fishbed', MiG-23 'Flogger', MiG-27 'Flogger' and possibly MiG-29 'Fulcrum', plus the Sukhoi Su-7/17/20 'Fitter', Su-24 'Fencer' and possibly Su-27 'Flanker'. Each of these aircraft can carry a maximum of two TN series weapons. Anti-submarine nuclear weapons are covered by the same basic designation, and it is known that two such bombs can be carried by the Beriev M-12 'Mail', Ilyushin Il-38 'May' and Tupolev Tu-142 'Bear', and one each by the Kamov Ka-25 'Hormone', Kamov Ka-27 'Helix' and Mil Mi-14 'Haze' anti-submarine helicopters.

ZAB series weapons: The ZAB prefix indicates low-drag incendiary bombs; the four known members of this tactically significant series are the 100-kg (220.5-lb) **ZAB-100**, 250-kg (551.1-lb) **ZAB-250**, 500-kg (1,102.3-lb) **ZAB-500** and 750-kg (1,653.4-lb) **ZAB-750**. A closely related variant is the 200-kg (440.9-kg) **ZAP-200** incendiary container weapon.

British conventional and nuclear bombs

UK

In full accord with the reticence typical of the British approach to all matters military, very little has ever been revealed about British free-fall bombs, either elderly conventional or more modern nuclear types. There are two sizes of these, though the only example of the 540-lb (244.9-kg) size is the 12.79-in (325-mm) diameter **Mk 1/2** intended for use by most British aircraft involved in attacks on battlefield targets. The other variety is the 1,000-lb (453.6-kg) size, and this is more than amply served by the 16.5-in (419-mm) diameter **Mk 2, Mk 6, Mk 6*, Mk 7, Mk 7*, Mk 9, Mk 10, Mk 11, Mk 11*, Mk 12, Mk 13, Mk 14, Mk 15, Mk 16** and **Mk 19** weapons intended for use by all British aircraft involved in battlefield and other tactical tasks. Given the need for modern aircraft to operate at low level over the modern battlefield the British have developed two types of retarder tail for use with these weapons.

For use with the Mk 1/2 (plus the American M64, Mk 81 and Mk 82 bombs) there is the **Hunting Bomb Retarder Tail Type 118**, which is 38.2 in (970 mm) long and has a span of 18.25 in (464 mm), producing a weapon with an overall length of 80.32 in (2040 mm). For use with the other British bombs (plus the American M65, M117 and Mk 83) there is the **Hunting Bomb Retarder Tail Type 117**, which is 40.75 in (1035 mm)

The Panavia Tornado can carry a very considerable weight of many weapon types on its underfuselage and underwing hardpoints, but is seen here sporting only a pair of ML Aviation CBLS 200 practice bomb carriers under the fuselage and two 330-Imp gal (1500-litre) drop tanks under the wings.

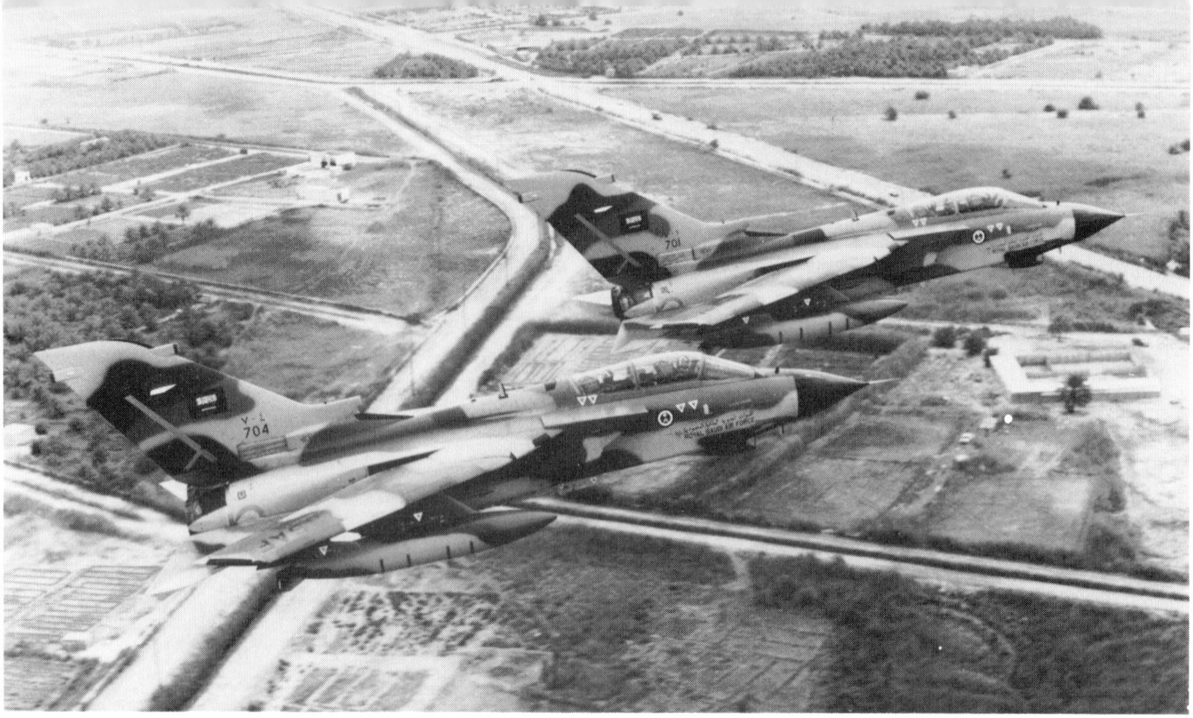

The Panavia Tornadoes of the Royal Saudi air force are fitted to carry the same weapon types as British Tornadoes, a key part of the aircraft fit to ensure accurate weapon delivery being the Ferranti laser ranger and marked-target seeker, fitted in a bulge fairing under the fuselage in line with the windscreen.

long to produce a weapon 91.39 in (2321 mm) in length. This latter comes in two versions, namely the **Type 117 Mk 3** for externally carried bombs and the **Type 117 Mk 4** for internally and externally carried bombs. The Type 117 Mk 3 has a span of 23.0 in (584 mm), and the Type 117 Mk 4 a span of 16.5 in (419 mm).

Both the Type 117 and Type 118 work on the same basic principle, with the four panels of the tail hinged about their rear ends to spring outward as airbrakes and also open the ribbon parachute segments between four arms. This system is highly effective, providing a great deal of drag after a precise interval from aircraft separation. The British bombs (especially the Mk 13/18) can also be fitted with the Texas Instruments Paveway guidance and control package to turn them into laser-homing 'smart' weapons.

Very little has ever been revealed about British nuclear weapons, but a slip by the Ministry of Defence in 1986 revealed that the designation of the UK's standard tactical nuclear weapon is **WE.177**. No information about this has been revealed, though it is believed to have both land and anti-submarine applications.

UK

Hunting Engineering HADES

Type: medium-weight cluster bomb
Dimensions: diameter 16.5 in (419 mm); length 96.5 in (2451 mm); fin span 22.25 in (565 mm)
Weight: total round 582 lb (264. 0 kg)
Warhead: 49 HB876 area-denial minelets

Variant and application

HADES: This **Hunting Area-DEnial System** cluster bomb combines the dispenser of the BL 755 dispenser bomb with the HB876 area-denial minelet of the JP233 anti-airfield dispenser system. The weapon works in exactly the same way as the BL755, but the payload comprises 49 5.5-lb (2.49-kg) HB876 minelets, which are each 100 mm (3.94 in)

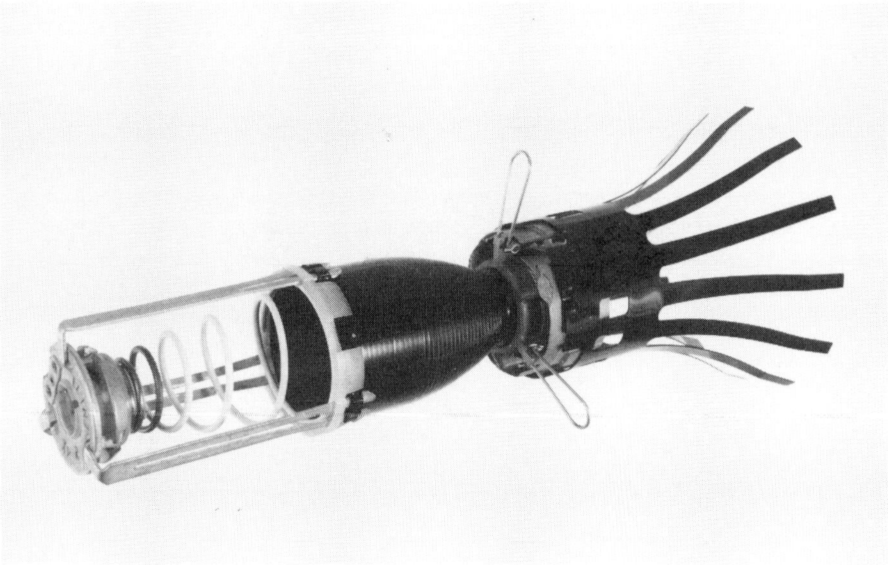

The bomblet of the BL755 anti-armour cluster bomb is seen in the fully-deployed condition in this excellent photograph: the tail-compressing mechanism has been discarded, and the spring-loaded nose/fuse section has been extended to ensure optimum stand-off detonation distance for the hollow-charge warhead.

in diameter and 150 mm (5.91 in) in length. These submunitions free-fall to the ground after being dispensed in the air, right themselves on their radial rings of sprung legs, and await detonation by any close-by disturbance. The individual minelets self-destruct after a variable but preset interval if not previously detonated.

Hunting Engineering BL755 UK

Type: medium-weight cluster bomb
Dimensions: diameter 16.5 in (419 mm); length 96.5 in (2451 mm); fin span 22.25 in (565 mm)
Weight: total round 582 lb (264.0 kg)
Warhead: 147 anti-tank/fragmentation bomblets

Variant and application
BL755: This dual-role weapon entered service in 1972, and is designed to provide the British air forces (plus those of selected export customers) a high 'kill' probability against a considerable range of hard and soft targets. The designers appreciated that high-speed low-altitude attacks cannot be made with the utmost accuracy (at least with the standard of avionics current at the time the weapon was developed), so the BL755 covers the target area with a bomblet pattern whose dimensions are proportional to anticipated aiming error.

The BL755 bomb can be hung on a single point, or on twin suspension lugs of the NATO standard 14-in (356-mm) or Warsaw Pact standard 250-mm (9.84-in) varieties. Before the aircraft takes off one of four time delays is set on the weapon to ensure adequate separation of the aircraft and weapon before the latter is fully activated. Thus, when the bomb is dropped the primary striker is armed and, after the set interval, this fires the primary cartridge which rapidly generates the gas pressure to blow off the two-part bomb body skins and then fires the main gas cartridge: this latter ejects the bomb-

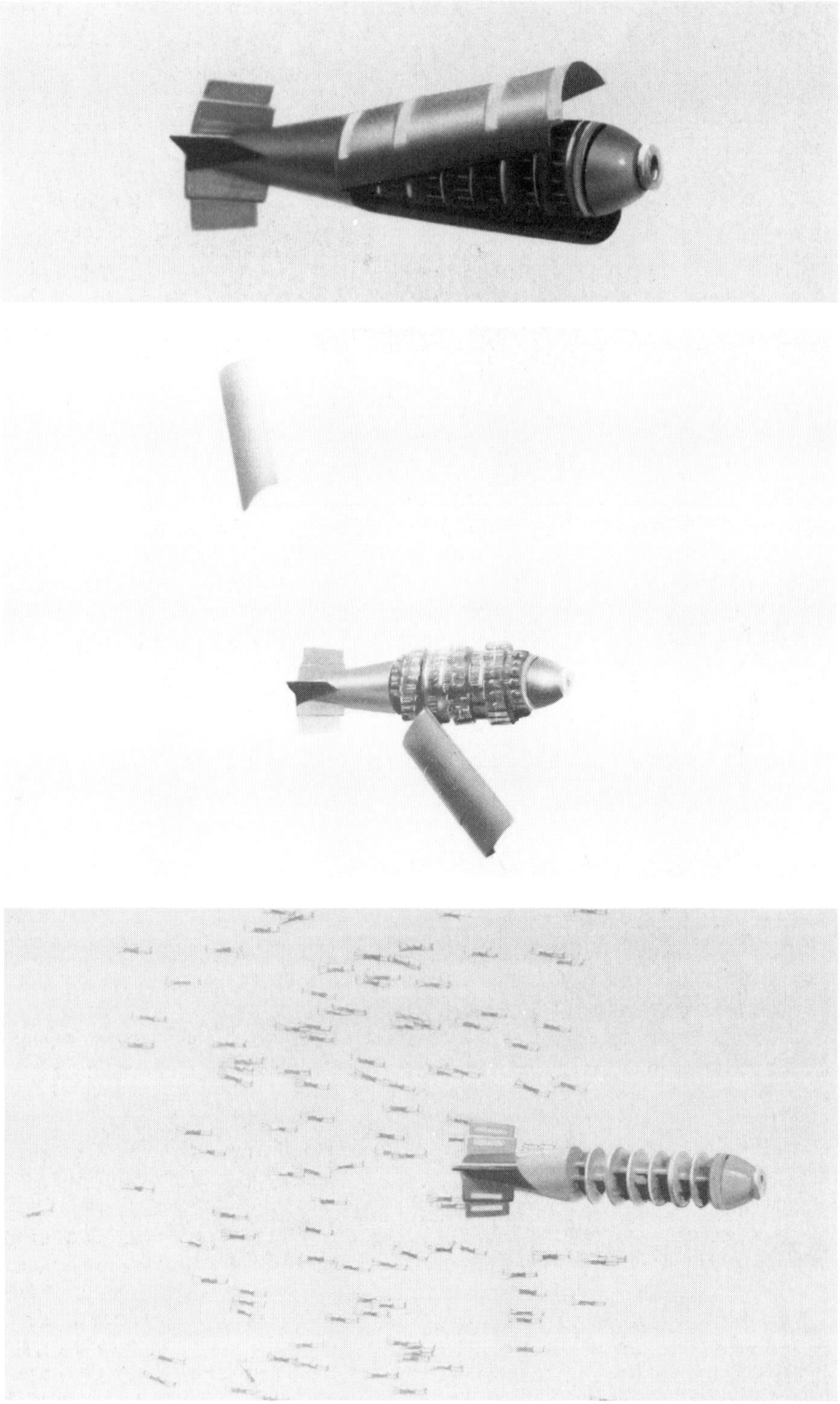

This manufacturer's artwork sequence explains the operating method used in the BL755 and, with variations, many other cluster bombs: as the weapon is released, the primary striker is armed and, after a preset interval, fires the primary cartridge; the gas pressure thus created releases the two bomb body skins and then activates the main gas cartridge to eject the 21 bomblets in each of the weapon's seven bays; each bomblet is armed shortly after ejection, and detonates on impact.

Above *A target tank receives the effect of just one of the 147 hollow-charge bomblets carried by the BL755 anti-tank cluster bomb, indicating the power of this simple and inexpensive weapon.*

Left *As part of the weapon kit for the Panavia Tornado, MBB has developed the Vorgeladener (pre-loaded) quadruple store-carrier for attachment to the lateral underfuselage hardpoints. This Vorgeladener is already carrying its four BL755 cluster bombs as it is raised towards the aircraft's belly.*

lets, which are accommodated seven to each of the weapon's 21 bays. Bomblet ejection from each of the bays is varied in velocity to ensure the attainment of the desired pattern on the ground. Each 68-mm (2.68-in) diameter bomblet is 150 mm (5.91 in) long as it is fired, lengthening to 356 mm (14.0 in) as the spring tail and nose probe deploy. The bomblet is armed as it falls, and detonates on impact with the shaped-charge warhead able to penetrate at least 250 mm (9.84 in) of armour as well as scattering a cloud of at least 2,000 lethal fragments. The weapon is used by the British forces under the designation **Cluster Bomb No.1**, the basic version described above being the Cluster Bomb No.1 Mk 1 and the three variants the Cluster Bomb No.1 Mk 2 with shorter time delays, the Cluster Bomb No.1 Mk 3 suitable for use on Warsaw pact suspension lugs, and the Cluster Bomb No.1 Mk 4 version of the Mk 3 with the time delays of the Mk 2.
Improved BL755: Entering service in 1987, this is a much developed version of the original BL755 (and possessing the same external dimensions and weights) but carrying considerably more powerful bomblets using the latest shaped-charge techniques. Each bomblet is still of 68-mm (2.68-in) diameter, but is 149 mm (5.87 in) long as it is fired, lengthening to 627 mm (24.69 in) complete with extending nose probe and retarding parachute. The system makes full use of increased launch aircraft attack accuracy, and the submunitions are retarded to increase their angle of attack, and thus their armour-penetrating capability.

UK

Hunting Engineering JP233

Type: heavy-weight airfield-attack and area-denial submunition dispenser
Dimensions: width 33.07 in (840 mm); height 23.6 in (599 mm) for underfuselage con-

One of the most impressive airfield attack weapons in the British inventory is the Hunting JP233 dispenser system. The aircraft (in this instance a Panavia Tornado GR. Mk 1) still has to overfly the target airfield and its defences, the underfuselage dispenser scattering 30 SG357 runway-cratering primary submunitions plus 215 HB876 area-denial secondary submunitions to prevent repair crews beginning immediate repairs.

figuration and for underwing configuration with SG357 submunitions, or 22 in (559 mm) for underwing configuration with HB876 submunitions; length 257.9 in (6551 mm) for underfuselage configuration, 223.5 in (5677 mm) for underwing configuration with SG357 submunitions, and 166.9 in (4240 mm) for underwing configuration with HB876 submunitions; fin span 63.07 in (1602 mm) for underwing configuration with SG357 submunitions, and 53.39 in (1356 mm) for underwing configuration with HB876 submunitions
Weight: total system 5,148 lb (2335.1 kg) for underfuselage configuration, 3,020 lb (1369.9 kg) for underwing configuration with SG357 submunitions, and 2,535 lb (1149.9 kg) for underwing configuration with HB876 submunitions
Warhead: (underfuselage twin configuration) 30 SG357 concrete-penetrating and 215 HB876 area-denial submunitions

Variant and application
JP233: Used only by the Royal Air Force (though qualified for use on the tactical aircraft of several other nations), the JP233 is a dispenser system designed to render inoperative for a substantial period hardened targets such as airfields, choke points, docks, headquarter areas and dumps. The system is available in different sizes to suit particular aircraft, the most effective being that on the Panavia Tornado comprising two side-by-side containers under the fuselage. These each contain 30 runway-catering and 215 area-denial submunitions (in the rear and front sections of the dispenser respectively), the laying of the two types simultaneously destroying the target and making it difficult for the enemy to move back into the area and effect repairs. The 57-lb (25.85-kg) SG357 is a substantial submunition comparable to the STABO used in the West German MW-1 dispenser system: the weapon is 35.0 in (889 mm) long with its impact sensor in the retracted position, and has a basic diameter of 7.1 in (180 mm) increasing to 10.0 in (254 mm) over the fins. It is a parachute-retarded submunition and thus falls nearly

A Panavia Tornado GR.Mk 1 lifts off with a useful airfield-attack load: two JP233 dispensers carried on the lateral underfuselage hardpoints, which are tandem arrangements so that forward 14-in (356-mm) suspension lugs can carry the containers for HB876 area-denial minelets and the rear 30-in (762-mm) lugs can support the containers for the larger SG357 runway-catering bomblets. The outboard ECM pod is a 'must' for so hazardous a mission.

The detonation of an SG357 runway-cratering submunition from a JP233 dispenser indicates the potentially decisive nature of the weapon, which has two charges and a rocket motor to ensure that the runway surface is broken before the penetrator is driven below the surface to explode and cause additional crater and heavy damage.

vertically to be the ground, where the Thorn EMI contact fuse system takes over to detonate the hollow-charge primary warhead and so open a hole through which the second charge is fired to penetrate into the earth and detonate, thus creating a large crater and a larger area of fracture and heave damage. The concomitant use of the HB876 area-denial minelet effectively slows enemy repair teams, for these 5.5-lb (2.49-kg) free-fall minelets are fitted with a disturbance fuse and a variable but preset self-destruct fuse. The HB876 is 100 mm (3.94 in) in diameter and 150 mm (5.91 in) in length, and free-falls to the ground, where it is turned upright by spring legs: its warhead is designed to produce both a high-velocity slug capable of destroying repair vehicles, and a hail of highly destructive steel fragments.

USA

B28 Mod 4 RE

Type: free-fall thermonuclear bomb (with option of parachute retarder)
Dimensions: diameter 20.0 in (508 mm); length 166.0 in (4216 mm)
Weight: total round 2,170 lb (984.3 kg), some sources giving a weight up to 2,540 lb (1152.1 kg) in unspecified variants
Warhead: this is about 36.0 in (0.91 m) long, and its components are plutonium (for fission), lithium-6 deuteride and tritium (for fusion), and PBX-9505 or Cyclotrol (for implosion leading to fission/fusion reaction)

Yield: each weapon can be set for only one yield (out of a possible five options), the known options being 70 kilotons, 350 kilotons, 1.1 megatons and 1.45 megatons, and the suggested fifth option being 29 megatons

Variants and application

B28 Mod 4 EX: The B28 operational- and strategic-level thermonuclear bomb was developed at the Los Alamos National Laboratory from 1955, and began to enter service with the US Air Force's Strategic Air Command in 1958. Production had begun in the previous year and continued up to about 1970. More than 1,200 examples were produced in a wide number of variants, all designed to be hung on NATO standard 30-in (762-mm) suspension lugs. The W28 warhead variants of the bomb payload equipped the TM-76 Mace ground-launched and AGM-28 Hound Dog air-launched missiles. The original B28 Mod 0 to B28 Mod 3 free-fall variants have now been withdrawn, and the retirement of surviving B28 Mod 4 and B28 Mod 5 bombs began in 1984 (their replacement being the more modern B83), stockpile numbers having fallen by 1987 to some 900 out of a production total of 1,200.

The weapons are being retired in the order 29 megatons (?), 1.1 megatons, 350 kilotons, 1.45 megatons and 70 kilotons. The B28 Mod 4 EX is 170.0 in (4.318 m) long, weighs between 2,027 and 2,040 lb (919.4 and 925.3 kg), and has cruciform stabilizing fins. The weapon is designed for external carriage by strategic aircraft such as the Boeing B-52 Stratofortress (12 bombs) and Rockwell B-1 (12 bombs), and by tactical aircraft such as the Grumman A-6 Intruder (three bombs), Lockheed F-104 Starfighter (one bomb), McDonnell Douglas F-4 Phantom II (one bomb) and Vought A-7 Corsair II (four bombs). The extra length of this variant makes for better streamlining (allowing carriage at supersonic speed), and it has an F28 Mod 1, 4, 6, 8 or 9 radar fuse system (set on the ground before take-off) allowing air- and ground-detonation capabilities after over-the-shoulder or loft delivery at altitudes as low as 300 ft (91 m).

B28 Mod 4 RE: This is another externally-carried version, with a three-fin tail and optional parachute retarder, and has the same fusing options as the B28 Mod 4 EX.

B28 Mod 5 FI: This is an internally-carried version, a cylindrical weapon of bluff lines and a small cruciform of rectangular stabilizing fins, and fitted with a parachute retarder. The standard installation in the B-52 is a quadruple package (two superimposed pairs). The weapon has F 29 Mod 3, 7 or 10 full fusing options for air or ground burst, and is the only in-service model to have a laydown capability. Surviving B28 Mod 5 FI weapons are undergoing a Stockpile Improvement Program update with new electrical equipment, improved safety measures and revised HE arrangements more resistant to crash- or fire-induced detonation.

B43 Mod 1

USA

Type: free-fall thermonuclear bomb (with option of parachute retarder)

Dimensions: diameter 18.0 in (457 mm); length varies with Mod between 144 and 165 in (3658 and 4191 mm), the relationship between Mod and specific length not having been revealed, though it is thought that later Mods are longer than earlier ones

Weight: total round 2,120 lb (961.6 kg)

Warhead: this is of unrevealed length, and the components are oralloy (for fission),

lithium-6 deuteride and tritium (for fusion), and HE (for implosion leading to fission/fusion reaction)
Yield: 1 megaton

Variants and application
B43 Mod 0: This weapon was designed from 1956 by the Los Alamos National Laboratory, entered production in 1959 (ending before 1970) and began to enter service in 1961 with the US Air Force's Strategic Air Command (the Boeing B-52 Stratofortress carrying 12 bombs, the General Dynamics FB-111 carrying two bombs and the Rockwell B-1 carrying 12 bombs internally and 14 bombs externally) and Tactical Air Command (the General Dynamics F-16 Fighting Falcon carrying one bomb, the General Dynamics F-111 carrying three bombs and the McDonnell Douglas F-4 Phantom carrying one bomb). In the same year the weapon was introduced by the US Navy and US Marine Corps (the Grumman A-6 Intruder carrying three bombs, the McDonnell Douglas F/A-18 Hornet carrying two bombs and the Vought A-7 Corsair II carrying four bombs). The B43 started its retirement phase in 1984, numbers declining from a 1983 peak of 2,000 towards a 1987 figure of about 1,000 weapons.

The weapon is designed to be hung from NATO standard 30-in (762-mm) suspension lugs, is highly streamlined and terminates at the rear in a cruciform of stabilizing fins. The B43 was designed for both strategic and operational-level deployments, and is being replaced in the former capacity by the B83 and in the latter by the B61. The B43 Mod 0 is the original version and weighs 2,060 lb (934.4 kg). The weapon can be delivered in the laydown, over-the-shoulder or low/medium-angle loft techniques, and the ground-set fusing options include retarded or free-fall ground burst, or retarded or free fall air burst.
B43 Mod 1: This variant has the same fusing options as the B43 Mod 0, but unlike the earlier weapon has a fusing radar set and a Permissive Action Link safety system. As with the earlier model, the lowest permissible delivery altitude is 300 ft (61 m). The training versions of the B43 are the **BDU-6, BDU-8, BDU-18** and **BDU-24.**

USA

B53

Type: free-fall thermonuclear bomb (with option of parachute retarder)
Dimensions: diameter 50.0 in (1270 mm); length 148.0 in (3759 mm)
Weight: total round 8,850 lb (4014.4 kg)
Warhead: this is of unrevealed length, and consists of oralloy (for fission), lithium-6 deuteride (for fusion) and Cyclotrol HE (for implosion leading to fission/fusion reaction)
Yield: 9 megatons

Variant and application
B53: This extremely powerful weapon was developed by the Los Alamos National Laboratory from 1958, entering production in 1961 and service in 1962 as the primary armament of the Boeing B-52 Stratofortress, which can carry 12 such weapons internally on NATO standard 30-in (762-mm) suspension lugs; the weapon can be delivered in the free-fall, laydown or retarded modes, with air or ground burst options set on the ground before take-off. The weapon is a blunt-end cylinder terminating aft in a drum section supporting the four wedge-profiled rectangular fins of the stabilizing cruciform, and carrying the assembly of five retarding parachutes. These last comprise one 5-ft

(1.52-m) diameter pilot parachute, one 16-ft (4.88-mm) drogue parachute, and three 48-ft (14.63-m) main parachutes.

Production of the B53 continued into the late 1960s, and by 1983 the stockpile had dropped to 150 such weapons, with numbers decreasing steadily with the advent of the B83 as the type's replacement. The same basic warhead was used as the W53 in the Mk 6 re-entry vehicle carried by the LGM-25C Titan II intercontinental ballistic missile.

B57 Mod 1 USA

Type: free-fall nuclear bomb (anti-submarine and operational-level land warfare applications)
Dimensions: diameter 14.75 in (374.65 mm); length less than 119.0 in (3023 mm)
Weight: total round 500 to 510 lb (226.8 to 231.3 kg)
Warhead: this is of unrevealed length, and is believed to be of the fission type with unrevealed components
Yield: this is a variable-yield weapon with options believed to lie between less than 1 kiloton and 20 kilotons, but often quoted as 5 to 10 kilotons

Variant and application
B57 Mod 1: This is one of the US forces' standard tactical- and operational-level nuclear bombs and many NATO air arms are trained in the type's use on maritime aircraft. The weapon was developed by the Los Alamos National Laboratory from 1960, entering production in 1963 and service as the B57 Mod 0 in 1964. Production continued throughout the 1960s, and aircraft qualified to carry the current B57 Mod 1 weapon (the B57 Mod 0 having been retired in 1967) on NATO standard 14- or 30-in (356- or 762-mm) suspension lugs are the General Dynamics F-111 (three bombs), Grumman A-6 Intruder (three bombs), Lockheed P-3 Orion (two bombs), Lockheed S-3 Viking (one bomb), McDonnell Douglas F-4 Phantom II (one bomb), McDonnell Douglas F/A-18 Hornet (two bombs), Panavia Tornado (one or two bombs), Sikorsky SH-3 Sea King (one bomb), Sikorsky SH-60F Seahawk (one bomb) and Vought A-7 Corsair II (four bombs).

The weapon has a highly streamlined configuration for external carriage, and terminates aft in a cruciform of moderately swept fins. The weapon can be delivered in laydown, over-the-shoulder or low/medium-angle loft modes down to altitudes of 300 ft (91 m). In its land warfare application the weapon possesses air- and surface-burst fusing options, and in its anti-submarine application a depth pressure fuse. The 1983 stockpile was about 1,000 such weapons, and while naval weapons are being retained for the present, the land warfare weapons are being phased out in favour of the B61. For training there are the **BDU-12** and **BDU-19** devices.

B61 USA

Type: free-fall lightweight thermonuclear bomb (with option of parachute retarder)
Dimensions: diameter 13.4 in (340.4 mm); length 142.0 in (3607 mm)
Weight: total round between 718 and 765 lb (325.7 and 347.0 kg) depending on variant

Warhead: this is of unrevealed length, and its components are oralloy (for fission) probably 'doped' with lithium-6 deuteride and tritium (for fusion), and (B61 Mods 0-2) PBX 9404 HE or (B61 Mods 3-5) PBX 9502 Improved HE (for implosion leading to fission/fusion reaction)
Yield: this is a variable-yield weapon with four options lying between 100 and 500 kilotons; some sources claim an additional low-yield option of 10 kilotons, and others claim that the B61 Mod 1 is a strategic version with a yield in the megaton range

Variants and application
B61 Mod 0: The B61 is the 'standard' nuclear weapon of the US forces, and in its multipurpose operational-level guise is also allocated (under US key) to the air forces of Belgium, West Germany, Greece, Italy and Turkey. The weapon was developed by the Los Alamos National Laboratory from 1963 and began production during 1966 or 1967. The B61 Mod 0 began to enter service in January 1968, and production was completed in June 1969. Some 3,000 B61 bombs had been stockpiled by 1983 and possibly 5,000 by 1987. Aircraft qualified to carry the type on NATO standard 30-in (762-mm) suspension lugs include the Boeing B-52 Stratofortress (12 bombs carried internally), General Dynamics F-16 Fighting Falcon (one bomb), General Dynamics F-111 (three bombs), General Dynamics FB-111 (two bombs), Grumman A-6 Intruder (three bombs), McDonnell Douglas F-4 Phantom II (one bomb), McDonnell Douglas F/A-18 Hornet (two bombs), Panavia Tornado (two bombs), Rockwell B-1 (38 bombs, 28 carried internally and 14 externally) and Vought A-7 Corsair II (four bombs).

The B61 Mod 0 has a Category B Permissive Action Link (allowing the arming of the weapon from the cockpit), but no command disabling mechanism or enhanced electrical safety. The weapon has an inflight fusing and yield selection capability by the adjustment of a dial in the cockpit. The B61 is a highly streamlined bomb with a rear cruciform of swept fins for internal or supersonic external carriage, and can be delivered from as low as 50 ft (15 m) using laydown, free-fall or retarded delivery for air- or ground-burst detonation. The retarding parachute is made of Kevlar, and can decelerate the bomb from 1,000 to 35 mph (1609 to 56 km/h) in a mere 2 seconds.
B61 Mod 1: This is believed to be the strategic version of the series, and is thus not fitted with a Permissive Action Link. Like the B61 Mod 0 the weapon lacks a command disabling capability and enhanced electrical safety. The weapon entered service in February 1969, and production was completed in April 1971.
B61 Mod 2: This is an improved tactical version, and entered service in June 1975 with production continuing up to June 1979. The weapon has a Category F Permissive Action Link and possesses an inertial command disabling capability, but lacks enhanced electrical safety.
B61 Mod 3: This is the initial definitive version, development having started in 1976 for service entry during 1979 and production continuing to the present. The weapon has a Category F Permissive Action Link and a command disabling capability by means of a weak/strong link drive by a unique signal generator. The B61 Mod 3 can be delivered in laydown or retarded modes, and its full fusing options include air- and ground-burst detonation and a delay-fuse capability. All the bombs of the B61 series are by current standards 'one-point safety' weapons, meaning that 'in the event of a detonation initiated in any one point in the high explosive system, the probability of achieving a nuclear yield greater than 4 pounds of TNT equivalent shall not exceed one in one million'.
B61 Mod 4: This variant entered development in 1976 and began production in 1981 for service entry late in that year or early in 1982. The type has the same Category F Permissive Action Link, command disabling facility and fusing options as the B61 Mod 3. From 1982 all surviving B61 Mod 0, 1, 2 and 5 weapons are being upgraded

to an approximate B61 Mod 3/4 standard. The training versions of the B61 are the **BDU-36, BDU-38** and **BDU-39E**.

B83 USA

Type: free-fall thermonuclear bomb (with option of parachute retarder)
Dimensions: diameter not available; length 144.0 in (3658 mm)
Weight: total round 2,408 lb (1092.2 kg)
Warhead: this is of unrevealed length, and its components are probably plutonium (for fission), oralloy (for fusion) and PBX-9502 Improved HE (for implosion leading to the fission/fusion reaction)
Yield: not revealed, but believed to be in the low megaton range

Variant and application
B83: This weapon was developed from January 1979 by the Livermore National Laboratory, and entered production in 1981 for initial deployment in 1984 as replacement for the B28, B43 and B53 thermonuclear weapons. In this capacity the B83 is the principal gravity weapon carried on NATO standard 30-in (762-mm) suspension lugs by the Boeing B-52 Stratofortress (eight bombs carried internally), General Dynamics FB-111 (two bombs carried internally) and Rockwell B-1 (38 bombs, of which 24 are carried internally and 14 externally) in the US Air Force's Strategic Air Command. Secondary carriers of this important new weapon (in replacement of the B28 and B43 weapons) are the tactical strike aircraft of the US Air Force and Navy, typically the General Dynamics F-16 Fighting Falcon, General Dynamics F-111, Grumman A-6 Intruder and Vought A-7 Corsair II.

A production total of some 2,500 B83 weapons is envisaged. The B83 has a Category D Permissive Action Link and a non-violent command disabling capability via a weak/strong link driven by a unique signal generator. The weapon was designed with the low-level, high-speed role of the B-1B principally in mind, and can thus withstand high-g evasive manoeuvres and an extremely rapid deceleration to 60 mph (97 km/h) after delivery at altitudes between 150 and 50,000 ft (45 and 15240 m) and speeds as high as Mach 2; the weapon has full fusing capability for air- and ground-burst detonation, but is intended mainly for ground burst to destroy silo-based missiles.

US bombs USA

Type	Diameter	Length	Weights	
			Nominal	Actual
BDU-27	data not available	data not available		
BDU-28				
BDU-30				
BDU-32			25 lb (11.3 kg)	
BDU-34				
BDU-37				
M30			100 lb (45.4 kg)	101 lb (45.8 kg)

Type	Diameter	Length	Weights Nominal	Weights Actual
M38			100 lb (45.4 kg)	100 lb (45.4 kg)
M46			100 lb (45.4 kg)	51.9 lb (23.5 kg)
M47			100 lb (45.4 kg)	100 lb (45.4 kg)
M52			1,000 lb (454 kg)	1,078 lb (489 kg)
M56			4,000 lb (1814 kg)	4,535 lb (2057.1 kg)
M57			250 lb (113.4 kg)	261 lb (118.4 kg)
M58			500 lb (226.8 kg)	537 lb (243.6 kg)
M59			1,000 lb (454 kg)	990 lb (449.1 kg)
M64			500 lb (226.8 kg)	535 lb (242.7 kg)
M65			1,000 lb (454 kg)	1,081 lb (490.3 kg)
M66			2,000 lb (907 kg)	2,053 lb (931.2 kg)
M70			115 lb (52.2 kg)	122.5 lb (55.6 kg)
M76			500 lb (226.8 kg)	475 lb (215.5 kg)
M78			500 lb (226.8 kg)	
M79			1,000 lb (454 kg)	960 lb (435.5 kg)
M81			260 lb (117.9 kg)	262 lb (118.8 kg)
M82			90 lb (40.8 kg)	88.5 lb (40.1 kg)
M84			100 lb (45.4 kg)	
M86			120 lb (54.4 kg)	
M88			220 lb (99.8 kg)	
M103			2,000 lb (907 kg)	2,039 lb (924.9 kg)
M104			100 lb (45.4 kg)	
M105			500 lb (226.8 kg)	
M109			12,000 lb (5443 kg)	
M110			22,000 lb (9979 kg)	
M113			125 lb (56.7 kg)	
M116	data not available		750 lb (340.2 kg)	
M117			750 lb (340.2 kg)	823 lb (373.3 kg)
M118			3,000 lb (1361 kg)	3,049 lb (1383 kg)
M120			150 lb (68.0 kg)	
M121			10,000 lb (4536 kg)	
M122			100 lb (45.4 kg)	
M124			250 lb (113.4 kg)	
M129			750 lb (340.2 kg)	
MC1			750 lb (340.2 kg)	725 lb (328.9 kg)
Mk 1			100 lb (45.4 kg)	112 lb (50.8 kg)
Mk 1			2,000 lb (907.2 kg)	2,030 lb (920.8 kg)
Mk 1			1,600 lb (726 kg)	1,590 lb (721.2 kg)
Mk 15			3 lb (1.36 kg)	
Mk 15			100 lb (45.4 kg)	
Mk 23			3 lb (1.36 kg)	
Mk 33			1,000 lb (454 kg)	1,008 lb (457.2 kg)
Mk 36			500 lb (226.8 kg)	576 lb (261.3 kg)
Mk 47			1,000 lb (454 kg)	
Mk 58			500 lb (226.8 kg)	
Mk 65			500 lb (226.8 kg)	
Mk 66			1,000 (454 kg)	
Mk 76			25 lb (11.34 kg)	
Mk 77 Mod 0			750 lb (340.2 kg)	760 lb (344.7 kg)
Mk 77 Mod 1			500 lb (226.8 kg)	520 lb (235.9 kg)
Mk 78			750 lb (340.2 kg)	760 lb (344.7 kg)
Mk 79			1,000 lb (454 kg)	912 lb (413.7 kg)

Type	Diameter	Length	Weights Nominal	Actual
Mk 81	9 in (229 m)	74.0 in (1880 mm)	250 lb (113.4 kg)	262 lb (118.8 kg)
Mk 82	11 in (279 mm)	87.4 in (2220 mm)	500 lb (226.8 kg)	531 lb (240.9 kg)
Mk 83	13.75 in (349 mm)	118.1 in (3000 mm)	1,000 lb (454 kg)	985 lb (446.8 kg)
Mk 84	13.0 in (330 mm)	147.6 in (3749 mm)	1,970 lb (893.6 kg)	1,970 lb (893.6 kg)
Mk 86			250 lb (113.4 kg)	
Mk 87			500 lb (226.8 kg)	
Mk 88			1,000 lb (454 kg)	
Mk 89			56 lb (25.4 kg)	
Mk 94			500 lb (226.8 kg)	441 lb (200 kg)
Mk 106			6 lb (2.72 kg)	
Mk 111			1,000 lb (454 kg)	1,130 lb (512.6 kg)
Mk 116			750 lb (340.2 kg)	701 lb (318 kg)
Mk 122				

Variants and applications

Armour-piercing bombs: The US forces use an extraordinarily wide assortment of free-fall bombs for a host of applications. These weapons fall into basic categories according to application, and are treated thus in this listing. The role of armour-piercing weapons is self-evident but, whereas their use was previously dedicated to attacks on tanks and warships, they have in general been replaced by specialist weapons such as cluster bombs and anti-ship missiles. The type nevertheless retains a capability against strongpoints and other hardened targets of the fixed variety, and the weapons of this type still in the US inventory are the **M52A1** armour-piercing (a converted artillery shell with 58 lb/26.3 kg of Explosive D), the **M58** semi-armour-piercing (a cast steel casing with 143.1 lb /64.9 kg of Picratol), the **M59** semi-armour-piercing (a cast steel casing with 313 lb/142 kg of Picratol), the **M103** semi-armour-piercing (a cast steel casing with 556.5 lb/252.4 kg of Picratol), the **Mk 1** armour-piercing (as cast steel casing with 209 lb/94.8 kg of Explosive D), the **Mk 33** armour-piercing (a cast steel casing with 140 lb/63.5 kg of Explosive D) and the **Mk 58A2** semi-armour-piercing (a cast steel casing with 143.1 lb/64.9 kg of Picratol).

Chemical bombs: Though thc USA bccamc a signatory to thc Gcncva Protocol of Junc 1925 only in 1977 (though retaining the right to stockpile chemical weapons for use in the event that other nations should attack the USA with gas or chemical weapons), the US forces have available to them a number of chemical warfare weapons. The primary fillings of these weapons are white phosphorus (which ignites spontaneously in contact with air, thereafter burning at 537°F/280.5°C), mustard gas (a persistent and often lethal agent attacking the skin and lungs to cause blisters, ulcers and inflammation), prussic acid agents (including less persistent but more lethal phosgene, cyanogen chloride and hydrocyanic acid attacking the lungs to cause coughing, asphyxia, pneumonia and death) and Sarin (a lethal nerve gas derived from phosphine oxide and attacking the central nervous system to cause motor ataxia and convulsions, coma, paralysis of the autonomic nervous system, and death.

During the 1960s and 1970s the USA developed a number of more 'advanced' nerve gases more difficult to treat than GB, but these are highly classified. Air-dropped weapons of the chemical-warfare type include the **M70** and **M70A1** with a 57-lb (25.9-kg) filling of white phosphorus, or alternatively or persistent mustard gas or gel, the **M78** with a filling of non-persistent AC (hydrocyanic gas), CK (cyanogen chloride) or CG (phosgene), the **M79** with a varying filling of non-persistent AC, CK or CG, the **M113** with a filling of unspecified persistent gas, the **M125** with a 2.6-lb (1.18-kg) filling of non-persistent GB, the **MC1** with a 220-lb (99.8-kg) filling of non-persistent GB, the **Mk 94** (converted Mk 82 general-purpose bomb) with a 108-lb (49-kg) filling of non-persistent GB (Sarin), the **Mk 116 'Weteye'** with a 403-lb (182.8-kg) filling of non-persistent GB and **'Bigeye'** whose two non-toxic agents are mixed on detonation to create a single highly lethal agent.

Two BAe Harriers show off part of the formidable modern inventory of battlefield attack weapons, in this instance fire bombs using jellied petroleum for devastating and far-reaching effect.

Destructor and demolition bombs: These are designed for maximum blast effect on or close to the surface, and are thus designed with a comparatively thin casing and as much HE filling as possible. Destructors are intended not to penetrate into the earth and are often fitted with a long nose probe to ensure detonation at the optimum height above the surface. Such weapons are frequently used for clearance of minefields or the creation of open spaces in wooded and forested regions. Demolition bombs are designed with a strengthened nose but thin walls to penetrate into the ground before detonation to create an earthquake effect. Apart from the M117D and Mk 83 Mod 2/3 mentioned separately (see general-purpose bombs), the main types in US service are the **M56** light-case demolition bomb (a sheet steel casing with 3,515 lb/ 1458.3 kg) of Tritonal), the **Mk 1** demolition bomb (a forged steel casing with 1,060 lb/ 480.8 kg of TNT), the **Mk 1 Mod 1** demolition bomb (a welded, cast or sheet steel casing with 55 lb/24.9 kg of Amatol or Lyconite, or 59 lb/27.8 kg of TNT), the **Mk 36** destructor (essentially the Mk 82 Mod 1 retarded bomb fitted with acoustic, infra-red or seismic fuses to detonate the weapon as a vehicle passes) and the **Mk 111** demolition bomb (a welded, cast or sheet steel casing with 632 lb/286.7 kg of TNT).

Fire bombs: Incendiary weapons have come to play a decisive part in tactical warfare since World War 2. Most common of the fillings is napalm (a mixture of 25% petrol, 25% benzene and 50% aluminium or polystyrene soap as the thickening agent). The most important of these incendiaries are submunitions such as the M50, M69, M74 and M126 carried in clusters (see separate BLU and CBU entry), and the US forces field a large assortment of different incendiary bombs such as the **M47** series (**M47A2, M47A3** and **M47A4**) with various incendiary or plasticized gel fillings (to a maximum of 100 lb/45.4 kg including 40 lb/18.1 kg of incendiary gel) or alternatively a smoke filling (either 100 lb/45.4 kg of white phosphorus or 74 lb/33.57 kg of plasticized white phosphorus), the **M76** incendiary with 180 lb (81.65 kg) of petrol gel and magnesium, the **M116** napalm bomb with 80 or 100 US gal

(302.8 or 378.5 litres) of thickened fuel according to variant (**M116A1** to **M116A23**), the **Mk 77** aluminium-cased fire bomb with 110 or 75 US gal (416.4 or 283.9 litres) of petroleum oil respectively in the **Mk 77 Mod 0** and **Mk 77 Mod 1** variants, the **Mk 78 Mod 1** and **Mk 78 Mod 2** steel-cased fire bombs with 110 US gal (416.4 litres) of petroleum oil, the **Mk 79 Mod 1** steel-cased incendiary bomb with 112 US gal (424 litres) of napalm and petroleum, and the **Mk 122 'Fireye'** fire bomb with a new filling of advanced composition and effect.

Fragmentation bombs: These were designed originally for use against personnel, but advances in the sciences of explosives (leading to higher fragment velocities) and metallurgy (leading to the development of precisely-formed fragments) have now given fragmentation weapons a very useful capability against light armour at close range. The casing of the bomb is pre-fragmented in manufacture to shatter in the optimum number of effective pieces propelled outward at high velocity from the detonation.

The most common types in US service are small submunitions such as the M40A1 fragmentation, the M41A1 (and M41A2) ribbon-stabilized dual-role type with shaped-charge/fragmentation warhead able to penetrate 100 mm (3.94 in) of armour, and M83 'Butterfly' fragmentation as used in cluster bombs (see separate BLU and CBU entry). Larger singleton types include the **M81** (a cast steel casing with a spiral winding of square-section steel wire over a 36-lb/16.3-kg filling of Composition B), the **M82** (a cast steel casing with a spiral winding of square-section steel wire of a 61.2-lb/27.8-kg filling of Composition B), the **M86** and the **M88**.

General-purpose bombs: General-purpose bombs are designed mainly for a combination of blast and fragmentation, and are thus based on a comparatively thick-walled casing with a substantial explosive filling to ensure penetration and considerable upheaval: typically a 2,000-lb (907-kg) general-purpose bomb can form a crater 50 ft (15.24 m) in diameter and 36 ft (10.97 m) in depth by blasting out some 8,500 cu ft (240.7 m^3) of earth. Depending on drop height the bomb can alternatively penetrate 15 in (380 mm) of metal or 11 ft (3.35 m) of concrete, and disperse shrapnel to a lethal radius of 400 yards (366 m). General-purpose bombs are carried as the primary offensive loads of tactical aircraft, and are generally of a low-drag type to disturb flight performance as little as possible.

The General Dynamics F-111 is generally associated with guided weapons of various types, but as illustrated here its four wing hardpoints can be fitted with tandem racks to carry a maximum of 24 750-lb (340-kg) free-fall bombs (four tandem triplets), providing the means of delivering a massive weight of high explosive on to an area target.

Observed by judges, a US Air Force armourer checks 250-lb (113-kg) Mk 81 general-purpose bombs before they are loaded on to a McDonnell Douglas F-4 Phantom II multi-role fighter during an exercise.

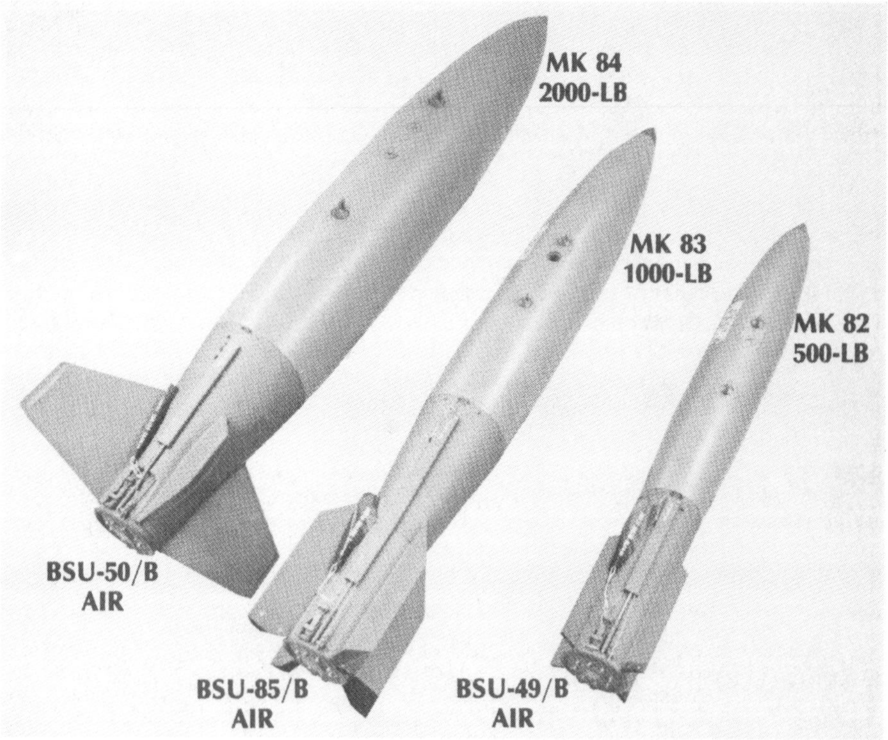

Goodyear retarders of the ballute type can be fitted to Mk 82, Mk 83 and Mk 84 general-purpose bombs, turning them into 'Snakeye' retarded bombs suitable for high-speed low-level delivery.

The most important types in service with the US forces are the **M30A1** (as cast steel casing with 54.2 lb/24.6 kg) of Amatol, TNT or Tritonal), the **M57** (a cast steel casing with 132 lb/59.9 kg of Amatol, TNT or Tritonal), the **M64** and **M64A1** (a cast steel casing with 262 lb/118.8 kg of Amatol, Composition B, TNT or Tritonal), the **M65A1** (a cast steel casing with 572 lb/259.5 kg of Amatol, Composition B, TNT or Tritonal), the **M66A1** and **M66A2** (a cast steel casing with 1,036 lb/469.9 kg of Amatol, TNT or Tritonal), the **M109**, the **M110**, the **M117A** and **M117A1E2** (a cast steel casing with 403 lb/182.8 kg of Minol 2 or Tritonal), the **M118E2** a cast steel casing with 1,975 lb/895.9 kg of Trinitrol), the **M121,** the **Mk81** (a cast steel casing with 96 lb/43.5 kg of H6, Minol 2 or Tritonal), the **Mk 82** (a cast steel casing with 192 lb/87.1 kg of H6, Minol 2 or Tritonal), the **Mk 83** (a cast steel casing with 445 lb/

Though planned as a long-range strategic bomber with thermonuclear weapons, the Boeing B-52 was pressed into tactical service during the Vietnam War with conventional ordnance carried in the bomb-bay and on the underwing pylons previously earmarked for AGM-69 Hound Dog and cancelled AGM-87 Skybolt cruise missiles. Here US Air Force armourers complete loading the four triplets of 750-lb (340-kg) 'slick' general-purpose bombs. Such free-fall weapons may lack the pinpoint accuracy of guided weapons, but dropped en masse *can fulfil a very significant interdiction role.*

201.9 kg of H6, Minol 2 or Tritonal) and the **Mk 84** (a cast steel casing with 945 lb/428.7 kg of H6, Minol 2 or Tritonal). Of these the most important and widely used are the M117A, M118, Mk 81, Mk 82, Mk 83 and Mk 84.

The M117A, Mk 81, Mk 82 and Mk 83 'slick' low-drag bombs can be adapted as retarded or 'Snakeye' bombs by the substitution of a different tail. The M117A takes an MAU-91A/B retarder to become the 863-lb (391.5-kg) **M117D** destructor, the Mk 81 takes the Mk 14 tail to become the 302-lb (137-kg) **Mk 81 Mod 1** retarded bomb, the Mk 82 takes the Mk 15 or Mk 115 tail to become the 571-lb (259-kg) **Mk 82 Mod 1** retarded bomb, and the Mk 83 takes the MAU-91A/B tail to become the 1,025-lb (464.9-kg) **Mk 83 Mod 2/3** destructor. Also available are three types of Goodyear ballute (balloon parachute) air-inflated retarder tail:

Right *The McDonnell Douglas AV-8B Harrier II was designed as a 'bomb truck' for US Marine Corps close support operations, and this capability is well exemplified here: 17,500-lb (227-kg) Mk82 'slick' general-purpose bombs carried as five triplets and two singletons.*

Far right *The F-15E is the long-range all-weather attack and interdiction version of the McDonnell Douglas Eagle air-superiority fighter, and is seen here with its conformal FAST (Fuel And Sensor Tactical) packs carrying 10,000-lb (4536 kg) of fuel and providing tangential carriage for bombs within the carried total of 22 500-lb (227-kg) Mk 82 general-purpose bombs. Bombs can be delivered with extreme accuracy thanks to the capabilities of the APG-63 synthetic-aperture radar, which can detect and display mobile targets in all weather conditions out to a range of 20 miles (32 km).*

the BSU-49/B is matched to the Mk 82, the BSU-50/B to the Mk 84 and the BSU-85/B to the Mk 83. The BSU-49/B tail unit with stabilizing fins and ballute weighs 56 lb (25.4 kg), and its dimensions include a length of 26 in (660 mm), a diameter of 8.75 in (222 mm) and a fin span of 15 in (381 mm). The BSU-50/B tail unit with stabilizing fins and ballute weighs 96 lb (43.55 kg), and its dimensions include a length of 30.5 in (775 mm), tail cone dimensions of 15.86 by 8.75 in (403 by 222 mm) and a fin span of 37 in (940 mm). The BSU-85/B tail unit with stabilizing fins and ballute weighs 90 lb (40/82 kg), and its dimensions include a length of 39.2 in (996 mm), tail cone dimensions of 12.85 by 8.75 in (326 by 222 mm) and a fin span of 20.32 in (516 mm).

Leaflet bombs: The least offensive type of bomb carried by US aircraft is the leaflet variety such as the **M104**, the **M105**, the **M129**, the M129E1 and the **M129E2.**

Photoflash bombs: These are designed to illuminate objectives for aerial photography, and the principal types in US service are the **M46** (a cast steel casing with 25 lb/11.3 kg of Photoflash compound producing a 500 million-candlepower light), the **M120** and **M120A1**, and the **M122.**

Practice bombs: The practice bombs are designed to simulate the ballistic qualities of live ordnance for training purposes, and are often considerably smaller than the live weapons they ape, though full-weight versions are available so that pilots can better experience the effect that a mass of ordnance can exert on his aircraft. The primary bombs in this category are the **BDU-27/B** fragmentation dummy, the **BDU-28/B** fragmentation dummy, the **BDU-30/B** practice, the **BDU-33/B** and **BDU-33/C** practice, the **BDU-34/B** practice, the **BDU-37/B** practice, the **M38A2** practice (a cast steel casing with 80 lb/36.3 kg of sand and 3 lb/1.36 kg of black powder as a spotting charge), the **M124** practice, the **Mk 5 Mod 2** and **Mk 5 Mod 3** practice miniature, the **Mk 15 Mod 2, Mk 15 Mod 3** and **Mk 15 Mod 4** practice, the **Mk 23 Mod 1** practice miniature, the **Mk 43 Mod 1** practice miniature, the **Mk 65 Mod 0** practice, the **Mk 66 Mod 0** practice, the **Mk 76** practice miniature, the **Mk 86 Mod 0** and **Mk 86 Mod 1** practice, the **Mk 87 Mod 0** practice, the **Mk 88 Mod 0, Mk 89 Mod 0** and **Mk 88 Mod 1** practice miniature, and the **Mk 106 Mod 0** practice miniature. Licensed copies of several

Right *One of today's most remarkable aircraft is the Grumman A-6E/TRAM Intruder, a carrierborne attack and strike platform able to deliver massive quantities of ordnance with pinpoint accuracy in all weather conditions. The Intruder has five hardpoints each able to carry tandem triplets of 250- or 500-lb (113- or 227-kg) general-purpose bombs, the former type being seen here.*

291

AE
155687
NAVY
500

US practice bombs are made by a number of countries, typical being the **PB-101** improved version of the Mk 87 made by Orlite Engineering in Israel: this is externally identical to the US original, but is made of glassfibre-reinforced polyester plastic; the type weighs 30 kg (66.1 lb) empty, and for use is filled with 60 kg (132.3 lb) of water; located in the centre of the tail is a housing for any type of marker chosen by the operator.
Target-identification bomb: The sole weapon of this nature in the US inventory is the M84A1.
Northrop Inertially-Aided Munition: In 1987 Northrop's Precision Products Division introduced an add-on kit to improve the accuracy of the USA's standard range of free-fall weapons: this kit comprises a tailcone (complete with steerable fins) containing a strapdown inertial navigation system to control the bomb's rear fins via a microprocessor. Before release, the Inertially-Aided Munition is primed with up-to-the-second navigation data from the launch aircraft, allowing the trajectory to be controlled after release to optimize the bomb's chances of striking the designated target at the best possible impact angle. The bomb can be released at a range of 17.25 miles (27.76 km) from the target by a high-altitude aircraft, or at a range of 5.75 miles (9.25 km) from the target by a low-altitude aircraft using the toss delivery technique.

USA

Bomb Live Unit (BLU) and Cluster Bomb Unit (CBU) weapons

Type	Diameter	Length	Weights	
			Nominal	Actual
BLU-1			750 lb (340.2 kg)	
BLU-3				1.75 lb (0.79 kg)
BLU-4				1.2 lb (0.54 kg)
BLU-6			data not available	
BLU-7				1.4 lb (0.635 kg)
BLU-10			250 lb (113.4 kg)	250 lb (113.4 kg)
BLU-11			500 lb (226.8 kg)	
BLU-14			data not available	
BLU-15			data not available	
BLU-16			data not available	1.6 lb (0.73 kg)
BLU-17	data not available			1.1 lb (0.5 kg)
BLU-18				0.4 lb (0.19 kg)
BLU-19			data not available	
BLU-20			data not available	
BLU-21			data not available	
BLU-22			data not available	
BLU-23			500 lb (226.8 kg)	490 lb (222.3 kg)
BLU-24			1 lb (0.45 kg)	1.2 lb (0.54 kg)
BLU-25			data not available	
BLU-26			1 lb (0.45 kg)	1.2 lb (0.54 kg)
BLU-27			750 lb (340.2 kg)	
BLU-28			data not available	

Type	Diameter	Length	Weights Nominal	Weights Actual
BLU-29			data not available	160 lb (72.58 kg)
BLU-30			data not available	
BLU-31			750 lb (340.2 kg)	
BLU-32			500 lb (226.8 kg)	589 lb (267.2 kg)
BLU-33			1,500 lb (680.4 kg)	
BLU-34			3,000 lb (1361 kg)	
BLU-35			250 kg (113.4 kg)	
BLU-36				0.95 lb (0.43 kg)
BLU-39				0.18 lb (0.08 kg)
BLU-40				1.7 lb (0.78 kg)
BLU-41			data not available	
BLU-46			data not available	
BLU-47			data not available	
BLU-48			data not available	
BLU-49				13.4 lb (6.1 kg)
BLU-50			data not available	
BLU-51			data not available	
BLU-52			700 lb (317.5 kg)	
BLU-53			20 lb (9.1 kg)	
BLU-57			500 lb (226.8 kg)	
BLU-58			500 lb (226.8 kg)	
BLU-59				0.95 lb (0.43 kg)
BLU-60				13 lb (5.9 kg)
BLU-61	data not available			2.2 lb (1.0 kg)
BLU-62				0.95 lb (0.43 kg)
BLU-63				0.95 lb (0.43 kg)
BLU-64				770 lb (349.3 kg)
BLU-65				280 lb (127.0 kg)
BLU-66			1 lb (0.45 kg)	1.1 lb (0.5 kg)
BLU-67				11 lb (5.0 kg)
BLU-68				0.93 lb (0.42 kg)
BLU-69				1.6 lb (0.73 kg)
BLU-70				0.88 lb (0.4 kg)
BLU-72			1,100 lb (497.0 kg)	
BLU-73				130 lb (59.0 kg)
BLU-74				235 lb (106.6 kg)
BLU-75			data not available	
BLU-76			2,650 lb (1202 kg)	
BLU-77			1 lb (0.45 kg)	1.02 lb (0.46 kg)
BLU-80			data not available	
BLU-82			15,000 lb (6804 kg)	
BLU-86				1 lb (0.45 kg)
BLU-89			data not available	
BLU-90			data not available	
BLU-91			data not available	
BLU-92			data not available	
BLU-94			data not available	
BLU-95			500 lb (226.8 kg)	
BLU-96			2,000 lb (907 kg)	

Type	Diameter	Length	Weight Nominal	Weight Actual
BLU-97				3.4 lb (1.54 kg)
BLU-106	4 in (101.6 mm)	53.5 in (1105 mm)	45 lb (20.4 kg)	43 lb (19.5 kg)
BLU-108			data not available	
BLU-109			2,000 lb (907 kg)	
M1			100 lb (45.4 kg)	122 lb (55.3 kg)
M26	data not available		500 lb (226.8 kg)	418 lb (189.6 kg)
M27			500 lb (226.8 kg)	585 lb (265.4 kg)
M28			100 lb (45.4 kg)	100 lb (45.4 kg)
M29			500 lb (226.8 kg)	415 lb (188.2 kg)
M36				722 lb (327.5 kg)
Mk 44			550 lb (249.5 kg)	560 lb (254 kg)

Variants and applications

BLU-1: The American BLU (Bomb Live Unit) designation covers a multitude of weapons and indeed of weapon types, ranging from small submunitions carried by CBU and SUU series dispensers to large special ordnance items. The series was originated in the 1960s as successor to earlier fragmentation weapons used in the anti-personnel and anti-armour roles, the concept being based on the premise that small BLUs could be developed comparatively easily for packaging in various numbers into Suspended Underwing Unit, which then became the specific Cluster Bomb Unit. The two primary types of SUU are the dropped type, designed to hinge open in the air and so release a concentrated cloud of BLUs which are self-dispersed by aerodynamic surfaces on the bomblets, and the fixed type (of which the SUU-7 was the first), designed to remain under the wing of the launch aircraft and use compressed air to dispense rearwards one or more streams of BLUs, dropping freely after all the BLUs have been launched. The BLU-1 is a fire bomb with 90 to 100 US gal (340.7 to 378.5 litres) of napalm; there are also **BLU-1B/B** and **BLU-1C/B** variants.

BLU-3 'Pineapple': This is a cylindrical bomblet with pop-out drag vanes, designed for use against unarmoured targets. The 0.35-lb (0.16-kg) Cyclotrol filling disperses 250 steel pellets at high velocity; there is also a **BLU-3/B** variant. A load of 360 BLU-3 bomblets in an SUU-7 dispenser constitutes the **CBU-2/A** cluster bomb, while 409 BLU-3 bomblets in the same dispenser constitute the **CBU-2B/A** variant. The combination of BLU-3/3 bomblets and SUU-7B/A dispenser is the **CBU-14/A** cluster bomb, while the combination of BLU-3/B bomblets and SUU-14A/A dispenser results in the **CBU-14A/A** cluster bomb.

BLU-4: This is an anti-materiel bomblet, on impact ejecting upwards a spherical fragmentation unit that explodes at a height of 10 ft (3 m). A load of 509 BLU-4 bomblets in an SUU-7 dispenser constitutes the **CBU-1** cluster bomb.

BLU-6: This is a smoke bomblet using the TMU-10 airborne smoke generator.

BLU-7/B: This is an anti-tank bomblet, using a hollow-charge warhead and armed by the deployment of its stabilizing parachute and available in **BLU-7/A** and **BLU-7A/B** live as well as **BLU-7A(D-1)/B** dummy variants. A load of 352 BLU-7/B bomblets in an SUU-10A/A dispenser constitutes the 660-lb (229.4-kg) **CBU-3A/A** cluster bomb, while 352 BLU-7A/B bomblets in the SUU-10/A dispenser constitute the **CBU-3/A** cluster bomb, and a load of 371 BLU-7A/B bomblets in the SUU-10B/A dispenser constitutes the **CBU-3B/A** cluster bomb.

BLU-10/B: This is a fire bomb based on an aluminium casing filled with 33 US gal (124.9 litres) of napalm weighing 211 lb (95.7 kg), and available in a **BLU-10A/B** variant.

BLU-11/B: This is a fire bomb filled with 65 US gal (246.1 litres) of napalm.

BLU-14: This is the designation of the **MLU10/B** mine for low-level skip bombing. The MLU-10/B weighs 766 lb (347.5 kg) and is filled with Destex explosive, which is deto-

nated by the weapon's advanced combination of infra-red, magnetic and seismic fuses.
BLU-15: Nothing is known of this other than it is an anti-materiel weapon.
BLU-16/B: This is a cylindrical smoke bomblet modified from an M8 grenade. A load of 261 such bomblets in an SUU-7B/A dispenser constitutes the **CBU-11/A** cluster bomb.
BLU-17/B: This is a cylindrical smoke bomblet modified from an M15 grenade with white phosphorus filling. A load of 261 such bomblets in an SUU-7B/A dispenser constitutes the 640-lb (290.3-kg) **CBU-12/A** cluster bomb, while the use of the SUU-7C/A dispenser turns this into the **CBU-12A/A** cluster bomb, and a mixture of 261 BLU-16/B and BLU-17/B smoke bomblets in the SUU-11B/A dispenser results in the **CBU-13/A** cluster bomb. A load of 72 BLU-17/B bomblets can also be carried in an SUU-14/A dispenser as the **CBU-22/A** cluster bomb.
BLU-18/B: This is a triangular anti-personnel fragmentation bomblet fitted with a delay fuse. A load of 1,200 of these bomblets in the SUU-31/A dispenser constitutes the **CBU-7/A** cluster bomb.
BLU-19/B23: This is a chemical warfare bomblet filled with explosively-dispersed GB

The Cessna A-37 Dragonfly was developed from the T-37 trainer, but notwithstanding its origins can carry a very comprehensive assortment of light attack weapons. This is an interim A-37A model, and amongst the stores illustrated are launchers for 2.75- and 5-in (69.85- and 127-mm) rockets, 250- and 500-lb (113- and 227-kg) general-purpose bombs, BLU-1/B napalm bombs, SUU-11B/A Minigun pods.

(Sarin) nerve gas. The combination of these bomblets with an SUU-13/A dispenser produces the **CBU-15/A** cluster bomb.
BLU-20/B23: This is a chemical warfare bomblet filled with thermally-dispersed BZ incapacitant gas.
BLU-21/B45: This is a biological warfare bomblet filled with a dry biological agent.
BLU-22/B45: This is a biological warfare bomblet filled with a wet biological agent.
BLU-23/B: This is a fire bomb based on an aluminiun casing filled with 67 US gal (253.6 litres) of Napalm-B, a heavier development of the basic napalm and weighing, in this application, 430 lb (195 kg).
BLU-24/B 'Orange': This is a spherical anti-personnel fragmentation bomblet with plastic fins and a special spin-delay fuse designed to ensure that the bomblet penetrates jungle canopy (reducing spin to less than 2,000 revolutions per minute) before detonation between the canopy and the ground. The type is based on a steel casing and has a Cyclotrol filling with 300 embedded steel fragments. There is also a **BLU-24A/B** variant. The combination of BLU-24/B bomblets and an SUU-14/A dispenser constitutes the **CBU-24/A** cluster bomb, the same bomblets in an SUU-14 dispenser constitute the **CBU-25** cluster bomb, and 264 BLU-24/Bs in an SUU-50 dispenser constitute the 750-lb (340-kg) **CBU-60/A** cluster bomb.
BLU-25/B: This is a cylindrical anti-personnel bomblet. Such bomblets in an SUU-13A/A dispenser constitute the **CBU-18/A** cluster bomb.
BLU-26/B 'Guava': This is a spherical anti-personnel and anti-materiel bomblet. The type is based on a cast steel casing 2.36 in (6.0 cm) in diameter and fitted with aerodynamic vanes, is filled with 0.7 lb (0.32 kg) of Cyclotrol or TNT, and on detonation spews out a rapidly-expanding sphere of 600 steel fragments (embedded in the explosive) which are lethal to a radius of 40 ft (12.2 m). The BLU-26/B can be fitted with an airburst, impact, or delay fuse: the airburst type detonates the weapon at a height of 30 ft (9.1 m), and the delay type can be set to provide detonation one minute or more after the weapon has hit the ground. These bomblets are carried in an SUU-31/B dispenser as the **CBU-23/B** cluster bomb, and 665 of the type (or alternatively 665 BLU-36/B APAMBs) in an SUU-30/B dispenser are the **CBU-24/B** cluster bomb. The SUU-30/B splits open in the air to release the bomblets, whose vanes spin them outward from the release point to ensure a wide scatter and whose three fuse types provide a highly confusing mix of air, ground and delay burst from a single delivery.
BLU-27/B: This is a fire bomb filled with 100 US gal (378.5 litres) of Napalm-B.
BLU-28/B: This is a biological warfare bomb with a filling of self-dispersing biological agent.
BLU-29/B: This is an anti-personnel and anti-materiel bomb (APAMB) using a flame agent canister.
BLU-30/B: This is a chemical warfare bomb with a filling of dry chemical agent.
BLU-31/B: This is a 750-lb (340-kg) demolition bomb based on the MLU-10 mine with a new acoustic and seismic sensor package tuned to the sound and vibration of tanks.
BLU-32/B: This is a fire bomb based on a welded aluminium casing filled with 67 US gal (253.6 litres) of Napalm-B weighing, in this application, 529 lb (240 kg); there is also a 595-lb (269.9-kg) **BLU-32A/B** version.
BLU-33/B: This is a demolition bomb of unrevealed type.
BLU-34/B: This is a demolition bomb for use against hard surface structures.
BLU-35/B: This is a fire bomb with a module to dispense 310 lb (140.6 kg) of napalm.
BLU-36/B: This is an anti-personnel and anti-materiel bomblet (APAMB) fitted with a random-delay fuse. A load of 665 of these bomblets can be carried in the SUU-30/B dispenser of the CBU-24/B cluster bomb as an alternative to the BLU-26/B bomblets, and 670 of these bomblets in an SUU-30/B dispenser constitute the 840-lb (381.0-kg) **CBU-29/B** cluster bomb.

BLU-39: This is a chemical warfare bomblet with the alternative designation **XM16**; the filling is CS gas, which is emitted as the bomblet bounces over the ground. The **CBU-19** is a twin 'strongback' dispenser for 16 modules each carrying an internal load of 33 BLU-39/B bomblets, and the **CBU-30** cluster bomb is an SUU-13 dispenser carrying 40 CDU-12 clusters each of 30 BLU-39/B bomblets.
BLU-39/B23: This is a dummy version of the BLU-39.
BLU-40/B: This is a fragmentation bomblet with a random-delay fuse.
BLU-41/B: This is a spherical fragmentation bomblet with spin arming and a spin-delay fuse.
BLU-42/B: This is a spherical anti-personnel minelet with a number of surface and trip-wire sensors for the 0.25-lb (0.11-kg) filling of Composition-B; the mine is also fitted with an anti-tamper device and a self-destruct system. These minelets are grouped in units of 54 to create CDU-18 or CDU-19 clusters, of which 10 in an SUU-38/A dispenser constitute the **CBU-34/A** cluster bomb.
BLU-43/B 'Short Dragontooth': This is a triangular anti personnel blast minelet; 120 of these minelets constitute a CDU-12 cluster unit, of which 40 can be loaded into an SUU-13 dispenser to create the **CBU-28** cluster bomb.
BLU-44/B 'Long Dragontooth': This is a triangular anti-personnel blast minelet; 120 of these minelets form the CDU-3 clusters, of which 40 are loaded into an SUU-13/A dispenser to create the **CBU-37/A** cluster bomb.
BLU-45/B: This is a shaped-charge anti-tank minelet with a timed self-destruct system. A load of 30 such mines in an SUU-36/A dispenser constitutes the 770-lb (349.3-kg) **CBU-33/A** cluster bomb.
BLU-46/B: This is an anti-personnel bomblet.
BLU-47/B: This is a general-purpose bomblet.
BLU-48/B: This is a spherical fragmentation bomblet and, grouped into CDU-37 clusters, such bomblets are carried in an SUU-37 dispenser to form the 840-lb (381-kg) **CBU-43** cluster bomb. The bomblets are also carried in an SUU-30 dispenser casing to create the **CBU-68** cluster bomb.
BLU-49/B: This is a ring tailed fragmentation bomblet with a 70/30 Cyclotrol filling. A load of 40 such bomblets in an SUU-13/A dispenser constitutes the **CBU-38/A** cluster bomb.
BLU-50/B: This is a chemical warfare bomb with a filling of BZ incapacitant gas.
BLU-51/B: This is a fire bomb.
BLU-52/B: This is a chemical warfare bomb with a 615-lb (279-kg) filling of CS gas.
BLU-53/B: This is a canister fire bomblet filled with Napalm-B. A load of 18 such bomblets in an SUU-51 dispenser constitutes the 485-lb (220-kg) **CBU-41/B** cluster bomb.
BLU-54/B: This is an anti-personnel minelet identical to the BLU-42/B but fitted with a long-life self-destruct system. A total of 54 such minelets is grouped in the CDU-20 or CDU-21 cluster, of which 10 in an SUU-38/A dispenser constitute the **CBU-42/A** cluster bomb.
BLU-55/B: This is an anti-personnel minelet. Such minelets are clustered in an SUU-13 dispenser casing to create the **CBU-47** cluster bomb.
BLU-56/B: This is an anti-personnel mine.
BLU-57/B: This is a fragmentation bomb.
BLU-58/B: This is a general-purpose bomb.
BLU-59/B: This is a fragmentation bomblet identical to the BLU-26 but fitted with a shorter-life random-delay fuse. A load of 670 such bomblets in an SUU-30/B dispenser constitutes the 840-lb (381-kg) **CBU-49/B** cluster bomb.
BLU-60/B: This is a fragmentation bomblet with a filling of Cyclotrol. A load of 40 such bomblets in an SUU-13 dispenser constitutes the **CBU-50/A** cluster bomb.

BLU-61/B: This is a spherical anti-materiel incendiary/fragmentation bomblet with a filling of zirconium and Composition-B. A load of 254 such bomblets in an SUU-30B/B dispenser constitutes the 770-lb (349.3-kg) **CBU-52/B** cluster bomb, while 290 of the bomblets in an SUU-51B/B dispenser casing constitute the **CBU-76/B** cluster bomb.
BLU-62/B: This is an anti-personnel and anti-materiel fragmentation bomblet fitted with a Cyclotrol-filled bounding unit.
BLU-63/B: This is a spherical anti-personnel and anti-materiel fragmentation bomblet filled with spin- and impact-fused Cyclotrol inside a pre-fragmented case. A load of 650 such bomblets in an SUU-30A/B dispenser constitutes the 815-lb (369.7-kg) **CBU-58A/B** cluster bomb, while 1,800 of the bomblets in an SUU-54A/B dispenser constitute the **CBU-75A/B** cluster bomb, and 790 of the same bomblets in an SUU-51B/B dispenser constitute the **CBU-77/B** cluster bomb.
BLU-64/B: This is a general-purpose fuel/air explosive bomb filled with 440 lb (199.6 kg) of hydrocarbon fuel. The weapon was combat tested in 1967 and 1968, but was not standardized for service.
BLU-65/B: This is a fire bomb.
BLU-66/B 'Pineapple': This is a spherical anti-personnel bomblet with plastic fins: the type is based on a cylindrical steel case filled with TNT and 260 embedded steel fragments, and is fitted with an impact fuse. A load of 640 such bomblets (or alternatively the same number of BLU-24/Bs) in an SUU-7C/A dispenser constitutes the 880-lb (399.2-kg) **CBU-46B/A** cluster bomb.
BLU-67/B: This is a cratering bomblet. A load of 40 such bomblets in an SUU-13/A dispenser constitutes the **CBU-51/A** cluster bomb.
BLU-68/B: This is a spherical incendiary anti-personnel and anti-materiel bomblet (APAMB) with a filling of zirconium sponge. A load of 670 such bomblets in an SUU-30B/B dispenser constitutes the 815-lb (369.7-kg) **CBU-54/B** cluster bomb.
BLU-69/B: This is a spherical incendiary anti-personnel and anti-materiel bomblet (APAMB). A load of 132 such bomblets in an SUU-14A/A dispenser constitutes the **CBU-57** cluster bomb.
BLU-70/B: This is a spherical incendiary anti-personnel and anti-materiel bomblet. A load of 670 of these bomblets in an SUU-30B/B dispenser constitutes the 815-lb (369.7-kg) **CBU-53/B** cluster bomb.
BLU-71/B: This is a land mine.
BLU-72/B 'Pave Pat I': This is a 2,500-lb (1134-kg) fuel/air explosive weapon based on a sheet steel container and designed for low-speed delivery with a retarding drogue and filled with 2,245 lb (1018.3 kg) of propane. The weapon was combat tested in 1967 and 1968, but was not standardized for service.
BLU-73/B: This is a fuel/air explosive canister bomb 21 in (533 mm) long and 13.75 in (349 mm) in diameter, and fitted with a retarding drogue. The weapon weighs 100 lb (45.4 g) and contains 72 lb (32.66 kg) of ethylene oxide, three such weapons in an SUU-49 dispenser constituting the nominal 500-lb (226.8-kg) but actual 460-lb (208.7-kg) **CBU-55/B** cluster bomb, which has a diameter of 13.75 in (349 mm) and length of 78.75 in (2000 mm). Three BLU-73/B bombs are also carried in the **CBU-72** cluster bomb: once dropped, the dispenser is split by a detonating cord to free the BLU-73s for their 90-ft (27.4-m) per second para-retarded fall. At an altitude of 30 ft (9 m) a proximity fuse detonates the charge in each BLU-73 to create an aerosol cloud of fuel and air some 60 ft (18.3 m) in diameter and 8 ft (2.44 m) thick, which sinks to the ground and is then ignited by another detonator to create a rapidly-expanding wave front whose 300-lb/sq in (21.98-kg/cm^2) overpressure flattens everything within a 60-ft (18.3-m) radius circle and cause very great damage beyond this circle.

The same BLU-73/B core is used in the US Army's **FAESHED** (Fuel/Air Explosive

HElicopter Delivered) weapon, which is in effect the US Navy's CBU-55 revised slightly for carriage by and delivery from a helicopter. As well as a high efficiency in clearing landing circles for helicopters in wooded and jungle terrain, the FAE weapon has also proved highly successful in minefield clearance by the simple expedient of detonating all pressure-fused mines within a radius of 29 ft (8.84 m) and all trip-fused mines within a radius of 85 ft (25.91 m). A similar weapon used by the US Marine Corps is the **MAD-FAE** (Mass Air Delivery Fuel/Air Explosive); this consists of fin-stabilized aluminium racks, each holding 12 136-lb (61.69-kg) FAE units, carried on the external hook of helicopters such as the Boeing Vertol CH-47 and Sikorsky CH-53 for individual or salvo release over minefields. Considerable research and development effort is continuing into FAE weapons using fuels such as butane, ethylene oxide, methyl acetylene, propadiene, propane and propylene oxide. Typical of this effort is the US Army's SLUFAE (Surface-Launched Unit, Fuel/Air Explosive), a mine-clearing system using the 5-in (127-mm) Zuni rocket fitted with a BLU-73/B warhead.

BLU-74/B: This is a modular fire bomb with a filling of Napalm-B.

BLU-75/B: This is a fire bomb.

BLU-76/B 'Pave Pat II': This is a fuel-air explosive weapon designed primarily for high-speed delivery but possessing the same low-speed capability as the BLU-72/B. The BLU-76/B contains a higher ethylene oxide payload, but requires some four seconds for the formation of the aerosol cloud before the second detonation can be started.

PLU-77/B (sic)**:** This is an anti-tank fragmentation bomblet with the advanced capability of discriminating between hard and soft targets; the type has a secondary anti-personnel function, the shaped-charge warhead having a bounding function in this role. A load of 717 such bomblets in a Mk 7 Mod 3 dispenser constitute the US Air Force's 750-lb (340-kg) **CBU-59/B** cluster bomb, otherwise known as the **'Rockeye II'** cluster bomb.

BLU-80 'Bigeye': This is a chemical warfare bomb the world's first air-delivered binary weapon in which two separately stored 'safe' agents are mixed by the bomb's explosion to create a single highly lethal agent.

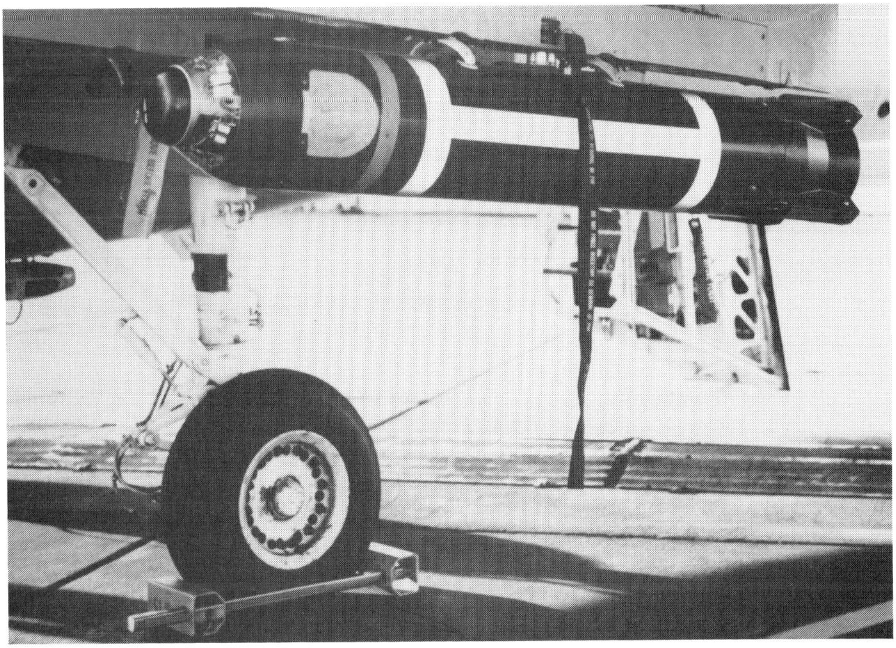

Though visually unimpressive, chemical weapons such as this 'Bigeye' are extremely important types. This joint US Air Force and US Navy project has produced in the 'Bigeye' the world's first air-dropped binary bomb, in which two individually non-toxic chemicals are mixed by the detonation of the bomb to create a persistent nerve agent.

BLU-81/B 'Grasshopper': This is a cancelled anti-vehicle minelet. Such weapons were to have been grouped in CDU-24/B canisters for carriage in an SUU-51/B dispenser as the **CBU-66** cluster bomb.
BLU-82/B: This is a 15,000-lb (6804-kg) general-purpose bomb used mainly for the clearing of helicopter landing zones in jungle. In the **BLU-82/B Mod 'Big Blue 82'** developed version the warhead comprises 12,600 lb (5715.4 kg) of GSX, a gelled slurry blast explosive (DBA-22M, an aqueous solution of ammonium nitrate, aluminium powder and polystyrene soap) detonated just above ground level by the 38-in (965-mm) nose probe to create a 1,000-lb/sq in (73.27-kg/cm^2) overpressure.
BLU-86A/B: This is a steel-cased fragmentation bomblet with a filling of Composition-B or Cyclotrol. A load of 670 such bomblets in an SUU-30A/B dispenser constitutes the 815-lb (369.7-kg) **CBU-71** cluster bomb, while 1,800 of the bomblets can be used as an alternative load for the BLU-63/Bs in the **CBU-75A/B** cluster bomb.
BLU-87(D-1)/B: This is a dummy fragmentation bomblet, of which the training version is the **BLU-87(T-1)/B.**
BLU-89/B: This is a general-purpose blast fragmentation bomb.
BLU-90/B: This is a general-purpose fragmentation bomb with a shaped-charge penetration nose.
BLU-91/B 'Gator': This is an anti-tank minelet, and such weapons can be carried in an SUU-58/B dispenser to create the **CBU-78/B** cluster bomb, in the SUU-58/B dispenser to create the **CBU-82/B** cluster bomb, and in the SUU-54A/B dispenser to create the **CBU-84/B** cluster bomb.
BLU-92/B 'Gator': This is an anti-personnel mine, and such weapons can be carried in an SUU-58/B dispenser to create **CBU-78/B** cluster bomb, in the SUU-58/B dispenser to create the **CBU-83/B** cluster bomb, as an alternative to BLU-91/Bs in the SUU-54A/B dispenser to create the **CBU-84/B** cluster bomb, and in the SUU-4A/B dispenser to create the **CBU-86/B** cluster bomb.

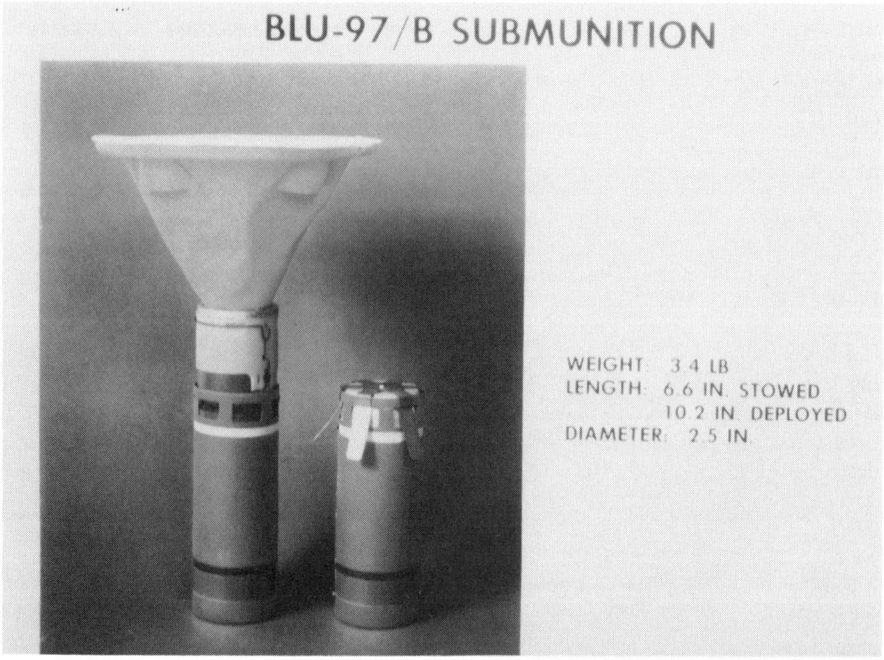

The BLU-97/B submunition is seen in stowed and deployed positions.

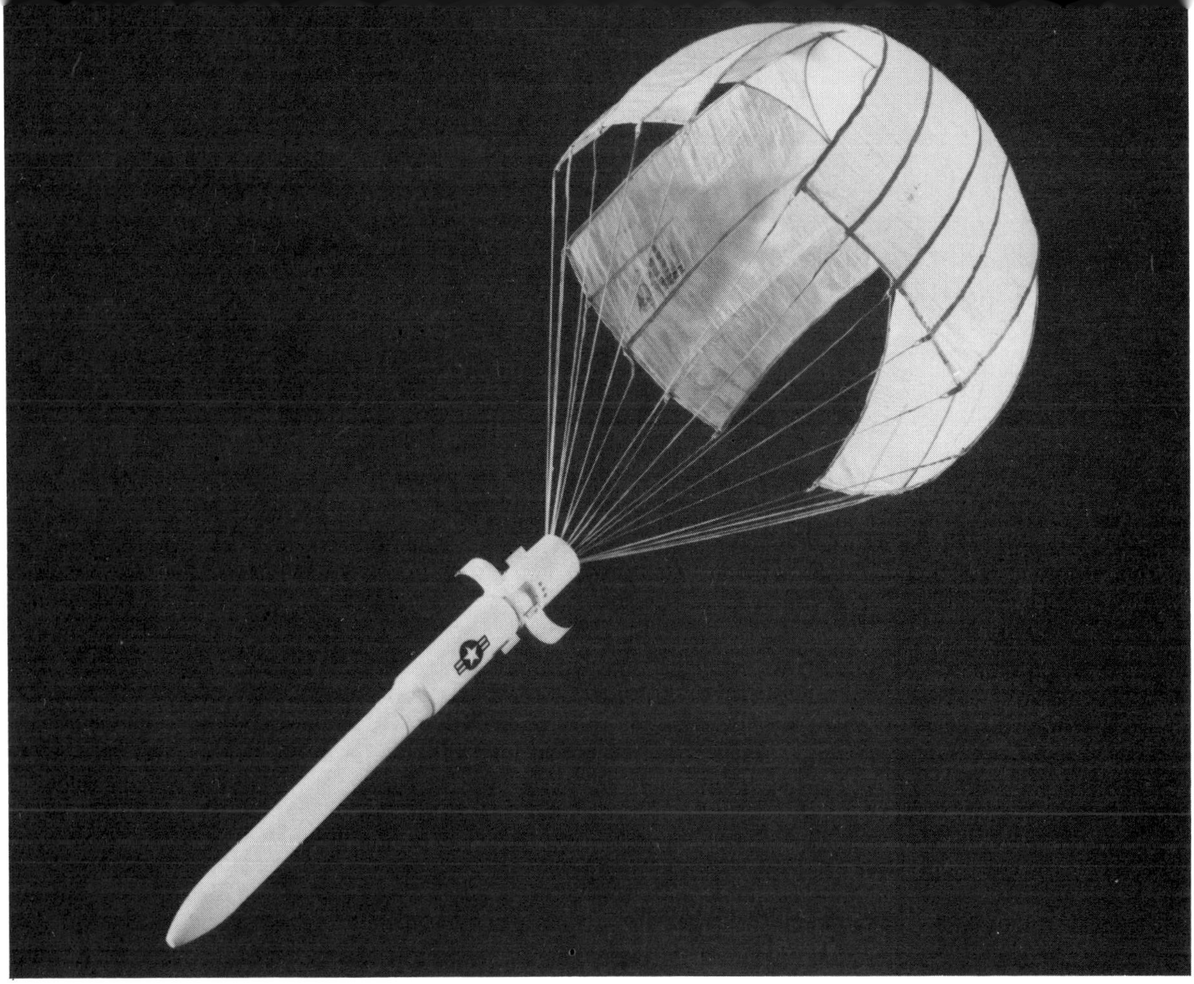

The Avco BLU-106/B BKEP in the orientation and stabilization phase of its delivery, with the parachute retarding the weapon and allowing the nose to pitch down.

BLU-94/B: This is a shallow-water anti-shipping mine with secondary anti-tank capability.
BLU-95/B: This is a 500-lb (226.8-kg) fuel/air explosive weapon containing 300 lb (136.1 kg) of propylene oxide fuel. Development of this weapon was undertaken for the US Air Force with the initial designation **HSF-1** (High-Speed FAE No.1).
BLU-96/B: This is a 2,000-lb (907-kg) fuel/air explosive weapon containing 1,400 lb (635 kg) of propylene oxide fuel. The BLU-96 is a payload carried by the GBU-15 guided glide bomb (see Air-to-Surface Missiles section), and development was undertaken for the US Air Force with the initial designation **HSF-11** (High-Speed FAE No.11).
BLU-97/B: This is the submunition carried by the CBU-87/B Combined Effects Munition (see below). The submunition has a conical stabilizing retarder and weighs 3.4 lb (1.54 kg), its dimensions including a diameter of 2.5 in (63.5 mm) and a length increasing from 6.6 in (167.6 mm) stowed to 10.2 in (401.6 mm) with its tail fully deployed.
BLU-106/B: This weapon was developed as the **Boosted Kinetic-Energy Penetrator** for use in the increasingly important tasks of runway and HAS (Hardened Aircraft Shelter) destruction, mainly by the Medium-Range Air-to-Surface Missile derivative of the BGM-109 Tomahawk cruise missile (28 submunitions in two bays), in the SUU-64/B dispenser carried by the Direct Airfield Attack Combined Munition (see below), and in the SUU-54 dispenser carried by the AGM-130B stand-off weapon.

Each BLU-106/B weighs 45 lb (20.4 kg), and is 43.5 in (1105 mm) long, with diameter increasing from 3.4 in (86 mm) in the case of the penetrator to 4 in (102 mm) in the case of the booster section. The BLU-106/Bs are dispensed with varying lateral ejection

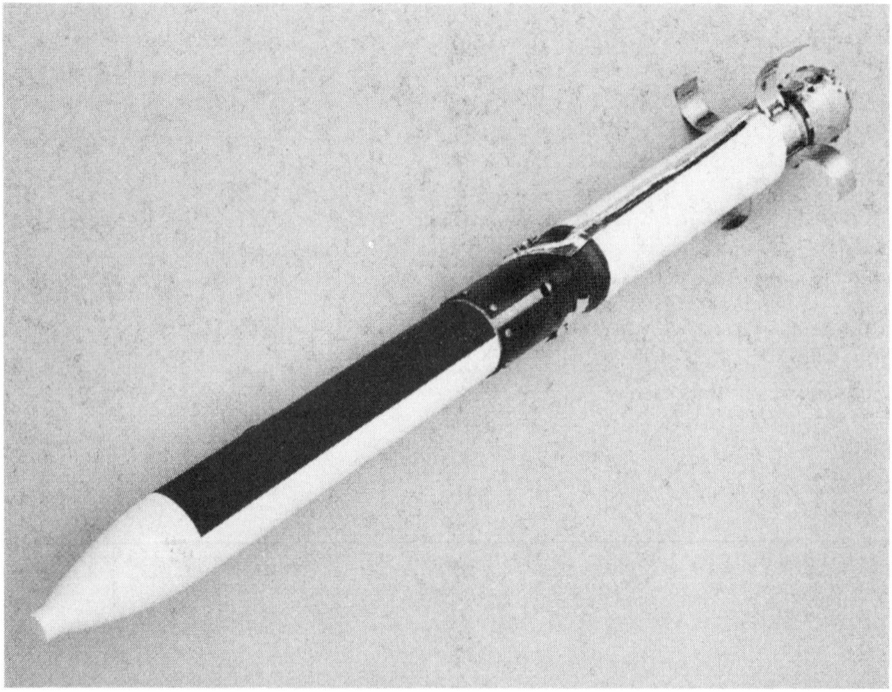

The BLU-106/B is an effective submunition, specially tailored to the dual tasks of cratering runways and piercing hardened aircraft shelters.

velocities and parachute-opening delays to create the designed target pattern, and each submunition works in the standard fashion of such weapons, being para-retarded immediately after release and then, as soon as the nose of the submunition has dropped to 60-65° below the horizontal, freed from the retarder and accelerated by a solid-propellant rocket. The weapon is stabilized by four flick-out fins, and the rocket burn imparts sufficient velocity for the warhead to penetrate the target surface and explode its 6.5 lb (2.95 kg) of HE only when deep under it, causing considerable heave and fracture damage.

BLU-108/B: This is the US Air Force submunition carrying as its payload four Avco Skeet devices. The Skeet is the world's first production weapon of the 'self-forging fragment' variety, wobbling through the air so that its onboard infra-red sensor can detect the heat of a tank, whereupon the microprocessor-based guidance system guides the Skeet directly over the target, and detonates a transverse explosive sheet that imparts very high velocity to a slab of copper, iron, tantalum and uranium alloy, at the same time turning it into a streamlined self-forging fragment that hits the tank's weakest armour with a speed of 9,000 ft (2743 m) per second. The BLU-108/B is used as the payload of the Sensor-Fused Weapon (see below).

BLU-109/B: This is the **Improved 2,000-lb Warhead**, designed to provide the tactical aircraft of the US Air Force with a capability against major hardened targets such as reinforced concrete bunkers, bridges, artillery sites, and hardened aircraft shelters. In this task the BLU-109/B succeeds the Mk 84 general-purpose bomb, which was designed in the 1950s and uses a low-grade steel case of very limited penetration capability. The BLU-109/B has a casing 1 in (25.4 mm) thick and made of high-grade steel forged as a single unit. The warhead has provision for a delay-action tail fuse to ensure detonation only after the bomb has penetrated the target. The warhead is 14.5 in (368 mm) in diameter and 95 in (2413 mm) in length, and is compatible with a number of guidance packages for increased accuracy: typical are the GBU-10 Paveway II laser system, the GBU-15 electro-optical or infra-red systems (including the extended-range capability of the AGM-130 rocket-boosted version), and the GBU-24 Paveway III low-level laser system.

CBU-5: This is an M30 cluster adaptor with 57 **M138** bomblets filled with BZ incapacitant gas.

CBU-8/A: This is an 860-lb (390-kg) cluster bomb based on an SUU-7A/A dispenser carrying 409 BDU-27/N spotting charges.

CBU-9/A: This is an 860-lb (390-kg) cluster bomb based on an SUU-7A/A dispenser carrying 406 BDU-28/B dummy bomblets.

CBU-10: This is a weapon based on the SUU-7 dispenser, but other details are lacking.

CBU-16/A: This is a cluster bomb based on an SUU-13A/A dispenser carrying CDU-9 clusters of bomblets filled with BZ incapacitant gas.

CBU-17/A: This is a cluster bomb based on an SUU-13/A dispenser carrying practice bomblets.

CBU-26/A: This is a cluster bomb based on an SUU-10/A dispenser carrying BDU-37/B dummy fragmentation bomblets.

CBU-44: This is a cluster bomb (dispenser) carrying anti-personnel fragmentation bomblets.

CBU-45/B: This is a 1,000-lb (454-kg) cluster bomb (dispenser) carrying anti-personnel and anti-materiel fragmentation bomblets.

CBU-59: This is the anti-personnel and anti-tank cluster munition better known as the 'Rockeye II' (see separate entry below).

CBU-61: This is the cluster bomb (dispenser) carrying bomblets filled with CS incapacitant gas.

The McDonnell Douglas F/A-18A Hornet was designed for the interchangeable fighter and attack roles, and this versatility is reflected in the type's very wide armament capability as exemplified in this layout view. Flanking the 570 rounds of 20-mm ammunition for the M61A1 Vulcan inbuilt cannon are two AGM-88A HARM anti-radar missiles plus (on the left) the ASQ-173 laser spot tracker and strike camera pod and (on the right) the complementary AAS-38 forward-looking infra-red pod. Behind these are LAU-68 seven- and LAU-61 19-tube rocket launcher pods, a drop tank, two AIM-7M Sparrow air-to-air missiles and two AIM-9L/M Sidewinder air-to-air missiles. Next to the rear come eight Mk 20/CBU-59/B Rockeye cluster bombs, eight Mk 82 500-lb (227-kg) bombs, nine Mk 83 1,000-lb (454-kg) bombs and four Mk 84 2,000 lb (907-kg) bombs; while right at the rear are five SUU-20 practice dispensers. It is worth noting that Paveway laser-guided bombs can replace the Mk 82, Mk 83 and Mk 84 free-fall weapons, and that other options include AGM-62 glide bombs, AGM-65 Maverick air-to-surface missiles, AGM-84 Harpoon anti-ship missiles, napalm bombs and nuclear weapons.

The components of the CBU-87/B Combined Effect Munition.

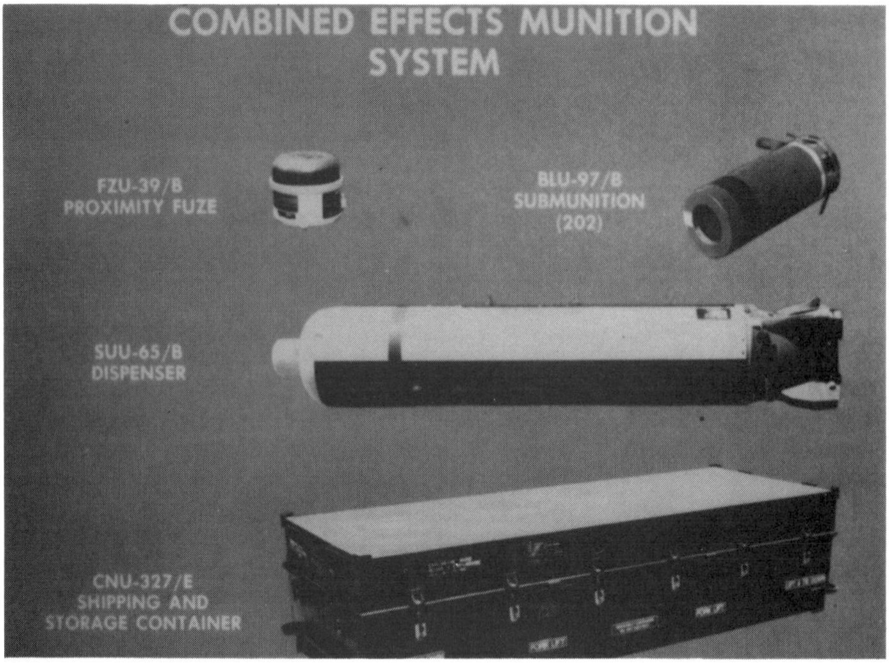

CBU-62: This is an 840-lb (381-kg) cluster bomb based on the SUU-30 dispenser carrying **M38** fragmentation grenades.

CBU-63: This is an 840-lb (381-kg) cluster bomb based on the SUU-30 dispenser carrying **M40A1** fragmentation bomblets each weighing 23 lb (10.43 kg).

CBU-70: This is a cluster bomb based on an SUU-30 dispenser, but other details are lacking.

CBU-87/B: Designed to replace the 'Rockeye' series and other elderly types such as the CBU-58, this is the 950-lb (430.9-kg) **Combined Effects Munition** designed to tackle all types of target (armour, materiel and personnel) with a single dispensed payload of bomblets used in the close air support, defence-suppression and interdiction roles. The CBU-87/B is based on the SUU-65/B dispenser and BLU-97/B submunition, and is designed for delivery at any altitude over 200 ft (60 m) and at any speed up to 805 mph (1295 km/h), the dispenser being spun by its fins at any chosen rate up to 2,500 revolutions per minute (to optimize the bomblet pattern) before its time or ground-proximity controller commands the dispenser to split open and so scatter the payload of 202 bomblets: these each deploy a cone-shaped decelerator and fall in the desired pattern with equal detonation capability on land, mud, water and snow. Each bomblet has a pre-fragmented anti-personnel case over a hollow-charge anti-tank warhead, and a disc of incendiary zirconium is also built into the charge to add to the weapon's destructive capability. The fragmentation section can disable motor transport at a range of 50 ft (15 m) and aircraft at 250 ft (76 m).

CBU-89/B: This is one of the US services' most important cluster munitions, an air-delivered scatterable mine system for which the US Air Force is the lead service with the US Army responsible for the BLU-91/B and BLU-92/B submunitions, and the US Navy for the mine/dispenser adaptor. The 700-lb (317.5-kg) weapon comprises the SUU-64/B Tactical Munitions Dispenser containing 72 BLU-91/B Gator anti-tank bomblets and 22 BLU-92/B Gator anti-personnel bomblets. The weapon can be dropped at any altitude between 200 and 40,000 ft (60 and 12190 m), the bomblets being dispensed on command of a time fuse or (when dropped from high altitude) an optional FZU-39/B proximity fuse. The bomblets are armed when they hit the ground, the BLU-91/B having a magnetic influence fuse and the BLU-92/B four tripwires.

CBU-92/B: This is the **Extended-Range Anti-armor Munition** currently under development for service entry with US Air Force tactical aircraft during 1993 as part of the impressive WAAM (Wide-Area Anti-armor Munitions) programme. The 950-lb (430.9-kg) weapon is based on the SUU-65/B dispenser filled with nine anti-tank and three unspecified area-denial submunitions, which are dispensed on command of a time fuse or (when dropped from high altitude) an optional FZU-39/B proximity fuse. The ERAM submunitions fall towards the ground under individual parachute retarders.

On the ground the submunition is stabilized by four legs, and its sensitive three-probe seismic/acoustic fuse system is activated. When triggered, the fuse classifies any target within a range of 500 ft (152 m), fixes its range and bearing from the submunition, and fires one of the submunition's two Avco Skeet warheads in the right direction. The Skeet's own infra-red fusing system then takes over to detonate the warhead as the Skeet passes over the target tank. The resultant self-forging fragment plunges straight down to pierce the tank's upper-surface armour. The manufacturer is also developing the ERAM submunition for use against taxiing aircraft as part of the US forces' increased effort in counter-air operations.

CBU-97/B: This is the 914-lb (414.6-kg) **Sensor-Fused Weapon** (SFW), currently under development within the context of the WAAM (Wide-Area Anti-armor Munitions) and Assault Breaker programmes to provide US tactical aircraft with the ability to make direct attacks on armoured formations. Each SFW is 92 in (2337 mm) long and 15.6 in (396 mm) in diameter, and comprises an SUU-64/B Tactical Munitions Dispenser carrying a load of 10 BLU-101/Bs: when released by the dispenser (under the control of a timer or of an FZU-39/B altitude fuse when the weapon is released from high altitude or in a toss manoeuvre), each of these deploys a parachute to retard and slow its descent to a pre-set altitude, where a retro-rocket is fired to drive the pack upwards again, spinning rapidly so that the BLU-108/B's four Skeet warheads are articulated outward on levers and are then released to fall individually, oscillating so that the infra-red sensor can search for a target within an area of about 4,800 sq yards (4015 m^2) immediately below it; if no target is detected the Skeet detonates just above the ground to cause blast and fragment damage to any soft-skinned target in the vicinity.

Direct Airfield Attack Combined Munition: The DAACM is currently under development as an 850-lb (385.6-kg) weapon based on the SUU-64/B Tactical Munitions carrying in its forward section eight BLU-106/B Boosted Kinetic-Energy Penetrator anti-runway

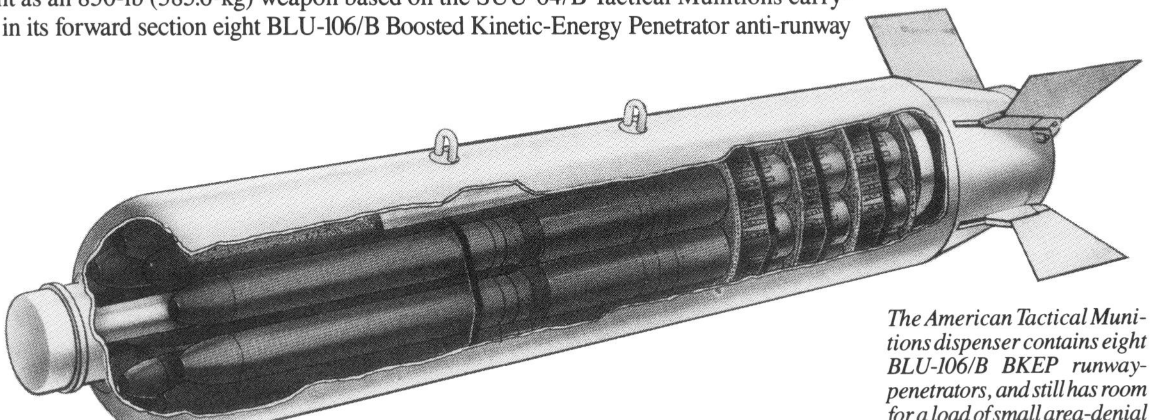

The American Tactical Munitions dispenser contains eight BLU-106/B BKEP runway-penetrators, and still has room for a load of small area-denial minelets to prevent repair crews getting to work quickly.

bomblets and in its rear section 24 Hunting HB876 area-denial mines. The DAACM is 15.6 in (396 mm) in diameter and 92 in (2337 mm) long, and is designed for delivery at any speed between 400 and 805 mph (644 and 1295 km/h) and at any altitude between 300 and 40,000 ft (90 and 12190 m). The weapon is designed to replace the French Durandal runway-cratering bomb in interim US service, with service entry scheduled for 1990.

The primary carriers for the DAACM in US service are the McDonnell Douglas F-4 Phantom II (four munitions), the McDonnell Douglas F-15E Strike Eagle (12 munitions), the General Dynamics F-16 Fighting Falcon (four munitions) and the General Dynamics F-111 (eight munitions).

M1A1: This is a 100-lb (45.4-kg) fragmentation cluster (a suspended rod and adaptor rings) comprising six **M41A1** fragmentation bomblet: these each weigh 19.8 lb (9 kg) and comprise a light sheet steel casing spirally wound with square-section steel wire over a 2.7-lb (1.2-kg) filling of TNT. There is also an **M1A2** variant, as well as the **M1A3** and **M1A4** derivatives each carrying six improved **M41A2** bomblets.

M4A2: This is a 100-lb (45.4-kg) fragmentation cluster bomb comprising three M40A1 bomblets each weighing 23 lb (10.43 kg).

M12: This is an incendiary cluster bomb.

M13: This is a 500-lb (226.8-kg) incendiary cluster bomb.

M17: This is a 500-lb (226.8-kg) incendiary cluster bomb.

M19: This is a 500-lb (226.8-kg) incendiary cluster bomb comprising 36 **M69** incendiary bomblets each weighing 6 lb (2.72 kg). There is also an **M19A2** version.

M20: This is a 500-lb (226.8-kg) incendiary cluster bomb. There is also an **M20A1** version.

M21: This is a 500-lb (226.8-kg) incendiary cluster bomb.

M22: This is a 500-lb (226.8-kg) incendiary cluster bomb; there is also an **M22A1** version.

M26A2: This is a 500-lb (226.8-kg) fragmentation cluster comprising an assembly or rods, braces and adaptors to carry 20 M41A1 fragmentation bomblets: these each weigh 19.8 lb (9 kg) and comprise a light sheet steel casing spirally wound with square-section steel wire over a 2.7-lb (1.2-kg) filling of TNT.

M27A1: This is a 500-lb (226.8-kg) fragmentation cluster comprising a structural assembly and tail unit for the carriage of six M82 fragmentation bombs (see separate US bombs entry).

M28A2: This is a 100-lb (45.4-kg) fragmentation cluster bomb comprising a sheet metal casing carrying 24 **M83** bomblets: this latter weighs 4 lb (1.81 kg), and is a cylindrical steel case filled with 1.2 lb (0.54 kg) of TNT and fitted with spring-opened case covers to retard the bomblet's fall.

M29A1: This is a 500-lb (226.8-kg) fragmentation cluster bomb comprising a sheet metal casing carrying 90 M83 bomblets: this latter weighs 4 lb (1.81 kg), and is a cylindrical steel case filled with 1.2 lb (0.54 kg) of TNT and fitted with spring-opened case covers to retard the bomblet's fall.

M31: This is a 500-lb (226.8-kg) incendiary cluster bomb comprising 38 **M74** incendiaries each weighing 10 lb (4.54 kg).

M32: This is a 1,000-lb (454-kg) non-persistent gas cluster bomb comprising 76 **M125** bomblets each weighing 10 lb (4.54 kg) and carrying 2.6 lb (1.18 kg) of GB.

M34A1: This is a 1,000-lb (454-kg) non-persistent gas cluster bomb comprising 76 **M125A1** bomblets each weighing 10 lb (4.54 kg) and carrying 2.6 lb (1.18 kg) of GB.

M35: This is a 750-lb (340-kg) incendiary cluster comprising 57 **M74A1** incendiaries each weighing 10 lb (4.54 kg).

M36: This is a 750-lb (340-kg) incendiary cluster comprising a sheet steel container for 182 **M50A3** incendiaries each weighing 3.7 lb (1.68 kg) and carrying 0.6 lb (0.27 kg)

of thermite (aluminium/magnesium) in a hexagonal sheet-steel casing. The container breaks open in the air, showering the incendiaries over an area determined by the release altitude of the container.
M43: This is a 750-lb (340-kg) gas cluster bomb comprising 57 M138 bomblets filled with BZ incapacitant gas.
M44: This is a 175-lb (79.4-kg) gas generator cluster bomb for BZ incapacitant gas.
Mk 12 'Padeye': This is a 600-lb (272.2-kg) dispenser weapon with smoke or BZ incapacitant gas.
Mk 17 'Gladeye': This is a Mk 4 dispenser with seven bomblet canisters.
Mk 21 'Sadeye': This is an 815-lb (369.7-kg) Mk 5 dispenser carrying an unspecified number of BLU-26/B fragmentation bomblets.
Mk 22 'Sadeye': This is an 815-lb (369.7-kg) Mk 5 dispenser carrying an unspecified number of BLU-26/B fragmentation bomblets.
Mk 44: This is a 550-lb (249.5-kg) shrapnel cluster bomb comprising an M16 cluster adaptor (a sheet steel case) with about 10,000 'Lazy Dog' 2-in (51-mm) shaped iron fragments. The weapon is designed round a 21.8-lb (9.9-kg) charge of TNT, which explodes in the air and drives the 'Lazy Dogs' downward at very high velocity into an area of about 100 sq yards (83.6 m^2): a simple comparison indicates that the effect is similar to that of several hundred 12-bore shotguns fired straight down from a height of 50 ft (15 m).

'Rockeye' Cluster Bomb Mk 20 Mod 0

USA

Type: medium-weight anti tank cluster bomb
Dimensions: diameter 13.2 in (335 mm); length 92 in (2337 mm); fin span 34 in (864 mm)
Weight: total round 476 lb (215.9 kg)
Warhead: 334 lb (151.5 kg) of bomblets

Variants and application
'Rockeye' Cluster Bomb Mk 20 Mod 0: This weapon was developed by the US Naval Weapons Center as an anti-tank cluster bomb for the full range of US Navy tactical aircraft with NATO standard 14-in (356-mm) suspension lugs. The weapon has as its core a Mk 7 Mod 2 dispenser filled with 247 **Mk 118 Mod 0** anti-tank bomblets. Each of these weighs 1.93 lb (0.875 kg) and contains 1.1-lb (0.5-kg) shaped-charged/fragmentation warhead with 6.4 oz (181 g) of Minol 2 or H6 explosive. The fragments of the hardened high-density steel case move at a minimum of 4,000 ft (1219 m) per second, and thus possess good capability against armoured as well as unarmoured targets. The tail unit has four flick-out stabilizing fins, and the nose is fitted with a Mk 339 Mod 0 mechanical time fuse to split the case open (and so dispense the bomblets) at a predetermined moment. The 'Rockeye' can be released at any altitude down to 250 ft (76 m) in level flight or 100 ft (30 m) in a pitch-up manoeuvre, and the saturation area from a 500-ft (150-m) release is a 3,333-sq yard (2785-m^2) oval. There is also a **Cluster Bomb Mk 2 Mod 2** which weighs 490 lb (222.3 kg).
CBU-59/B 'Rockeye II': This is the 766-lb (347.5-kg) 'Rockeye' variant for US Air

The versatility of the McDonnell Douglas F-15 Eagle is considerable, and though this aircraft was designed as an air-superiority fighter it can double in the attack role. This is the F-15E Strike Eagle demonstrator (a modified F-15B two-seater combat-capable trainer) carrying AIM-9L/M Sidewinder self-defence missiles and 22 CBU-59/B Rockeye anti-tank cluster bombs.

Force tactical aircraft, and comprises 717 PLU-77/B anti-personnel and anti-materiel bomblets in a Mk 7 Mod 3 dispenser to create this dual role cluster bomb.
ISCB-1: Developed on the basis of the 'Rockeye' cluster bomb by ISC Technologies, the main manufacturer of the 'Rockeye' series, the ISCB-1 is the US forces' latest area-denial cluster weapon. The weapon comprises a Mk 7 dispenser, a Mk 339 nose fuse, a tail cone assembly and a filling of 160 electronically-timed operational mines plus 65 dummy mines. Up to the time the bomblets are dispensed the ISCB-1 functions just like the 'Rockeye': thereafter the individual delay of the bomblets is controlled by the programmable bomb panel located in the tail cone, set before take-off for any delay between 0 seconds and 24 hours by 10-second increments. At the appropriate moment the Mk 339 fuse initiates the linear shaped charge that splits open the dispenser to release the payload over a denial area of between 3,600 and 5,625 sq yards (3010 and 4705 m^2).

West Germany

MBB Modular Dispenser System

Type: modular submunition dispenser
Dimensions: width 630 mm (24.8 in); height varies from 340 to 660 mm (13.39 to 25.98 in) according to exact configuration; length varies from 2540 to 4500 mm (100 to 177.17 in) according to exact configuration
Weight: total dispenser varies from 600 to 2000 kg (1,322.75 to 4,409.2 lb) according to exact configuration
Warhead: submunitions of various sizes and weights up to a maximum weight determined by exact dispenser configuration and launch aircraft loadings

Variant and application

MDS: Under development for the West German air force and possibly for export, the **Modular Dispenser System** is an advanced weapon based conceptually on the Messerschmitt-Bölkow-Blohm MW-1 but with physical affinities to the Apache/CWS stand-off dispenser to offer far lower drag and the facility for configuration to meet the loading requirements of most tactical aircraft with NATO standard 762-mm (30-in) sus-

The increasing capability of ground-based air defences is making it ever more hazardous to overfly the land battlefield for the delivery of conventional ordnance. The 1980s have thus witnessed considerable emphasis on the development of dispenser weapons that allow the aircraft to make a single pass strewing a number of submunitions designed principally to defeat armour and matériel concentrations. One of the most advanced of these systems is the MBB Modular Dispenser System, here seen under a prototype of the Northrop F-20A Tigershark. This weapon is designed to succeed the bulky MW-1 system, and amongst its many attractions is modular design so that the weapon can be configured to suit any type of tactical aircraft.

A cloud of submunitions scatters to each side of the Modular Dispenser System carried by a McDonnell Douglas F-4 Phantom II in the system's initial firing trials.

pension lugs. The system is designed for reloading and re-use and thus, unlike the MW-1, is not as a rule jettisoned after firing. The weapon is based on the use of two standard submunition-launcher units (each possessing four 132-mm/5.2-in diameter double-ended tubes, in one type of unit arranged as a rectilinear unit with side-by-side tubes, and in the other as a square unit with two superimposed pairs of tubes) that can be built up into dispensers of varying weights, loadings and sizes (typically 40 or 44 tubes to a maximum dispenser depth of six single- or three twin-row modules), and then faired into a neat streamlined package by the addition of sloped nose and tail sections.

The tail section contains the dispenser's inbuilt safe-and-fire unit as well as the intervalometer to sequence the lateral firing of the submunitions to the left and right of aircraft ground track. The limits on launch are a minimum height of 50 m (165 ft) and an aircraft speed between 735 and 1100 km/h (457 and 684 mph). The pattern of dispersed munitions embraces a width of 350 m (385 yards) and a length varying between 200 and 2000 m (220 and 2,185 yards) depending on intervalometer setting and tube number. As presently planned, the MDS is to use the same range of submunitions as the MW-1 (see separate entry), though more advanced submunitions are also envisaged. One of the system's many attractions is its ability to work adequately with the avionic standards of the launch aircraft, which can vary from the simple (optical sight and air-data computer) to the most advanced (head-up display, stores-management system and main computer with various sensor inputs), though of course the more advanced avionic suite provides the launch aircraft with greater first-pass attack accuracy.

West Germany

MBB MW-1

Type: submunition dispenser
Dimensions: width 1500 mm (59.06 in); height 700 mm (27.56 in); length 5300 mm (208.66 in)
Weight: total dispenser about 5000 kg (11,022.9 lb)
Warhead: about 4000 kg (8,818.3 lb) of submunitions

Variant and application
MW-1: The MW-1 (**Mehrzweckwaffe Nr 1**, or multiple dispenser weapon no 1) was developed by Messerschmitt-Bölkow-Blohm especially for the Panavia Tornado multi-role aircraft of the West German air force, with which it began to enter service in 1984. The MW-1 is a very substantial dispenser that fits onto the Tornado's underfuselage NATO standard 762-mm (30-in) suspension lugs. The complete dispenser is made up of four 28-tube sub-units (two rectilinear sections and two modestly streamlined nose and tail sections) for a total of 112 132-mm (5.2-in) diameter double-ended launcher tubes.

The submunitions can be loaded with four different ejection charges to provide different lateral ranges as a means of creating the desired ground patterns, which are designed primarily for attacks on armoured formations and airfields depending on the mix of submunitions. These are produced by Raketen Technik Gesellschaft (a subsidiary of

The first-generation MW-1 dispenser weapon is costly and, after its submunitions have been dispensed, must generally be jettisoned because of its high aerodynamic drag. The Modular Dispenser System has far lower drag and can thus often be brought back for reloading. This is a 42-tube prototype weapon during the first firing trials under a McDonnell Douglas F-4 Phantom II.

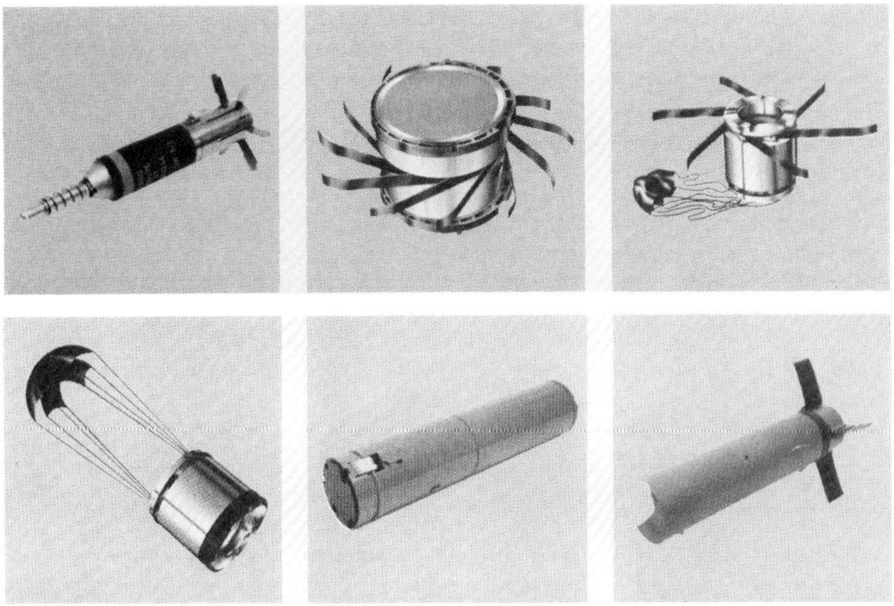

The six types of submunitions fired by the MW-1 dispenser weapon are the KB44 anti-tank bomblet (top left) *the MIFF anti-tank minelet* (top centre) *the MUSA area-denial minelet* (top right) *the MUSPA delayed-action area-denial minelet* (bottom left), *the STABO runway-cratering bomblet* (bottom centre), *and the ASW hardened shelter destroyer* (bottom right).

MBB and the ordnance manufacturer Diehl), and come in six varieties. The ASW is intended for attacks on hardened aircraft shelters, and is a full-calibre 132-mm (5.2-in) munition that operates in three phases; after launch the type descends vertically under its parachute retarder until it hits the earthen covering of the shelter. This initial impact triggers the type's booster rocket, which drives the body of the weapon through the earth until it meets the concrete of the shelter; this second impact in turn triggers the shaped-charge warhead in the nose, which blasts a hole through the concrete to allow a secondary projectile to be fired into the interior of the shelter, where it detonates with catastrophic results.

The KB44 is the standard anti-armour submunition, and is a fin-stabilized sub-calibre bomblet with a diameter of 44 mm (1.73 in), weight of 0.6 kg (21.16 oz) and an impact fused shaped-charge warhead able to defeat the top and side armour of all current tanks. The type is packed in clusters of seven, and three such clusters can be loaded into each half of the launch tube, making for a theoretical maximum of 4,704 such free-fall bomblets, which can all be ejected in 0.6 second to cover an area 185 m (605 ft) wide and 500 m (1,640 ft) long in a low-altitude attack. The MIFF is a full-calibre 132-mm (5.2-in) cylindrical anti-tank minelet weighing 3.4 kg (7.5 lb); seven or eight such minelets can be loaded into each launch tube, giving a theoretical maximum of 872 per fully loaded dispenser. After launch the minelets free-fall to the ground, where they are stabilized with a flat side upwards by the action of radially-arranged spring legs to await the detonation of the two flat-plate charges as a result of the vibration or magnetic influence of a nearby tank. If not otherwise detonated, the MIFF self-destructs after a pre-set time.

The MUSA is another full-calibre 132-mm (5.2-in) submunition, in this instance a 4.2-kg (9.26-lb) cylindrical fragmentation type for use against personnel and soft-skinned vehicles; after launch the minelet falls under its retarding parachute to the ground, where its radial legs right the weapon before it detonates and scatters fragments to a lethal radius of 100 m (330 ft). Six such minelets can be loaded into each launch tube, making a theoretical maximum of 672 though 648 is a more usual load. The MUSPA is a close relative of the MUSA, and is thus a full-calibre 132-mm (5.2-in) weapon again weighing 4.2 kg (9.26 lb); the type is used in the airfield-denial role with a delay-action acoustic fuse triggered by any passing aircraft; the submunition also possesses a random-delay self-destruct fuse.

Finally there is the STABO runway-cratering submunition, a full-calibre 132-mm (5.2-in) weapon that descends under its parachute after ejection. As it hits the runway, the weapon detonates a first shaped-charge warhead to pierce the runway concrete, whereupon a secondary projectile is fired through the hole to penetrate into the earth before detonating to cause considerable heave and fracture damage; the MW-1 dispenser can carry a maximum of 224 (or more commonly 200) such munitions. With all its various submunitions dispensed, the empty MW-1 would in war be jettisoned (despite the fact that it constitutes some 10% of the cost of each of the 1,400 MW-1 systems currently envisaged) as its high drag would impose too severe a tactical limitation on the egressing Tornado.

West Germany

MBB VEBAL/Syndrom

Type: submunition dispenser
Dimensions: data not available
Weight: data not available
Warhead: probably 30 submunitions

Variant and application

VEBAL/Syndrom: This is currently the most advanced of the fixed dispenser systems developed by Messerschmitt-Bölkow-Blohm, and can be regarded as a modern version of the concept embodied in the French Giboulée system, making full use of the latest advances in sensor technology. Dispensers of an earlier concept were designed to saturate an extensive area as the means of providing an adequate 'kill' probability: West German estimates for the MW-1, for instance, indicate that the release of a full load of 4,704 KB44 bomblets against ten tanks will knock out only four of them. The VEBAL/Syndrom weapon is designed to overcome this inherent wastefulness by increasing the single-shot 'kill' probability of individual submunitions by using a sensor pod in conjunction with the dispenser pod to increase accuracy.

The Syndrom pod includes an infra-red linescanner, laser, radar and radiometer (the last to assess the potential target's thermal signature and so ensure positive identification). Once activated by the aircraft pilot or weapons operator, the system is fully automatic: as the aircraft overflies a hostile area at an altitude of 75 m (245 ft) the sensor pod watches a strip 30 m (100 ft) wide under the aircraft's flight path and uses a high-speed microprocessor to pass fire-control information about any detected target to the VEBAL (VErtical BALlistic) dispenser. This part of the complete system, also known as the **VBW** (**VertikalBordWaffe**, or vertical attack weapon), contains some 20 to 30 tubes for large and extremely powerful anti-tank submunitions, which are fired individually downwards and rearward with a very high likelihood of hitting and destroying the particular target under attack. This reduces waste and so makes the VEBAL/Syndrom system highly cost-effective, whilst also increasing the number of tanks likely to be destroyed. Current thinking is devoted to the development of integrated VBW dispenser containing about 18 launch tubes and their associated fire-control system (sensors and computer). At a height of 75 m (245 ft) the sensors scan a swathe about 30 m (98.5 ft) wide.

ROCKETS

EDESA CBAS-1 Albatros

Argentina

Type: air-launched rocket (air-to-surface)
Dimensions: diameter 70 mm (2.76 in); length and fin span data not available
Weight: total round between 8.4 and 10.3 kg (18.52 and 22.71 lb) depending on specific warhead type
Warhead: varies according to type (see below)
Propulsion: 5.5-kg (12.13-lb) extruded double-base solid-propellant rocket delivering unknown thrust
Performance: range 8000 m (8,750 yards)

Variant and application
CBAS-1 Albatros: This Argentine rocket was introduced in 1973 after development by CITEFA (Scientific and Technical Research Institute of the Armed Forces), and is associated mainly with naval aircraft. The rocket is produced by EDESA, which is the industrial contractor responsible for the rocket as its integrated weapon system (launcher unit and fire-control system). The casing of the 5.5-kg (12.13-lb) core section is made of aluminium, the propellant weighs 2.8 kg (6.17 lb) and the rocket is stabilized in flight by four wrap-around fins that are deployed mechanically as the rocket leaves its launch tube. The warhead types currently available are the 2.9-kg (6.39-lb) smoke, 4.45-kg (9.81-lb) anti-tank/anti-personnel, 4.65-kg (10.14-lb) fragmentation, and 4.8-kg (10.58-lb) anti-matériel/anti-personnel.

Forges de Zeebrugge 2.75-in FFAR

Belgium

Type: air-launched rocket (air-to-surface and air-to-air)
Dimensions: diameter 2.75 in (69.85 in); length 46.85 in (1190 mm); fin span 10.0 in (254 mm)
Weight: 8.6 to 11.9 kg (18.96 to 26.23 lb) depending on warhead type
Warhead: 2.9-kg (6.39-lb) FZ-32 smoke, 3.0-kg (6.61-lb) FZ-49 armour-piercing, 3.935-kg (8.675-lb) LXT-454-69 chaff, 4.15-kg (9.15-lb) FZ-63 white phosphorus, 4.30-kg (9.48-lb) FZ-71 HE fragmentation, 4.30-kg (9.48-lb) FZ-85 anti-vehicle and anti-materiel, 4.30-kg (9.48-lb) FZ-86 smoke, 4.52-kg (9.96-lb) FZ-58 penetrating HE, 4.90-kg (10.80

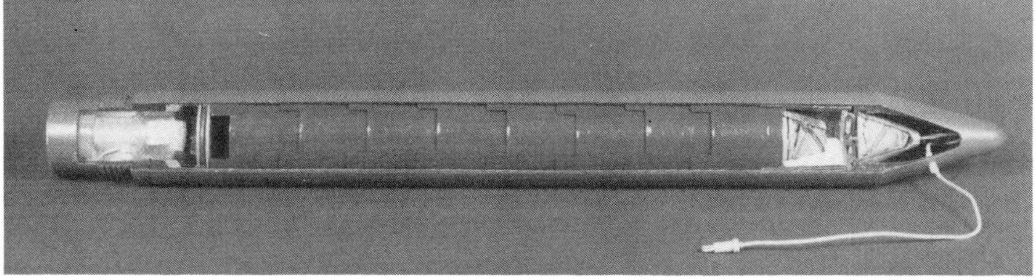

Latest of the 2.75-in (6.85-mm) FFAR developments produced by Forges de Zeebrugge is the FZ-100. Submunitions are ejected forwards at predetermined intervals to create a dense saturation on the ground under the rocket's flightpath.

Even the lightest aircraft, such as this SIAI-Marchetti SM.1019 observation type, can be turned into useful low-intensity attack platforms with specialized armament: seen on the four underwing hardpoints are two seven-tube launchers for Forges de Zeebrugge 2.75-in (69.85-in) rockets and two machine-gun pods.

lb) M257 illuminating, 6.2-kg (13.67-lb) FZ-100 anti-vehicle and anti-personnel submunition, and WDU-4A/A flechette
Propulsion: one solid-propellant rocket delivering 750-lb (348-kg) thrust for 1.2 to 1.4 seconds (Mk 4 and Mk 40 types), or 650-kg (1,433-lb) thrust for 0.85 second (FZ-67 and FZ-68 types), or 675-kg (1,488-lb) thrust for 2.3 seconds (NRZ-96 type)
Performance: range 500/6000 m (545/6,560 yards) for Mk 4 and Mk 40 types, 500/8800 m (545/9,625 yards) for FZ-67 and FZ-68 types, or 50/9000 m (545/9,845 yards) for NRZ-96 type

Variants and applications
FZ 2.75-in FFAR: This very widely used tactical rocket with four folding tail fins was developed in the USA as the 'Mighty Mouse' series in the air-to-air role with a 6-lb (2.8-kg) warhead, developing in the 1960s to a 10-lb (4.54-kg) warhead and ultimately to a 17-lb (7.7-kg) type. The rocket has been produced in huge numbers by American and overseas contractors, but since 1958 the Belgian company Les Forges de Zeebrugge has been the chief source of FFAR weapons, and has developed the type into a highly capable air-to-surface type with a number of motors and warheads.

The motor of the original Mk 4 and Mk 40 versions had 5.75 lb (2.61 kg) of propellant and offered modest thrust for between 1.2 and 1.4 seconds, providing the rocket with adequate range but leaving it highly susceptible to gravity and crosswind effects. Then the FZ-67 (0.8-second burn) and FZ-69 (0.9-second burn) motors were developed from the US Mk 66 type to provide higher specific thrust with all the advantages this offered in terms of range and comparative immunity to gravity and crosswind. The latest NRZ-96 motor (used in the air- and surface-launched FFAR) was developed in conjunction with the French company SNPE, and is 5.7 kg (12.57 lb) in weight with 3.0 kg (6.61 lb) of composite propellant. It is 1030 mm (40.55 in) in length. The external motor shape remains unaltered, allowing the existing fin assembly and warhead types to be added, but its spin-inducing nozzle, high specific thrust and longer burn time provide greater velocity. Compared with that of the Mk 4 version the velocity increases from 600 to 715 m (1,969 to 2,346 ft) per second when carrying the FZ-71 warhead. Other advantages are improved accuracy and greater payload (heavier warheads).

There are ten main warhead types associated with the Forges de Zeebrugge 2.75-in rockets, giving the type very operational flexibility. The 2.9-kg (6.4-lb) FZ-32 warhead generates a 10/15-second smoke cloud after impact, and is used mainly for training purposes. The 3.0-kg (6.61-lb) FZ-49 is a hollow-charge anti-tank warhead to penetrate 400 mm (15.75 in) of armour. The 4.52-kg (9.96-lb) FZ-58 warhead is an HE penetration type for use against runways, hardened aircraft shelters, fortifications, bunkers and the like. The 4.15-kg (9.15-lb) FZ-63 warhead is a white phosphorus type used for smoke and incendiary purposes. The 4.3-kg (9.48-lb) FZ-71 warhead is an anti-personnel type whose detonation produces more than 8,000 fragments. The 4.3-kg (9.48-lb) FZ-85 is an area attack warhead with a pre-fragmented casing that shatters into some 3,000 pieces on detonation. The 4.3-kg (9.48-lb) FZ-86 warhead generates a 4.5-minute smoke screen. The 6.7-kg (14.77-lb) FZ-100 is a submunition (or cargo) carrier with nine 0.48-kg (1.06-lb) submunitions possessing anti-personnel (10.5-m/34.4-ft lethal radius) and anti-tank (105-mm/4.13-in penetration capability) capabilities, and each fitted with an aerodynamic stabilizer for concentrated vertical descent after launch under control of the sophisticated GRCP28 fire-control system to create an 'airburst wall' of submunitions that fall in a concentrated pattern onto the target area. The 3.935-kg (8.68-lb) LXT-454-69 is a chaff warhead designed to produce a 600-m^2 (6,460-sq ft) radar-impenetrable cloud. Finally the 4.9-kg (10.8-lb) M257 is an illuminating warhead able to generate a 1-million candlepower light for two minutes. There is also the WDU-4A/A anti-personnel warhead with a load of 2,200 1.3-g (0.06-oz) flechettes.

The only limitation to the use of these warheads is for the M257, which cannot be launched at aircraft speeds higher than 740 km/h (460 mph). The FZ-100 submunition warhead is perhaps the most interesting of the rocket's payloads, and comes in three variants: the FZ-100 Mod 1 has a pyrotechnic fuse with a fixed delay, the FZ-100 Mod 2 has an electronic fuse that can be remotely set from the cockpit, and the FZ-100 Mod 3 has a variable-time fuse that has to be set on the ground before the rocket is loaded into its launcher. All in all the FZ-100 is similar to the almost identical M261 warhead for the US Hydra 70 rocket, and allows the launch aircraft to deliver a concentration

The increasingly militant role of the helicopter over the battlefield has led to the development of armament kits for many types, generally on the basis of a transverse beam fitted with two or four hardpoints. This Sikorsky H-76 Eagle sports two Forges de Zeebrugge launcher each fitted for 12 2.75-in (69.85-in) rockets.

of small but effective warheads right over the target. The submunitions each contain 110 g (3.88 oz) of compressed Hexogen explosive, and are ejected forward when the fuse ignites the burster charge to descend under its special stabilizer.

Brazil

Avibras SBAT-37

Type: air-launched rocket (air-to-surface)
Dimensions: diameter 70 mm (2.76 in); length and fin span data not available
Weight: total round data not available
Warhead: data not available
Propulsion: one solid-propellant rocket delivering unrevealed thrust
Performance: not revealed

Variant and application
SBAT-37: This is a light spin-stabilized weapon designed for use mainly by helicopters and light tactical aircraft able to accommodate the associated seven-tube **LM-37/7** launcher.

Brazil

Avibras SBAT-70

Type: air-launched rocket (air-to-surface)
Dimensions: diameter 70 mm (2.76 in); length 1220 mm (48.03 in); fin span data not available
Weight: total round data not available
Warhead: 4-kg (8.82-lb) AVC-70/AC (HEAT anti-armour type), AVC-70/AP (HE-FRAG anti-personnel type), AVC-70/AC/AP (combined anti-armour and anti-personnel type) AVC-70/F (flechette anti-materiel and anti-personnel type), AVC-70/FB (white phosphorus smoke type), AVC-70/EF (practice smoke type) and AVC-70/E (inert practice type)
Propulsion: one rocket with either 4.98 kg (10.98 lb) of BD-108 double-base solid propellant delivering 470-kg (1,036-lb) thrust for 1.2 seconds, or 5.15 kg (11.35 lb) of CV-7 composite solid propellant delivering 525-kg (1,157-lb) thrust for 1.2 seconds
Performance: data not available

Variants and application
SBAT-70: This Brazilian rocket is in the same basic class as the US FFAR type, and

Items of unguided ordnance such as rockets are well within the industrial capabilities of emergent countries, and their manufacture provides employment, reduces imports and, on occasion, generates foreign currency through exports. Good examples of this tendency are the Brazilian SBAT rocket, modelled on the American FFAR and Zuni. This Northrop F-5E Tiger II of the Brazilian air force is carrying two camouflaged LM-70/19 launchers each for 19 SBAT-70 rockets.

The armament of light and medium helicopters can be particularly useful in low-intensity warfare, and for this role Avibras has developed a trim armament fit that can accommodate an LM-70/7 seven-tube launcher for 2.75-in (69.85-mm) SBAT-70 rockets as well as a pintle-mounted 7.62-mm (0.3-in) MAG machine-gun.

is a folding-fin weapon optimized for the air-to-surface role in two subvariants: the **SBAT-70 M1** does not spin and is designed for high-speed launch, while the **SBAT-70 M2** spins and is designed for either low-speed launch by helicopters and light tactical aircraft, or high-speed launch by faster aircraft needing a weapon of greater accuracy and range than the SBAT-70 M1. The SBAT-70 rocket is associated with four reusable launchers, namely the two-round **LM-70/2,** four-round **LM-70/4,** seven-round **LM-70/7** and 19-round **LM-70/19.** There is also a single-use 19-round launcher, and the LM-70/7 can also be used as part of the Avibras helicopter armament system, which features a fixed rocket launcher surmounted by a flexibly mounted 7.62 mm (0.3 in) FN MAG machine-gun.

Avibras SBAT-127

Brazil

Type: air-launched rocket (air-to-surface)
Dimensions: diameter 127 mm (5 in); length and fin span data not available
Weight: total round data not available
Warhead: 20-kg (44.09-lb) HE
Propulsion: one solid-propellant rocket delivering unrevealed thrust
Performance: data not available

Variant and application
SBAT-127: This is a powerful tactical rocket in the same basic class as the American Zuni. The type is stabilized by a cruciform arrangement of four fins at the tail, and is associated with two- and four-round launchers.

Canada

Bristol Aerospace CRV7

Type: air-launched rocket (air-to-surface and air-to-air)
Dimensions: diameter 2.75 in (69.85-mm); length 41.0 in (1041 mm) without warhead; fin span data not available
Weight: round excluding warhead 14.55 lb (6.60 kg)
Warhead: varies according to specific type (see below)
Propulsion: one composite solid-propellant rocket in the form of either the RLU-5001/B (C14) delivering 2,310-lb (1048-kg) thrust for 1.4 seconds (designed for use on rockets carried by fixed-wing launch aircraft), or the RLU-5002/B (C15) based on the C14 but omitting the aluminium powder for reduced smoke emission and delivering 2,200-lb (998-kg) thrust for 2.2 seconds (designed for use on rockets carried by rotary-wing launch aircraft)
Performance: speed 2,800 mph (4506 km/h) with a 10-lb (4.5-kg) warhead after launch from an aircraft flying at 620 mph (998 km/h), or 3,350 mph (5391 km/h) with an 8-lb (3.3-kg) warhead after launch from an aircraft at 620 mph (998 km/h); range typically 7,100 yards (6490 m) in the air-to-surface role

Variant and applications

CRV7: This Canadian unguided rocket was designed for full compatibility with the 2.75-in (69.85-mm) FFAR type, especially in warheads and launchers. The rocket is spin-stabilized by a fluted exhaust nozzle, has three spring-deployed wrap-round fins at the tail, and is notable for its combination of multiple warhead options and good range plus low dispersion.

The 9.4-lb (4.3-kg) M151 HE blast fragmentation warhead is designed for use against vehicles, light armour, soft ground targets, personnel, light naval craft and helicopters. The 9.7-lb (4.4-kg) M156 white phosphorus/incendiary warhead is designed for target-marking. The 17-lb (7.7-kg) M229 HE blast fragmentation warhead is designed for use against vehicles, light armour, soft ground targets, personnel, light naval craft and helicopters. The 9.4-lb (4.3-kg) M247 HEDP shaped-charge blast fragmentation warhead is designed for use against vehicles, light armour, soft ground targets, personnel, ships and helicopters. The 10.8-lb (4.9-kg) M257 flare warhead is designed for night illumination and target marking. The 9.3-lb (4.2-kg) WDU-4A/A warhead contains 0.046-oz (1.3-g) flechettes for use against vehicles, materiel and personnel in the area-saturation role. The 8.7-lb (4.0-kg) WDU-13/A warhead contains 0.137-oz (3.9-g) flechettes for use against vehicles, materiel and personnel in the area-saturation role. The 9.4-lb (4.3-kg) WTU-1/B and WTU-5001/B warheads are practice types. The 8-lb (3.6-kg) WAU-5001/B warhead contains a plastic-cased tungsten rod for use against heavy tanks. The 8-lb (3.6-kg) BAH-002 warhead contains a plastic-cased steel rod for use against light and medium tanks. The 12.4-lb (5.6-kg) RA-79 warhead is an HE anti-ship type with a kinetic-energy penetrator combined with a delayed-action blast fragmentation HE charge for attacks on ships up to 3,000 tons. The 5-lb (2.3-kg) BAH-004 warhead is under development with a plastic-cased tungsten penetrator for long-range use against heavy tanks. The 16-lb (7.3-kg) WDU-5001/B warhead is under development with a hardened steel casing and zirconium liner to penetrate hardened aircraft shelters and then cause fires inside them. The 10-lb (4.5-kg) WDU-5002/B warhead is under development with six low-drag tungsten rod flechettes ejected at motor burn-out for use against armour.
LAU-5003A/A: This is the designation of the CRV7's primary combat launcher, a 19-tube disposable unit characterized by a length of 49.8 in (1265 mm) without nose and tail fairings, a diameter of 15.7 in (399 mm), an empty weight of 76.4 lb (34.7 kg) with

A Canadair CF-5 of the Canadian Armed Forces fires a salvo of Bristol Aerospace CRV7 unguided rockets. These are the world's highest-performance rockets, offering enhanced kinetic energy on the target, together with a flatter and therefore more accurate flight trajectory.

the rear fairing, and a loaded weight of 530 lb (240.5 kg) with 19 rockets using the C14 motor and 10-lb (4.5-kg) warhead. Bristol Aerospace is also developing a reusable 19-tube launcher for use by fixed- and rotary-wing aircraft.

LAU-5002A/A: This is the designation of the CRV7's primary helicopter and training launcher, a six-tube reloadable unit characterized by a length of 59.8 in (1519 mm) without fairings, a diameter of 9.9 in (251 mm), an empty weight of 64.3 lb (29.2 kg) with the rear fairing, and a loaded weight of 206.2 lb (93.5 kg) with six rockets using the C14 motor and 10-lb (4.54-kg) warhead. Bristol Aerospace is also developing seven-, 12- and 19-round lightweight launchers suitable for helicopter use.

SUU-5003: This is the designation of a combined rocket-launcher and bomb dispenser unit. The unit is 122.0 in (3099 mm) long, its elliptical frontal dimensions are 19.3 by 11.5 (490 by 292 mm), and it weighs 402 lb (182.5 kg) without bombs or its four CRV7 rockets.

The 19-tube LAU-5003A/A launcher allows even light tactical aircraft to carry a substantial load of the devastating CRV7 rockets, which can be fitted with an increasingly sophisticated range of conventional and submunition warheads, suiting the rocket to a host of operational tasks.

China

NORINCO 55-mm Unguided Rocket Type 1

Type: air-launched rocket (air-to-surface)
Dimensions: diameter 55 mm (2.17 in); length 915 mm (36.02 in); fin span 230 mm (9.06 in)
Weight: total round 3.992 kg (8.8 lb)
Warhead: 1.372-kg (3.02-lb) HE
Propulsion: one solid-propellant rocket delivering unrevealed thrust
Performance: data not available

Variant and application
57-mm Unguided Rocket Type 1: This is a general-purpose air-to-surface rocket modelled on the Soviet S-5 and, like the Soviet rocket, often considered a 57-mm (2.24-in) weapon. The Chinese rocket is generally carried in an eight-round underwing launcher by tactical attack aircraft. The weapon has capability against most types of ground targets (with the exception of heavy armour and bunkers), and can be fired singly or in salvoes.

China

NORINCO 90-mm Unguided Rocket Type 1

Type: air-launched rocket (air-to-surface)
Dimensions: diameter 90 mm (3.54 in); length 1267 mm (49.88 in); fin span 341 mm (13.43 in)
Weight: total round 14.6 kg (32.19 lb)

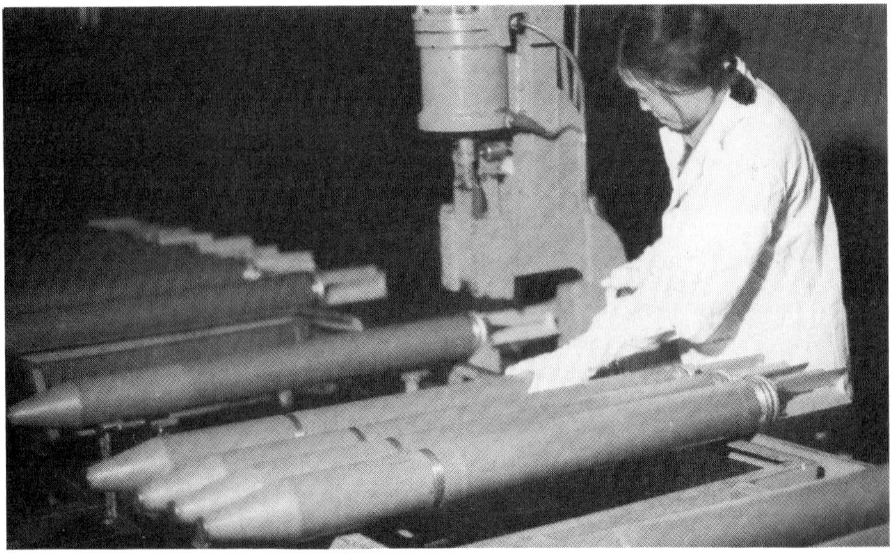

China's defence forces and their supporting industries are beginning to emerge on the world stage as operators and producers of advanced weapons, but the emphasis is still on total reliability and effectiveness under austere conditions. Typical of this long-established approach is the 90-mm (3.54-in) Unguided Rocket Type 1, fired from the Type HF-7 seven-tube launcher.

Warhead: 5.58-kg (12.3-lb) hollow-charge HE
Propulsion: one solid-propellant rocket delivering unknown thrust
Performance: data not available

Variant and application
90-mm Unguided Rocket Type 1: This is the standard anti-armour rocket of the Chinese air force, and its hollow-charge warhead (carrying 0.863 kg/1.9 lb of explosive) can penetrate 120 mm (4.72 in) of armour at impact angles up to 65°. The type is associated with the **Type HF-7** launcher carried by tactical aircraft such as the Nanchang Q-5.

Thomson Brandt SNEB 68

France

Type: air-launched rocket (air-to-surface and air-to-air)
Dimensions: diameter nominal 68 mm (2.68 in) and actual 71.8 mm (2.83 in); length varies from 847 to 1165 mm (33.35 to 45.87 in) depending on warhead type; fin span data not available
Weight: total round 4.3 to 6.26 kg (9.48 to 13.8 lb) depending on warhead
Warhead: 0.8-kg (1.76-lb) inert warhead on Type 252 rocket, 1.05-kg (2.31-lb) blast warhead on Type 251P rocket, 1.8-kg (3.97-lb) smoke warhead on Type 250 rocket, 1.8-kg (3.97-lb) hollow-charge and fragmentation warhead on Type 253 rocket, and 3.0-kg (6.61-lb) fragmentation on Type 246 rocket
Propulsion: one solid-propellant Type 25 or Type 68 F.1B rocket delivering unrevealed thrust
Performance: range 1000/4000 m (1,095/4,375 yards) from fixed-wing aircraft and 1000/2000 m (1,095/2,185 yards) from helicopters

Variants and applications
SNEB 68 Type 250: The SNEB 68 is amongst the most widely used air-launched rockets in the world, and has been produced in a number of forms to suit particular tactical purposes. The configuration of the rocket is standard, with a core section (rocket motor and tail-mounted stabilizer unit) to which is added any of several warheads. The rocket is stabilized by eight fins deployed by propellant gas as the weapon leaves its launch tube, and the leading edge of each fin in chamfered to impart spin-stabilization. The Type 250 is a practice round with a smoke warhead to allow accurate assessment of the impact point. The rocket has an overall length of 911 mm (35.87 in) and a weight of 5.05 kg (11.13 lb), and the velocity increment of the type (to be added to the speed of the launch aircraft) is 600 m (1,969 ft) per second. The SNEB 68 rocket series has recently been redesignated **TBA 68**.
SNEB 68 Type 251P EAA: This is the air-to-air variant of the SNEB 68 series, and carries an HE blast warhead. The rocket has an overall length of 847 mm (33.35 in) and a weight of 4.29 kg (9.46 lb). Its lighter weight gives this somewhat rare version a velocity increment of 800 m (2,625 ft) per second, and that means that when launched at a speed of 925 km/h (575 mph) at an altitude of 9000 m (29,530 ft), in just 2 seconds the rocket covers a distance of 1900 m (2,080 yards) compared with a typical distance for air-to-surface rockets (such as the Type 253 ECC and Type 256P EAP) of only some 1150 to 1300 m (1,260 to 1,420 yards) in the same time.
SNEB 68 Type 252 XF4: This is an air-to-surface practice round with a marker head and an overall length of 910 mm (35.83 in). The rocket comes as the **SNEB 68 Type**

Classic illustration of a BAe Harrier unleashing a salvo of 68-mm (2.68-in) SNEB rockets from its four underwing launcher pods.

252-4 and **SNEB 68 Type 252-5** subvariants with weights of 6.2 and 5.0 kg (13.67 and 11.02 lb) respectively.

SNEB 68 Type 253 ECC: This is the air-to-surface variant optimized for use against motor transport, matériel and armour. The model is 924 mm (36.38 in) long and weighs 5.06 kg (11.15 lb), and is fitted with a hollow-charge warhead which can pierce about 400 mm (15.75 in) of armour. The velocity increment of the variant is 600 m (1,969 ft) per second.

SNEB 68 Type 253-3 XF2: This is the practice version of the Type 253 ECC, and has a marker warhead.

SNEB 68 Type 256P EAP: This is the air-to-surface variant optimized for use against personnel and light equipment. The model is 924 mm (36.38 in) long and weighs 6.26 kg (13.8 lb), and it is fitted with a blast fragmentation warhead, which generates some 440 fragments on detonation. The velocity increment of the variant is 450 m (1,476 ft) per second.

SNEB 68 Type 259L LEM: This is the self-defence decoy variant of the series, and has an overall length of 1165 mm (45.87 in) and a weight of 6.2 kg (13.67 lb). The warhead is filled with chaff.

Multi-Dart 261-8 ABL: If the SNEB 68 can be considered the Mk 1 version of this important series with an HE warhead type, the Multi-Dart 260 series is the Mk 2 version specifically tailored to the anti-light armour role with a submunition warhead delivered from a low-level launch from impact angles of between 2° and 4° at velocity declining from 760 m (2,493 ft) per second at a range of 1500 m (1,640 yards) to

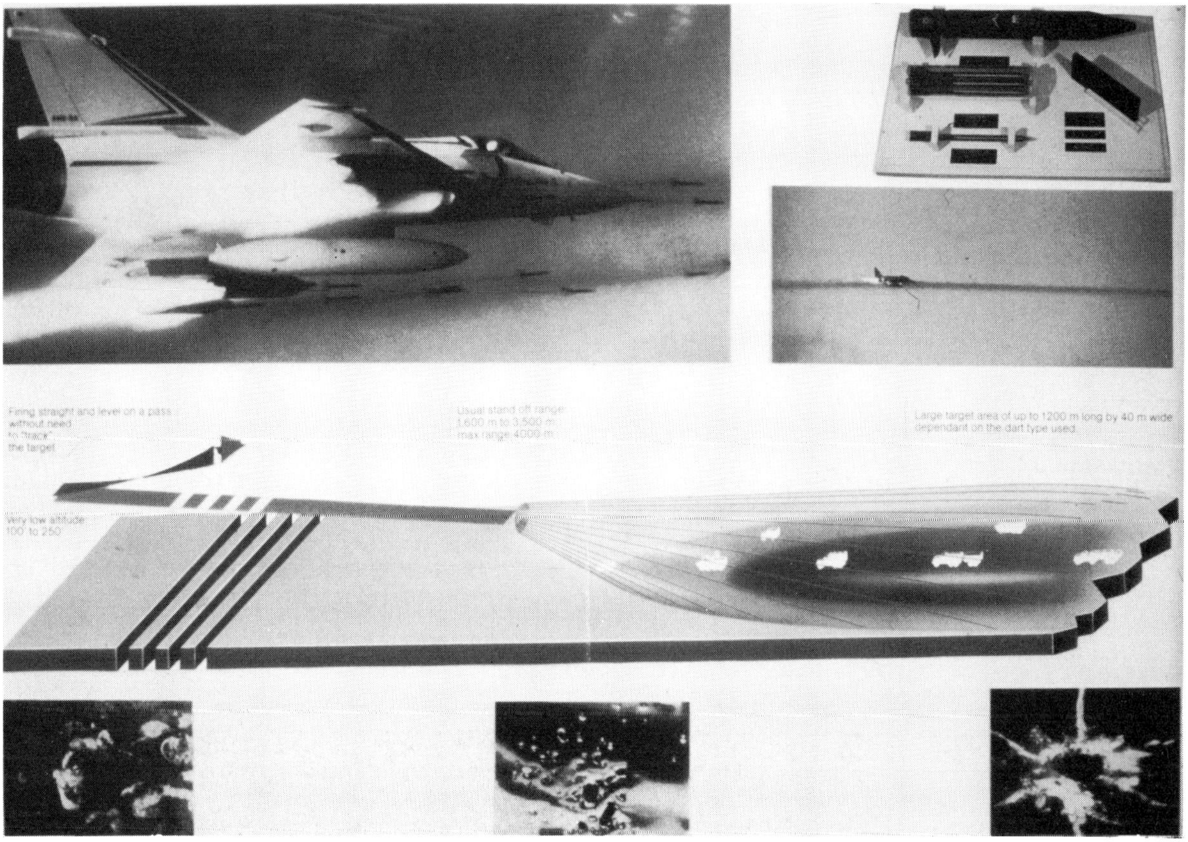

A composite illustration of the Thomson Brandt Multi Dart 68 provides graphic evidence of the capabilities of modern unguided rockets. Running clockwise from top left, the sequence shows a Dassault-Breguet Mirage 2000 ripple-firing a salvo of Multi Dart 68 rockets, the sub-projectiles of the AMV warhead, the flat trajectory of the rocket, a typical tactical employment diagram, and two types of damage.

330 m (1,083 ft) per second at 2200 m (2,405 yards). The type is fitted with the improved Type 26 or Type 68F2 motor, which contains 2.46 kg (5.42 lb) of Butalite solid propellant providing a thrust of 430 kg (948 lb) for 1.32 seconds for ranges between 1200 and 4000 m (1,315 and 4,375 yards) when launched from fixed-wing aircraft, and between 1200 and 3000 m (1,315 and 3,280 yards) when launched from helicopters. The Multi-Dart 261-8 ABL is the baseline model: possessing an overall length of 1380 mm (54.33 in) and weight of 8.2 kg (18.08 lb), this model carries a 3-kg (6.61-lb) submunition warhead. This latter contains eight flechettes each weighing 190 g (6.7 oz) and having dimensions that include a diameter of 13.5 mm (0.53 in) and a length of 270 mm (10.63 in). The flechettes are ejected over a 1.7-second period at a distance of 1000 m (1,095 yards) from launch to strike the target at 500 m (1,640 ft) per second at a range of 1800 m (1,970 yards). The flechettes can penetrate 15 mm (0.59 in) of armour.

Multi-Dart 261-36 AMV: This is the Multi-Dart model designed for use against armoured personnel carriers and similar matériel. The rocked has an overall length of 1380 mm (54.33 in) and a weight of 8.3 kg (18.3 lb), and carries a submunition warhead. This weighs 3 kg (6.61 lb), is 530 mm (20.87 in) long and contains 36 flechettes each weighing 35 g (1.23 oz) and having dimensions that include a diameter of 9 mm (0.35 in) and a length of 132 mm (5.2 in). The flechettes are ejected over a 1.7-second period at a distance of 1000 m (1,095 yards) from launch to strike the target at 400 m (1,312 ft) per second at a range of 1800 m (1,970 yards). The flechettes can penetrate 8 mm (0.315 in) of armour. The 1,296 flechettes of a typical 36-rocket salvo fired from a fixed-wing aircraft flying at 925 km/h (575 mph) cover a distance of 2000 m (2,185 yards) in 3 seconds and impact an oval area measuring some 1400 m (1,530 yards) long and 40 m (45 yards) wide with a velocity of 700 m (2,297 ft) per second. An equivalent figure for the 792 flechettes of a typical 22-rocket salvo fired from a helicopter are a distance of 2500 m (2,735 yards) covered in 4.5 seconds to blanket an oval area measuring 1000 m (1,095 yards) in length and 70 m (76.5 yards) in width with an impact velocity of 400 m (1,312 ft) per second.

Multi-Dart 262-4 XF4: This is a practice version of the Multi-Dart 262 rocket, and carries a marker warhead. The type's overall length is 1150 mm (45.29 in) and its weight 8.2 kg (18.08 lb).

Multi-Dart 262-6 XF1: This is a practice version of the Multi-Dart 262 rocket, and carries an inert warhead. The type's overall length is 1150 mm (45.29 in) and its weight 8.2 kg (18.08 lb).

Multi-Dart 269E ECL: This is the target-marking version of the Multi-Dart type, and has the same dimensions and weight as the Multi-Dart 261-8. The warhead has a brightness of 1-million candlepower, and lasts for 1 minute.

Matra RL F1: This is the Matra disposable launcher pod for the SNEB 68 rocket, of which 36 can be carried. The pod has a length of 2128 mm (83.78 in) and diameter of 565 mm (22.24 in), and its loaded weight is 270 kg (595.24 lb).

Matra RL F2: This is a reusable launcher pod for six SNEB 68 rockets. The pod's dimensions include a length of 1450 mm (57.09 in) and diameter of 246 mm (9.69 in), and the loaded weight is 58 kg (127.87 lb).

An Aérospatiale SA 342M Gazelle shows off the type of armament fit becoming increasingly popular for light battlefield helicopters: the seven-tube rocket launcher pod on the port hardpoint provides capability against an area target, while the fixed M621 cannon on the starboard hardpoint provides a point target capability and also the possibility of engaging other helicopters.

Among the Thomson Brandt range of rocket launchers for helicopter use is the 68.12, the designation revealing that this type carries 12 68-mm (2.68-in) rockets. The launcher with its SNEB rockets are mounted here on the starboard fuselage hardpoint of an Aérospatiale SA 355M Ecureuil 2.

Matra RL 155 116 MA: Otherwise known as the **RL F4,** this is the reusable 18-tube launcher pod of the Matra rocket-launcher series, and can fire both the SNEB 68 and Multi Dart types. The pod has an overall length of 2340 mm (92.13 in) and diameter of 410 mm (16.14 in), and weighs 185 kg (407.85 lb) fully loaded.
Matra RL HALMA: This carries 10 SNEB rockets.
Thomson Brandt 68.7: The Matra rocket-launchers are well streamlined units designed for use primarily on fixed-wing tactical aircraft, whereas Thomson Brandt produces two different series of launchers, one being a relatively unstreamlined type intended for use mainly with helicopters, and the other a lower-drag type intended for use with light tactical aircraft. The Thomson Brandt 68.7 is the smaller of the two launchers designed for fixed-wing aircraft, and carries seven SNEB 68 rockets. The launcher is 1270 m (50.0 in) long and 240 mm (9.45 in) in diameter, and its weight increases from an empty 30 kg (66.14 lb) to a loaded figure of between 60 and 73 kg (132.28 and 160.93 lb) depending on warhead type. The notional rate of fire is 1,800 rounds per minute.
Thomson Brandt 68.12: This is the smallest of the company's helicopter-mounted launchers, and comes in two subvariants each able to carry 12 rockets. The **Thomson Brandt 68.12C** subvariant is 1200 m (47.24 in) long, 310 mm (12.2 in) wide and 255 mm (10.04 in) high, and its weight increases from an empty 250 kg (55.1 lb) to a loaded figure of between 75 and 98 kg (165.34 and 216.05 lb) depending on warhead type; and the **Thomson Brandt 68.12L** subvariant is 1400 mm (55.1 in) long, 310 mm (12.2 in) wide and 255 mm (10.04 in) high, and its weight increases from an empty 30 kg (66.1 lb) to a loaded figure of between 80 and 103 kg (176.4 and 227.1 lb) depending on warhead type. Each launcher is carried on NATO standard 356-mm (14-in) suspension lugs, and the notional rate of fire is 300 rounds per minute. The Thomson Brandt 68.12C can carry only the SNEB rocket, while the Thomson Brandt 68.12L can carry either the SNEB or Multi-Dart rocket.
Thomson Brandt 68.18: This is the larger of the launchers designed for fixed-wing aircraft, and carries 18 rockets. The unit is 1345 mm (52.95 in) long and 394 mm (15.51 in) in diameter, and its weight increases from an empty 70 kg (154.32 lb) to a loaded figure of between 154 and 186 kg (339.51 and 410.06 lb) depending on warhead type. The notional rate of fire is 1,800 rounds per minute.
Thomson Brandt 68.22: This is the intermediate size of helicopter-mounted launcher and comes in two subvariants, each of them able to carry 22 rockets: the **Thomson Brandt 68.22C** subvariant is 1200 mm (47.24 in) long, 450 mm (17.72 in) wide and 325 mm (12.80 in) high, and its weight increases from an empty 43 kg (94.8 lb) to a loaded figure of between 145 and 190 kg (319.66 and 418.87 lb) depending on warhead type; and the **Thomson Brandt 68.22L** subvariant is 1400 mm (55.1 in) long, 450 mm (17.72 in) wide and 350 mm (13.78 in) high, and its weight increases from an empty

52 kg (114.6 lb) to a loaded figure of between 190 and 235 kg (418.9 and 518.1 lb) depending on warhead type. Each launcher is carried on NATO standard 356-mm (14-in) suspension lugs, and the notional rate of fire is 300 rounds per minute. The Thomson Brandt 68.22C can carry only the SNEB rocket, while the Thomson Brandt 68.22L can carry either the SNEB or Multi-Dart rocket.

Thomson Brandt 68.36: This is the largest size of helicopter-mounted launcher, and carries 36 rockets. The launcher is 1400 mm (55.12 in) long, 600 mm (23.62 in) wide and 400 mm (15.75 in) high, and its weight increases from an empty 83 kg (182.98 lb) to a loaded figure of between 310 and 390 kg (683.42 and 859.79 lb) depending on warhead type. The notional rate of fire is 300 rounds per minute.

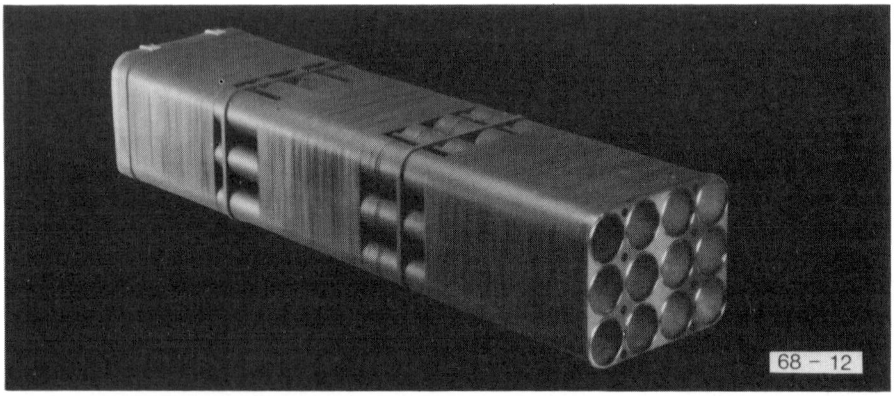

The Thomson Brandt range of 68-mm (2.68-in) rocket launchers for helicopters comprises the 12-tube 68.12, the 22-tube 68.22 and the very powerful though weighty 36-tube 68.36.

JL.100: This is an extremely useful underwing store combining fuel and rocket capability: the former is accommodated in the rear of the cylindrical store and amounts to 250 litres (55 Imp gal), while the latter comprises 18 68-mm (2.68-in) SNEB or Multi-Dart rockets.
RPK: This is a further development of the JL.100's concept, in this instance combining a tank and 18-tube launcher with the ability to carry four small drop loads (usually light bombs) supported laterally on the sides of the fuel/tank/ rocket launcher assembly supported by the aircraft hardpoint.

Thomson Brandt TBA 100 — France

Type: air-launched rocket (air-to-surface)
Dimensions: diameter nominal 100 mm (3.94 in) and actual 102.8 mm (4.05 in); length 2480 mm (97.64 in); fin span 700 mm (27.56 in)
Weight: total round between 36.9 and 42.6 kg (81.35 and 93.91 lb) depending on warhead
Warhead: 12.5 kg (27.56 lb) Type LUM illuminating, 14 kg (30.86 lb) Type ESP semi-armour-piercing, 14.4 kg (31.75 lb) Type ECC hollow-charge armour-piercing, 17.6 kg (38.80 lb) Type EEG general-purpose fragmentation, 17.6 kg (38.80 lb) Type IN inert practice, and 18.2 kg (40.12 lb) Type DEM demolition
Propulsion: one solid-propellant Type 100 F.1 rocket delivering unrevealed thrust
Performance: range 1500/4000 m (1,640/4,375 yards)

Variants and applications
TBA 100: Though similar to the SNEB 68 in configuration (and thus having eight folding fins deployed at the tail by propellant gas), the TBA 100 rocket packs a far heavier punch at comparable effective range, though only at the expense of higher weight and thus reduced numbers that can be carried by any particular launch aircraft. The basic rocket (minus warhead but including nose plug and stabilizing section) weighs some 24 kg (52.91 lb), and the choice of several warhead types adds significantly to the TBA 100's tactical flexibility. The rocket is launched at an aircraft speed of about 925 km/h (575 mph), and has a velocity increment (to be added to launcher speed) of 760 m (2,493 ft) per second. This translates to a flight time of 3 seconds to a range of 2500 m (2,735 yards) or of 6 seconds to 4000 m (4,375 yards).

All the warhead types have a length of 716 mm (28.19 in). The Type LUM warhead has a 750,000-candlepower flare that burns for 1 minute. The Type ESP semi-armour-piercing warhead has a hard steel core surrounded by 0.7 kg (1.54 lb) of HE and can penetrate 150 mm (5.91 in) of armour. The Type ECC hollow-charge armour-piercing and anti-personnel warhead possesses an anti-personnel sleeve (breaking into 1,300 fragments) over the hollow charge that can penetrate 500 mm (19.69 in) of armour. The Type EEG general-purpose and anti-personnel warhead shatters into some 2,000 1-g (0.035-oz) fragments on detonation. And the Type DEM demolition warhead is designed for penetration and destruction of hardened aircraft shelters with a performance that includes the ability to travel through 3.0 m (9.84 ft) of earth before piercing 300 mm (11.8 in) of concrete.
Multi-Dart 100: Like the SNEB 68, the TBA 100 has spawned a Mk 2 version in the form of the Multi-Dart 100, introduced in 1983 for the low-level attack of armoured targets with impact angles ranging from 3° to 5° and velocity dropping from 560 m (1,837 ft) per second at 1200 m (1,310 yards) to 430 (1,410 ft) per second at 4000 m (4,375 yards). The Type 100 F.1 solid-propellant rocket motor is used, and the Multi-

Dart 100 is available with three types of submunition warhead, each of them 885 mm (34.84 in) long. The Type AB 24 warhead is designed for use against tanks and other heavily armoured targets, and at a weight of 15.5 kg (34.17 lb) contains six 1650-g (58.2-oz) flechettes, each of 24 mm (0.94 in) diameter and 548 mm (21.57 in) length, able to penetrate 80 mm (3.15 in) of armour.

The Type ABL warhead is designed for use against armoured personnel carriers, and at a weight of 14.5 kg (31.97 lb) contains 36 190-g (6.70-oz) flechettes, each of 13.5 mm (0.53 in) diameter and 270 mm (10.63 in) length, able to penetrate 15 mm (0.59 in) of armour. The Type AMV warhead is designed for use against light armoured vehicles and matériel, and at a weight of 14 kg (30.86 lb) contains 192 35-g (1.23-oz) flechettes, each of 9 mm (0.35 in) diameter and 132 mm (5.2 in) length, able to penetrate 8 mm (0.315 in) of armour.

Matra RL C6: This is the Matra 100 mm (3.94 in) rocket-launcher pod designed for use on high-performance tactical aircraft. The type can carry six rockets, and its dimensions include a length of 3000 mm (118.1 in) and a diameter of 410 mm (16.14 in). The loaded weight varies between 370 and 420 kg (815.7 and 925.93 lb) depending on the warhead type.

Thomson Brandt 100.6: Also known as the **RL F3**, this pod can carry six 100 mm (3.94 in) rockets, and its dimensions include a length of 3000 mm (118.1 in) and a diameter of 340 mm (13.39 in). The empty weight is 120 kg (264.55 lb) and the loaded weight

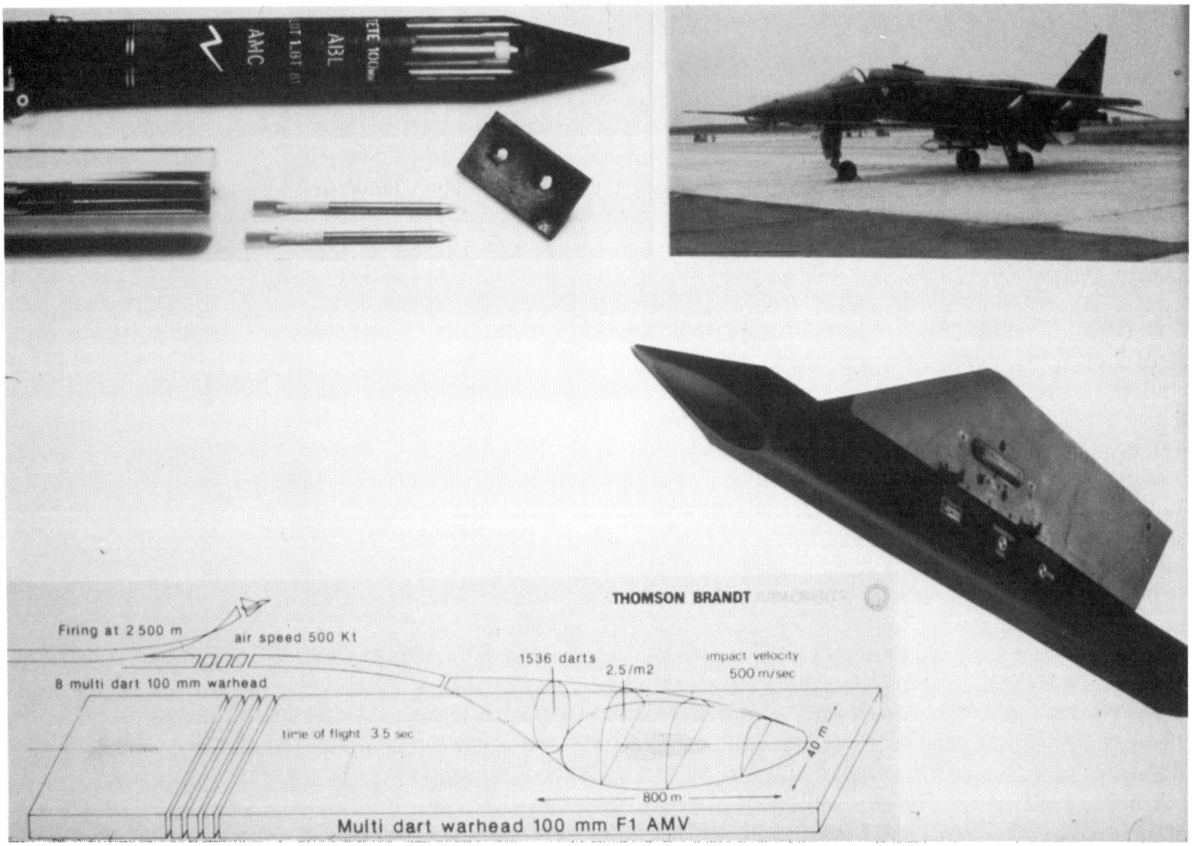

The Multi-Dart 100 is larger brother to the Multi-Dart 68, and offers greater kinetic energy delivered on to the target, though only at the expense of reduced projectile numbers. This composite illustration shows the nature and effect of the Multi-Dart 100: running clockwise from top left, the sequence shows the ABL warhead and its sub-projectiles (plus their penetration capabilities), a SEPECAT Jaguar A carrying Multi-Dart 100 launchers, the launcher and a typical tactical employment diagram.

between 320 and 375 kg (705.47 and 826.72 lb) depending on warhead type. The notional rate of fire is 400 rounds per minute.
Thomson Brandt 100.4: This launcher is also known as the RL F3, and can carry four 100 mm (3.94 in) rockets. The launcher's dimensions include a length of 2900 mm (114.17 in), a width of 230 mm (9.06 in) and a height of 240 mm (9.45 in). The empty weight is 70 kg (154.32 lb), and the loaded weight varies from 206 to 240 kg (454.14 to 529.10 lb) depending on warhead type.

SNIA BPD ARF/8M2

Italy

Type: air-launched rocket (air-to-surface)
Dimensions: diameter 51 mm (2.008 in); length 1030 mm (40.55 in) with API warhead or 1048 mm (41.25 in) with other warheads; fin span data not available
Weight: total round 4.0 kg (8.82 lb) with API warhead, or 4.8 kg (10.58 lb) with other warheads
Warhead: 1.4 kg (3.09 lb) API, or 2.2 kg (4.85 lb) others
Propulsion: one double-base solid-propellant rocket delivering 202 kg (445 lb) thrust for 1.1 seconds
Performance: burn-out velocity 515 or 470 m (1,670 1,540 ft) per second with API or other warhead respectively

Variants and applications
ARF/8M2: This is a light air-to-surface rocket designed for use by helicopters and light fixed-wing aircraft, yet provides good capabilities against a wide range of battlefield targets as a result of its light weight (allowing the carriage of a large number of rounds for use against unhardened area targets) and considerable accuracy (resulting from the use of a freely-rotating quadruple wrap-round fin assembly at the tail). The type is currently available with the 1.4 kg (3.09 lb) armour-piercing incendiary warhead, or with a number of alternative 2.2 kg (4.85 lb) warheads such as HE incendiary, pre-formed fragmentation, anti-tank/anti-personnel, smoke, spotting, practice smoke, practice flash, target practice, and illuminating.

The manufacturer is also improving the type with a new range of 3 kg (6.61 lb) warheads such as HE incendiary, pre-formed fragmentation, anti-tank/anti-personnel, practice smoke, practice flash and target practice. These latter increase overall length to 1150 mm (45.28 in) and weight to 6.6 kg (14.55 lb), but decrease burn-out velocity to 400 m (1,310 ft) per second. More powerful 3 kg (6.61 lb) warheads are under development.
ACR/MR: This is a variant of the ARF/8M3 with same warhead options.
Aerea AL-18-50: This is a lightweight launcher (for use on fixed-wing aircraft) with tubes for 18 50 mm (2 in) rockets. The launcher is 1740 mm (68.5 in) long and 305 mm (12.01 in) in diameter, and its weights are 30 kg (66.14 lb) empty and 116 or 131 kg (255.73 or 288.80 lb) loaded with 2.2 or 3.0 kg (4.85 or 6.61 lb) warheads respectively. The notional rate of fire is 600 rockets per minute.
Aerea AL-25-50: This is a lightweight launcher (for use on fixed-wing aircraft) with tubes for 25 50 mm (2 in) rockets. The launcher is 2000 mm (78.74 in) long and 392 mm (15.43 in) in diameter, and its weights are 47 kg (103.62 lb) empty and 167 or 187 kg (368.17 or 412.26 lb) loaded with 2.2 or 3.0 kg (4.85 or 6.61 lb) warheads respectively. The notional rate of fire is 600 rockets per minute.
Aerea AL-29-50: This is a lightweight launcher (for use on fixed-wing aircraft) with

tubes for 29 50 mm (2 in) rockets. The launcher is 1680 mm (66.14 in) long and 396 mm (15.59 in) in diameter, and its weights are 45 kg (99.21 lb) empty and 184 or 207 kg (405.64 or 456.35 lb) loaded with 2.2 or 3.0 kg (4.85 or 6.61 lb) warheads respectively. The notional rate of fire is 600 rockets per minute.

Aerea HL-14-50: This is a lightweight launcher (for use on rotary-wing aircraft) with tubes for 14 50 mm (2 in) rockets. The launcher is 1123 mm (44.21 in) long and 270 mm (10.63 in) in diameter, and its weights are 15 kg (33.07 lb) empty and 82 or 93 kg (180.78 or 205.03 lb) loaded with 2.2 or 3.0 kg (4.85 or 6.61 lb) warheads respectively. The notional rate of fire is 600 rockets per minute.

Aerea HL-18-50: This is a lightweight launcher (for use on rotary-wing aircraft) with tubes for 18 50 mm (2 in) rockets. The launcher is 1400 mm (55.12 in) long and 305 mm (12.01 in) in diameter, and its weights are 27 kg (59.52 lb) empty and 113 or 128 kg (249.12 or 282.19 lb) loaded with 2.2 or 3.0 kg (4.85 or 6.61 lb) warheads respectively. The notional rate of fire is 600 rockets per minute.

SNIA BPD's 50-mm (1.97-in) rockets are small in calibre, but come in a variety of warhead types and can thus provide light tactical aircraft as well as helicopters with a capable punch over short ranges.

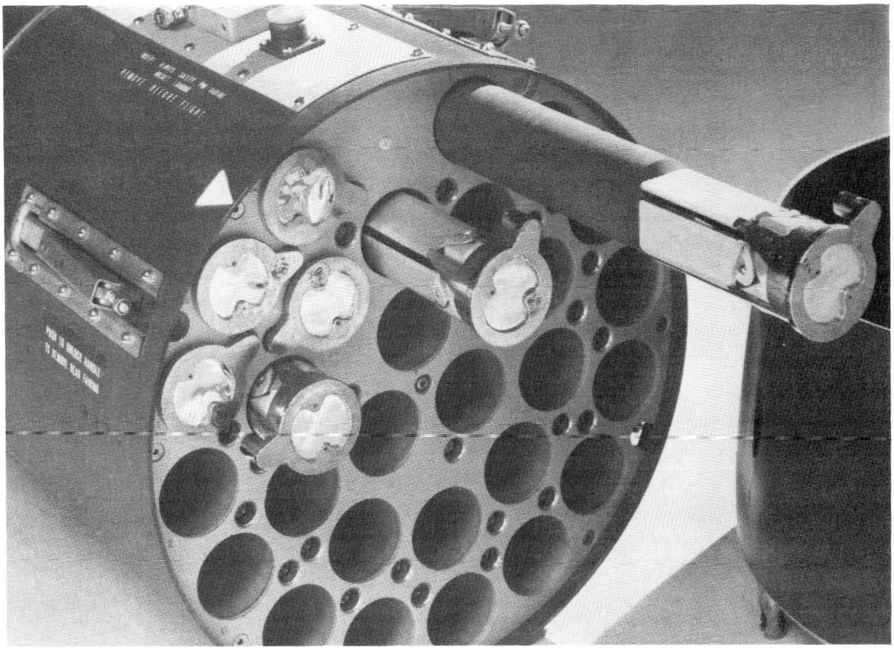

The rear face of an Aerea 28-tube rocket launcher pod with a partial load of 50-mm (1.97-in) rockets. Note the way in which the fins are folded, and the forward-facing lateral projection for the electrical firing connection.

Aerea HL-28-50: This is a lightweight launcher (for use on rotary-wing aircraft) with tubes for 28 50 mm (2 in) rockets. The launcher is 1400 mm (55.12 in) long and 380 mm (14.96 in) in diameter, and its weights are 40 kg (88.18 lb) empty and 174 or 197 kg (383.60 or 434.30 lb) loaded with 2.2 or 3.0 kg (4.85 or 6.61 lb) warheads respectively. The notional rate of fire is 600 rockets per minute.
Aerea HL-29-50: This is a lightweight launcher (for use on rotary-wing aircraft) with tubes for 29 50 mm (2 in) rockets. The launcher is 1680 mm (66.14 in) long and 396 mm (15.59 in) in diameter, and its weights are 45 kg (99.21 lb) empty and 184 or 207 kg (405.64 or 456.35 lb) loaded with 2.2 or 3.0 kg (4.85 or 6.61 lb) warheads respectively. The notional rate of fire is 600 rockets per minute.

SNIA BPD Medusa rocket

Italy

Type: air-launched rocket (air-to-surface)
Dimensions: diameter 81 mm (3.19 in); length 1570 mm (61;81 in) with unitary warheads or 1850 mm (72.83 in) with submunition warhead; fin span data not available
Weight: total round 15 kg (33.07 lb) with unitary warheads or 18 kg (39.68 lb) with submunition warhead
Warhead: 7 kg (15.43 lb) HE, pre-formed fragmentation, spotting and hollow-charge HE, or 10 kg (22.05 lb) multi-purpose with 11 AT-2 anti-tank submunitions
Propulsion: one double-base solid-propellant rocket delivering 917 kg (2,021 lb) thrust for 0.85 second
Performance: burn-out velocity 650 m (2,133 ft) per second with initary warheads or 550 m (1,084 ft) per second with submunition warhead; range 1400 to 6000 m (1,530 to 6,560 yards)

SNIA BPD is Italy's primary manufacturer of solid-propellant rocket motors and unguided rocket projectiles, the company's range covering today's basic calibres in a number of types characterized by different warheads.

Variants and applications

Medusa: This is a powerful medium-calibre air-to-surface rocket, and for stabilization uses a combination of spin (imparted by thrust deflectors in the rocket motor's exhaust nozzle) and triple wrap-round fins at the tail. After launch at 8° from a helicopter hovering at 20 m (65 ft), the rocket impacts the target at 8.5° with a velocity of 280 m (919 ft) per second at a range of 6000 m (6,560 yards) after a 15.4-second flight. The equivalent

The HL-7-80 rocket launcher pod, as carried by this Agusta A 109A, can give even medium helicopters an effective area attack capability at modest stand-off ranges.

data for an aircraft launch at -20° at 500 m (1,640 ft) at 720 km/h (447 mph) are a range of 1400 m (1,530 yards) for an impact velocity of 748 m (2,454 ft) per second after a 2.05-second flight. A warhead containing 11 anti-tank and anti-personnel submunitions is under development for this important rocket system.

Aerea AL-6-80: This is the launcher (for use on fixed-wing aircraft) with tubes for six Medusa rockets. The launcher's dimensions include a length of 2214 mm (87.17 in) and a diameter of 296 mm (11.65 in), and the weights are 38 kg (83.77 lb) empty and 134 or 142 kg (295.41 or 313.05 lb) with standard or submunition warhead. The nominal rate of fire is 600 rounds per minute.

Aerea SAL-6-80: This is the launcher (for use on supersonic fixed-wing aircraft) with tubes for six Medusa rockets. The launcher's dimensions include a length of 2600 mm (102.36 in) and a diameter of 296 mm (11.65 in), and the weights are 48 kg (105.82 lb) empty and 144 or 162 kg (317.46 or 357.14 lb) with standard or submunition warhead. The nominal rate of fire is 600 rounds per minute.

Aerea AL-12-80: This is the launcher (for use on fixed-wing aircraft) with tubes for 12 Medusa rockets. The launcher's dimensions include a length of 2600 mm (102.36 in) and a diameter of 396 mm (15.59 in), and the weights are 66 kg (145.50 lb) empty and 257 or 293 kg (566.58 or 645.94 lb) with standard or submunition warhead. The nominal rate of fire is 600 rounds per minute.

Aerea HLA-12-80: This is the launcher (for use on fixed- and rotary-wing aircraft) with tubes for 12 Medusa rockets. The launcher's dimensions include a length of 2300 mm (90.55 in) and a diameter of 396 mm (15.59 in), and the weights are 60 kg (132.28 lb) empty and 251 or 287 kg (553.35 or 632.72 lb) with standard or submunition warhead. The nominal rate of fire is 600 rounds per minute.

Aerea HL-7-80: This is the launcher (for use on rotary-wing aircraft) with tubes for seven Medusa rockets. The launcher's dimensions include a length of 1900 mm (74.80 in) and a diameter of 296 mm (11.65 in), and the weights are 31 kg (68.34 lb) empty and 142 or 163 kg (313.05 or 359.35 lb) with standard or submunition warhead. The nominal rate of fire is 600 rounds per minute.

SNIA BPD Falco

Type: air-launched rocket (air-to-surface)
Dimensions: diameter 122 mm (4.8 in); length (122/A) 2535 mm (99.8 in) with standard warheads or 3041 mm (119.72 in) with submunition warheads, and (122/H) 2716 mm (106.93 in) with standard warheads or 3344 mm (131.65 in) with submunition warheads; fin span data not available
Weight: total round (122/A) 59.9 kg (132.05 lb) with standard warheads or 65.9 kg (145.28 lb) with submunition warheads, and (122/H) 58.4 kg (128.75 lb) with standard warheads or 71.4 kg (157.41 lb) with submunition warheads
Warhead: (122/A) 26 kg (57.32 lb) standard types (HE and hollow-charge HE), or 32 kg (70.55 lb) submunition types (APM with 66 anti-personnel mines, or ATM with seven anti-tank mines or APAMB with 77 anti-personnel and anti-materiel bomblets), and (122/H) 19 kg (41.89 lb) standard types (HE, white phosphorus, smoke and target practice), or 32 kg (70.55 lb) submunition types (APM with 66 anti-personnel mines, or ATM with seven anti-tank mines or APAMB with 77 anti-personnel and anti-materiel bomblets)
Propulsion: one double-base solid-propellant rocket (short type for aircraft-launched 122/A and long type for helicopter-launched 122/H) delivering unknown thrust for 1.21 seconds
Performance: burn-out velocity 750 m (2,461 ft) per second with standard warhead or 670 m (2,198 ft) per second with submunition warhead; range not revealed

Variants and applications
Falco 122/A: This is a very powerful air-to-surface rocket system developed from the same company's FIROS field artillery rocket system and notable for its range and wide assortment of warhead types. The Falco 122/A aircraft-launched version uses a shorter motor, and is optimized for stand-off attacks against area targets or shorter-range attacks on point targets or the armoured and unarmoured types.
Falco 122/H: This is the helicopter-launched variant of the family using a longer motor to offset the lower launch speed imposed by helicopter operations.
Aerea SAL-4-122: This is the launcher (for use on supersonic fixed-wing aircraft) with tubes for four Falco rockets. The launcher's dimensions include a length of 3450 mm (135.83 in) and a diameter of 360 mm (14.17 in), and the weights are 69 kg (152.11 lb) empty and 308.6 or 332.6 kg (680.34 or 733.25 lb) with standard or submunition warheads.

The SAL-4-122 is a four-tube launcher for 122-mm (4.8-in) SNIA BPD rockets, and is designed for carriage by supersonic aircraft.

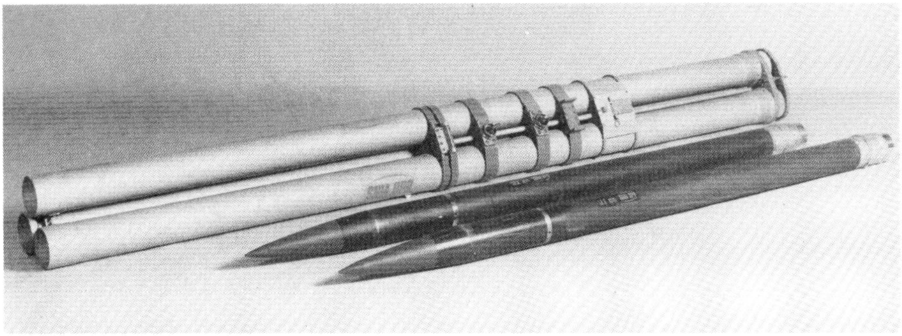

For helicopter use SNIA BPD has developed three- and six-tube tube assemblies. This is the three-tube variety, seen with a pair of the extremely effective 122-mm (4.8-in) rockets fired by the system.

Aerea HL-3-122: This is the launcher (for use on rotary-wing aircraft) with tubes for three Falco rockets. The launcher's dimensions include a length of 3450 mm (135.83 in) and a diameter of 276 mm (10.87 in), and the weights are 59 kg (130.07 lb) empty and 234.2 or 273.2 kg (516.31 or 602.29 lb) with standard or submunition warheads.
Aerea HL-6-122: This is the launcher (for use on rotary-wing aircraft) with tubes for six Falco rockets. The launcher's dimensions include a length of 3450 mm (135.83 in) and a diameter of 410 mm (16.14 in), and the weights are 101 kg (222.66 lb) empty and 451.4 or 529.4 kg (995.15 or 1,167.11 lb) with standard or submunition warhead.

INTA S-9

Spain

Type: air-launched rocket (air-to-surface)
Dimensions: diameter 37 mm (1.46 in); length and fin span data not available
Weight: total round data not available
Warhead: data not available
Propulsion: one solid-propellant rocket delivering unrevealed thrust
Performance: data not available

Variant and applications
S-9: This is a conventional lightweight rocket designed for the smaller types of tactical aircraft, and also valuable as a training type.
CASA Type 18.037: This is a reusable 18-tube launcher pod for the S-9 rocket, and is available in standard and low-drag models for use by light tactical aircraft and higher-performance machines respectively. The standard model has a length of 1240 mm (48.82 in) and a diameter of 250 mm (9.84 in), and its weights are 15.7 kg (34.61 lb) empty and 35 kg (77.16 lb) loaded. The low-drag model measures 1900 mm (74.8 in) in length and 250 mm (9.94 in) in diameter, and its weights are 19.3 kg (42.55 lb) empty and 38.6 kg (85.1 lb) loaded. With either type of pod the rockets can be fired singly or in a ripple at 0.3-second intervals.
CASA Type 54.037: This is a reusable 54-tube launcher pod for the S-9 rocket, and is available in standard and low-drag models for use by light tactical aircraft and higher-performance machines respectively. The standard model has a length of 1240 mm (48.82 in) and a diameter of 400 mm (15.75 in), and its weights are 39.5 kg (87.1 lb) empty and 97.5 kg (214.95 lb) loaded. The low-drag model measures 2270 mm (89.37 in) in length and 400 mm (15.75 in) in diameter, and its weights are 49 kg (108.02 lb) empty

and 107 kg (235.9 lb) loaded; the rockets can be fired singly or in a ripple at 0.2-second intervals.

Spain

INTA S-11

Type: air-launched rocket (air-to-surface)
Dimensions: diameter 69.85 mm (2.75 in); length and fin span data not available
Weight: data not available
Warhead: data not available
Propulsion: one solid-propellant rocket delivering unknown thrust
Performance: data not available

Variant and applications
S-11: This is a considerably larger and heavier weapon than the S-9, and able to deliver a considerable punch over moderate ranges.
CASA Type 06.070: This is a reusable six-tube launcher pod for the S-11 or 2.75 in (69.85 mm) FFAR rocket, and is available in standard and low-drag models for use by light tactical aircaft and higher-performance machines respectively. The standard model has a length of 1240 mm (48.82 in) and a diameter of 250 mm (9.84 in), and its weights are 14.8 kg (32.63 lb) empty and 63.6 kg (140.2 lb) loaded. The low-drag model measures 1900 mm (74.80 in) in length and 250 mm (9.84 in) in diameter, and its weights are 17.8 kg (39.24 lb) empty and 66.6 kg (146.83 lb) loaded. The rockets can be fired singly or in a ripple at 0.3-second intervals.
CASA Type 18.070: This is a reusable 18-tube launcher pod for the S-11 or 2.75 in (69.85 mm) FFAR rocket, and is available in standard and low-drag models for use by light tactical aircraft and higher-performance machines respectively. The standard model has a length of 1240 mm (48.82 in) and a diameter of 400 mm (15.75 in), and its weights are 37.3 kg (82.23 lb) empty and 183.8 kg (405.2 lb) loaded. The low-drag model measures 2270 mm (89.37 in) in length and 400 mm (15.75 in) in diameter, and its weights are 46.8 kg (103.17 lb) empty and 193.3 kg (426.15 lb) loaded. The rockets can be fired singly or in a ripple at 0.3-second intervals.

Spain

INTA S-12

Type: air-launched rocket (air-to-surface)
Dimensions: diameter 100 mm (3.94 in); length and fin span data not available
Weight: data not available
Warhead: data not available
Propulsion: one solid-propellant rocket of unrevealed thrust
Performance: data not available

Variant and application
S-12: This is the largest of the INTA series of rockets designed for the Spanish air force, and packs a very heavy punch.
CASA Lanco RC-06-100: This is a reusable underwing pod designed to carry and launch

six S-12 rockets. The launcher is 3400 mm (133.86 in) long, and its weights are 100 kg (220.45 lb) empty and 460 kg (1,014.11 lb) loaded. The rockets can be fired singly or in a ripple at intervals variable between 0.2 and 0.7 seconds.

Bofors rockets

Sweden

Sweden has been an adherent of the unguided rocket for many years. Such weapons offer very useful capabilities at modest cost to an economy already pressed by the results of political determination to ensure neutrality through armed strength, not reliant on imports for its primary weapons and weapon systems. The **Bofors M57** was developed specifically for Swedish tactical aircraft. The 75 mm (2.95 in) rocket can be fired from a single mounting or, more effectively, from a 19-tube launcher, and while the basic M57 is designed for general-purpose use with its 7 kg (15.43 lb) HE warhead, the **M57AC** is the anti-tank variant with a hollow-charge warhead, and the **M57B** the anti-personnel variant with a fragmentation warhead. The **M56** and **M60** are both elderly but still extremely powerful weapons. Each type can be carried on a single launcher or fired from a six-tube pod, but it has not been revealed how these 135 mm (5.315 in) rockets differ from each other. The **M49** is an old but very powerful 145 mm (5.71 in) anti-tank rocket used in small numbers as a singly-launched weapon.

Bofors M70

Sweden

Type: air-launched rocket (air-to-surface)
Dimensions: diameter 135 mm (5.315 in); length 2105 mm (82.87 in) with general-purpose warhead, or 2165 mm (85.24 in) with armour-piercing warhead; fin span data not available

Seen in one of the many possible attack configurations, with a centreline drop tank and four M70X pods for 135-mm (5.31-in) rockets, is a Saab AJ 37 Viggen of the Swedish air force.

Weight: total round 45.8 kg (100.97 lb) with general-purpose warhead, or 44.6 kg (98.325 lb) with armour-piercing warhead
Warhead: 19.6 kg (43.21 lb) hollow-charge armour-piercing, 20.8 kg (45.86 lb) general-purpose and inert practice
Propulsion: one 25 kg (55.1 lb) double-base solid-propellant rocket delivering 1050 kg (2,315 lb) thrust for 2.3 seconds
Performance: burn-out velocity 600 m (1,969 ft) per second; range 3000 m (3,280 yards)

Variant and application
M70: This rocket was designed as part of the weapon fit for the Saab 37 Viggen multi-role aircraft, and is an extremely powerful type that can be fitted with either of two operational warheads carrying a direct-action or direct-action delay fuse. The flight time to the typical range of 2000 m (2,185 yards) is 2.5 seconds and to the maximum of 3000 m (3,280 yards), 5.5 seconds. The rockets can be fired singly or in a ripple at 0.1-second intervals. The anti-tank warhead contains a 5 kg (11.02 lb) Hexotol charge, and the general-purpose warhead contains a 3.7 kg (8.16 lb) TNT charge.
M70X: This is the six-tube launcher pod for the M70 rocket, a low-drag type designed for carriage on a single point or on NATO standard 356 mm (14 in) twin suspension lugs by high-performance aircraft. The pod's overall dimensions include a length of 3226 mm (127.0 in) and a diameter of 486 mm (19.13 in), and the weights are 125 kg (275.57 lb) empty and 392.6 or 399.8 kg (863.7 or 879.5 lb) loaded with armour-piercing or general-purpose rockets respectively.

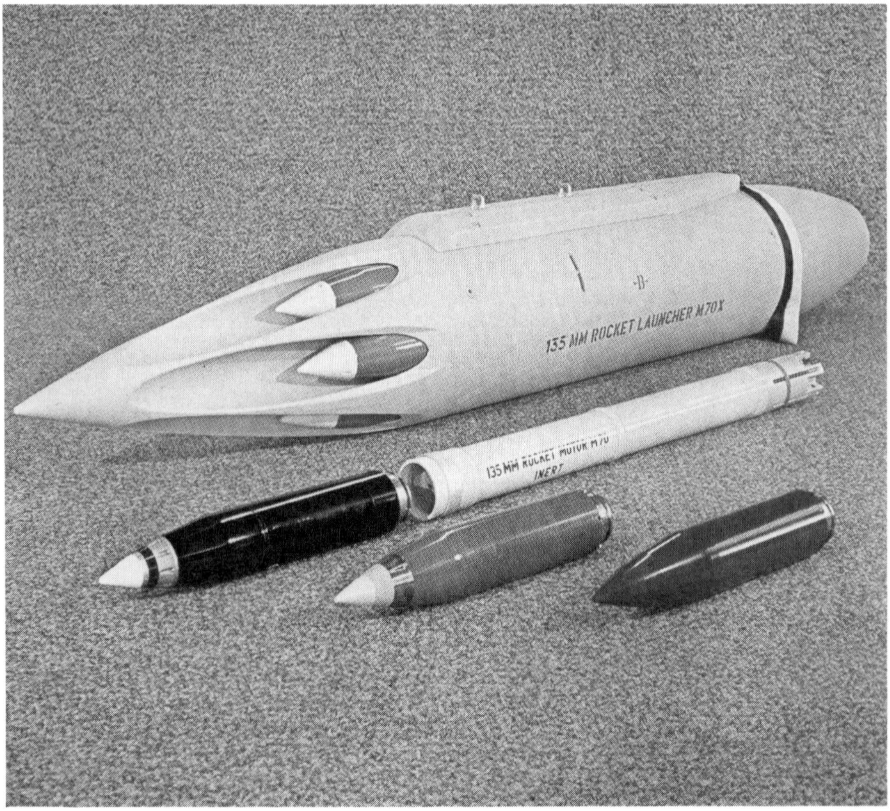

The M70X system is tailored exactly to the requirements of the Swedish air force, and in common with other unguided rockets, the massive projectile has a choice of warheads, in this instance three types.

Oerlikon-Bührle/SNIA BPD SNORA

Switzerland/ Italy

Type: air-launched rocket (air-to-surface)
Dimensions: diameter 81 mm (3.19 in); length 1416 to 1785 mm (55.75 to 70.28 in) depending on warhead type: fin span data not available
Weight: 13.2 to 19.7 kg (29.1 to 43.43 lb) depending on warhead type
Warhead: 4 kg (8.82 lb) target practice, 4.5 kg (9.92 lb) hollow-charge anti-tank and HE pre-formed fragmentation, 5 kg (11.02 lb) target practice, 7 kg (15.43 lb) target practice and HE pre-formed fragmentation, and 11 kg (24.25 lb) target practice and HE pre-formed fragmentation
Propulsion: one TWK 006 solid-propellant rocket delivering 1120 kg (2,469 lb) thrust for 0.85 second
Performance: range about 2500 m (2,735 yards)

Variants and applications
SNORA Rakete 022: The SNORA series of rockets was developed jointly by Oerlikon-Bührle and SNIA BPD, and is amongst the most widely used weapons of its type in the world. The basic rocket is powered by the TWK 006 motor section, which weighs 8.7 kg (19.18 lb) and is 1033 mm (40.67 in) long complete with the wrap-round fins that help to stabilize this spinning rocket. To this core is added the warhead/fuse section to make up the individual type. The Rakete 022 has the UGK-028 target practice warhead, which weighs 7 kg (15.43 lb) and is 515 mm (20.28 in) long for an overall rocket weight of 15.7 kg (34.61 kg) and length of 1548 mm (60.94 in). The type is capable of a velocity increment (added to launcher speed) of 670 m (2,198 ft) per second.
SNORA Rakete 023: This carries the SSK-031 HE pre-formed fragmentation warhead, which weighs 4.5 kg (9.92 lb) and is 383 mm (15.08 in) long for an overall rocket weight of 13.2 kg (29.1 lb) and length of 1416 mm (55.75 in). The type has a velocity increment of 820 m (2,690 ft) per second, and on impact the mechanical fuse detonates 1.7 kg (3.75 lb) of HE to produce some 1,215 fragments
SNORA Rakete 025: This carries the SSK-032 HE pre-formed fragmentation warhead, which weighs 11 kg (124.25 lb) and is 750 mm (29.53 in) long for an overall rocket weight of 19.7 kg (43.43 lb) and length of 1785 mm (70.28 in). The type has a velocity increment of about 670 m (2,198 ft) per second, and on impact the mechanical fuse detonates 2.8 kg (6.17 lb) of HE to produce some 2,133 fragments.
SNORA Rakete 026: This carries the PHK-030 hollow-change anti-tank warhead, which weighs 4.5 kg (9.92 lb) and is 477 mm (18.78 in) long for an overall rocket weight of 13.1 kg (28.88 lb) and length of 1510 mm (59.45 in). The type has a velocity increment of 820 m (2,690 ft) per second, and on impact the electro-mechanical fuse detonates 1 kg (2,205 lb) of HE to produce a jet of hot gas and vaporized metal able to penetrate 350 mm (13.78 in) of armour.
SNORA Rakete 027: This carries the UGK-034 target practice warhead, which weighs 11 kg (24.25 lb) and is 751 mm (29.57 in) long for an overall rocket weight of 19.7 kg (43.43 lb) and length of 1784 mm (70.24 in). The type has a velocity increment of 520 m (1,706 ft) per second.
SNORA Rakete 028: This carries the 4.5 kg (9.92 lb) target practice warhead, which is 393 mm (15.47 in) long for an overall rocket weight of 13.2 kg (29.1 lb) and length of 1426 mm (56.14 in). The type has a velocity increment of 820 m (2,690 ft) per second.
SNORA Rakete 029: This carries the USK-035 target practice warhead, which weighs

The Oerlikon-Bührle SNORA 81-mm (3.19-in) rocket is one of the most important weapons of its type in the world. The weapon is based on the TWK 006 motor section with four spring-opened fins (vertical left), to which are added any of the warhead sections. Seen here are the UGK-028 practice and USK-036 marker types (upper and lower horizontal), and the SSK-032 fragmentation, SSK-029 fragmentation, SSK-031 fragmentation and PHK hollow-charge types (left to right vertical). An example of the complete weapon, in this instance the RAK-024 with the SSK-029 warhead, is seen at the bottom.

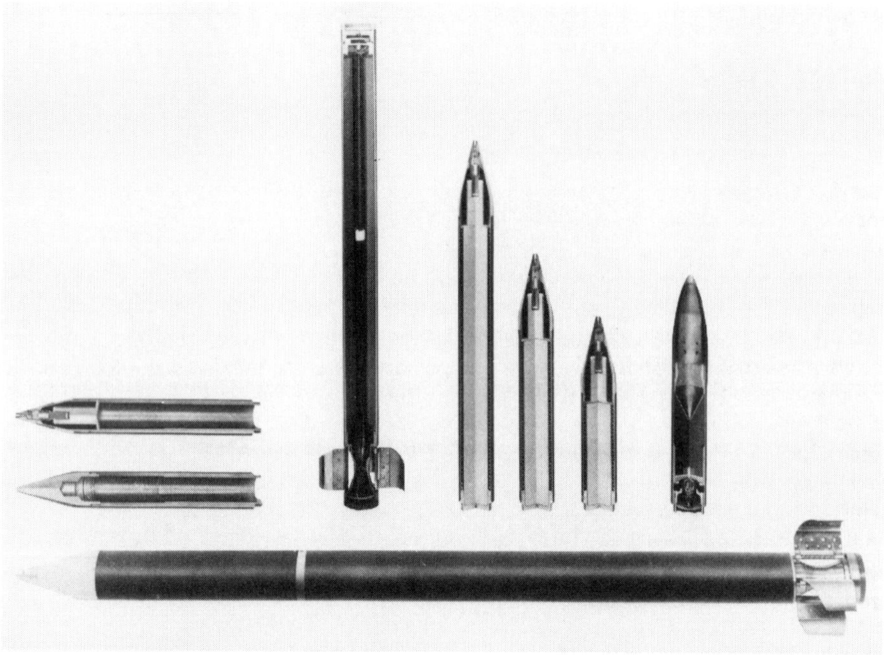

4.5 kg (9.92 lb) and is 393 mm (15.47 in) long for an overall rocket weight of 13.2 kg (29.1 lb) and length of 1426 mm (56.14 in). The type has a velocity increment of 820 m (2,690 ft) per second.

SNORA Rakete 030: This carries the USK-036 target practice warhead, which weighs 7 kg (15.43 lb) and is 515 mm (20.28 in) long for an overall rocket weight of 15.7 kg (34.61 lb) and length of 1548 mm (60.94 in). The type has a velocity increment of 670 m (2,198 ft) per second.

SNORA Rakete 031: This carries the USK-037 target practice warhead, which weighs 11 kg (24.25 lb) and is 751 mm (29.57 in) long for an overall rocket weight of 19.7 kg (43.43 lb) and length of 1784 mm (70.24 in). The type has a velocity increment of 520 m (1,706 ft) per second.

SNORA Rakete 054: This carries the PHK-042 hollow-charge anti-tank warhead, which weighs 4.5 kg (9.92 lb) and is 468 mm (18.43 in) long for an overall rocket weight of 13.2 kg (29.1 lb) and length of 1500 mm (59.06 in). The type has a velocity increment of 820 m (2,690 ft) per second, and on impact the electro-mechanical fuse detonates 1 kg (2.205 lb) of HE to produce a jet of gas and vaporized metal able to penetrate 450 mm (17.72 in) of armour.

Aerea AL-6-80: This is the Italian-produced launcher pod with tubes for six SNORA rockets. The type is desgned for use on subsonic aircraft, and the pod's primary data include a length of 2214 mm (87.17 in), a diameter of 296 mm (11.65 in), an empty weight of 38 kg (83.77 lb) and loaded weights varying from 116 to 156.2 kg (255.73 to 344.36 lb) depending on the warhead type.

Aerea SAL-6-80: This is the Italian-produced launcher pod with tubes for six SNORA rockets. The type is designed for use on supersonic aircraft, and the pod's primary data include a length of 2600 mm (102.36 in), a diameter of 296 mm (11.65 in), an empty weight of 48 kg (105.82 lb) and loaded weights varying from 124 to 163 kg (273.37 to 359.35 lb) depending on the warhead type. The nominal rate of fire is 10 rounds per second.

Aerea SAL-12-80: This is the Italian-produced launcher pod with tubes for 12 SNORA rockets. The type is designed for use on supersonic aircraft, and the pod's primary data include a length of 2600 mm (102.36 in), a diameter of 396 mm (15.59 in), an empty weight of 66 kg (145.50 lb) and loaded weights varying from 218 to 297 kg (480.60 to 654.76 lb) depending on the warhead type. The nominal rate of fire is 10 rounds per second.

Aerea HL-7-80: This is the Italian-produced launcher pod with tubes for seven SNORA rockets. The type is designed for use on helicopters, and the pod's primary data include a length of 1800 mm (70.87 in), a diameter of 296 mm (11.65 in), an empty weight of 33 kg (72.75 lb) and loaded weights varying from 123 to 143 kg (271.16 to 315.26 lb) depending on the warhead type. The nominal rate of fire is 10 rounds per second.

CASA Type 04.080: This is the Spanish-produced launcher rack able to accept four SNORA rockets, and designed for use on light tactical aircraft and helicopters. A variable intervalometer allows the rockets to be ripple-fired at intervals ranging from 0.2 to 0.5 seconds, and there is also a **Type 04.057/080** variant of the rack that can accept four 57 or 81 mm rockets.

Oerlikon-Bührle RWK 020 (SAL-12-80): This is the Swiss-produced launcher pod with tubes for 12 SNORA rockets; the type is designed for use on supersonic aircraft, and the pod's primary data include a length of 2550 mm (100.39 in), a diameter of 396 mm (15.59 in), an empty weight of 69 kg (152.12 lb) and loaded weights varying from 227 to 305 kg (500.44 to 672.40 lb) depending on the warhead type. The nominal

To complement its 81-mm (3.19-in) rockets, Oerlikon-Bührle makes a range of pods suitable for tactical aircraft ranging from slow helicopters to supersonic fixed-wing types. This is the RWK 020 type, otherwise designated SAL-12-180, for 12 SNORA rockets. The type is designed for carriage by supersonic aircraft, and is seen here without its aerodynamic nose fairing.

Oerlikon-Bührle SAL-12-80 (RWK 020) rocket launcher pod under the starboard wing of a Dassault-Breguet Mirage IIIS fighter. In place over the nose of the 12-tube launcher is the aerodynamic fairing that reduces drag until it is destroyed as the 81-mm (3.19-in) rockets are fired.

rate of fire is 12 rounds per second, and the rockets can be fired singly, or in 2-, 6- or 12-round salvoes.

Oerlikon-Bührle RWK 021 (SAL-6-80): This is the Swiss-produced launcher pod with tubes for six SNORA rockets. The type is designed for use on supersonic aircraft, and the pod's primary data include a length of 2600 mm (102.36 in), a diameter of 296 mm (11.65 in), an empty weight of 48 kg (105.82 lb) and loaded weights varying from 127 to 166 kg (280.0 to 366.0 lb) depending on the warhead type. The nominal rate of fire is 10 rounds per second, and the rockets can be fired singly, or in 2-, 3- or 6-round salvoes.

Oerlikon-Bührle RWK 022 (HL-7-80): This is the Swiss-produced launcher pod with tubes for seven SNORA rockets. The type is designed for use on helicopters, and the pod's primary data include a length of 1800 mm (70.87 in), a diameter of 296 mm (11.65 in), an empty weight of 33 kg (72.75 lb) and loaded weights varying from 125 to 143 kg (275.57 to 315.26 lb) depending on the warhead type. The nominal rate of fire is 10 rounds per second, and the rockets can be fired singly, or in 2-, 3- or 7-round salvoes.

Oerlikon-Bührle RWK 024 (AL-6-80): This is the Swiss-produced launcher pod with tubes for 12 SNORA rockets. The type is designed for use on subsonic aircraft, and the pod's primary data include a length of 2214 mm (87.17 in), a diameter of 296 mm (11.65 in), an empty weight of 38 kg (83.77 lb) and loaded weights varying from 117 to 132 kg (257.94 to 291.00 lb) depending on the warhead type. The nominal rate of fire is 10 rounds per second, and the rockets can be fired singly, or in 2-, 3- or 6-round salvoes.

Oerlikon-Bührle SURA-D

Switzerland

Type: air-launched rocket (air-to-surface)
Dimensions: diameter 81 mm (3.19 in); length 1077 to 1212 mm (42.40 to 47.42 in) depending on warhead type; fin span data not available
Weight: 12.7 to 14.2 kg (28.00 to 31.31 lb) depending on warhead type
Warhead: 3 kg (6.61 lb) target practice, 3 kg (6.61 lb) HE pre-formed fragmentation, 3 kg (6.61 lb) HE incendiary, 3 kg (6.61 lb) hollow-charge anti-tank, 4.5 kg (9.92 lb) target practice and 4.5 kg (9.92 lb) HE pre-formed fragmentation
Propulsion: one 8.4 kg (18.52 lb) TWK 007 double-base solid-propellant rocket delivering 745 kg (1,642 lb) thrust for 0.9 second
Performance: burn-out velocity varies from 530 to 595 m (1,739 to 1,952 ft) per second depending on variant; range 2500 m (2,735 yards)

Variants and applications

SURA-D Rakete 047: The SURA-D is in essence an upgraded version of the SURA-FL (the latter is no longer in service) designed for use by a wide variety of tactical fixed- and rotary-wing aircraft tasked with the engagement of point or area targets. The series is based on the TWK 007 motor section, which provides the rocket with spin stabilization (with the aid of exhaust deflectors), weighs 8.4 kg (18.52 lb), has a length of 819 mm (32.24 in) and can accommodate at its front any of seven warhead types to create a specific weapon.

A particular advantage of the SURA-D rocket is that it needs no special launcher pod. Before launch the cruciform fixed-fin assembly serves as the forward suspension (being attached to the front suspension lug on any NATO-standard hardpoint), and below this uppermost rocket can be attached in similar manner a tier of more SURA-D rockets limited in number only by considerations of the launch platform's weight and ground-

Simplest of the Oerlikon-Bührle family of 81-mm (3.19-in) unguided rockets, the SURA-D variety is designed for use on helicopters and low-performance tactical aircraft, the rockets being locked together in tiers by their sliding fin assemblies until the ignition of each rocket from the bottom upward moves it forward to pick up its fin section and free it from the bottom of the tier. These are RAK 052 rockets, comprising a TWK 007 core (motor and fin section) with a UIS 3 blast incendiary warhead.

clearance. The rockets are launched electrically from the bottom of the tier upwards, the sliding fin assembly serving as a guide until the rear of the rocket cone connects with it, whereupon the fin assembly is detached from the tier to serve as the rocket's tail-mounted stabilizing unit. The SURA-D Rakete 047 is the anti-tank variant of the series, and is fitted with the PI-3 hollow-charge anti-tank warhead. This weighs 3 kg (6.61 lb) and is 380 mm (14.96 in) long for an overall rocket weight of 11.4 kg (25.13 lb) and length of 1199 mm (47.20 in).

SURA-D Rakete 048: This is a target practice variant of the series, and is fitted with the UGK-033 warhead. This weighs 4.5 kg (9.92 lb) and is 393 mm (15.47 in) long for an overall rocket weight of 12.4 kg (27.34 lb) and length of 1212 mm (47.72 in).

SURA-D Rakete 049: This is a target practice variant of the series, and is fitted with the USK-035 warhead. This weighs 4.5 kg (9.92 lb) and is 393 mm (15.47 in) long for an overall rocket weight of 12.4 kg (27.34 lb) and length of 1212 mm (47.72 in).

SURA-D Rakete 050: This is a general-purpose and anti-personnel variant of the series, and is fitted with the SSK-031 HE pre-formed fragmentation warhead. This weighs 4.5 kg (9.92 lb) and is 383 mm (15.08 in) long for an overall rocket weight of 12.4 kg (27.34 lb) and length of 1202 mm (47.32 in).

SURA-D Rakete 051: This is a target practice variant of the series, and is fitted with the USE-3 warhead. This weighs 3 kg (6.61 lb) and is 258 mm (10.16 in) long for an overall rocket weight of 11.4 kg (25.13 lb) and length of 1077 mm (42.40 in).

SURA-D Rakete 052: This is a general-purpose and anti-personnel variant of the series, and is fitted with the UIS-3 HE pre-formed fragmentation warhead. This weighs 3 kg (6.61 lb) and is 326 mm (12.83 in) long for an overall rocket weight of 11.4 kg (25.13 lb) and length of 1145 mm (45.08 in).

SURA-D Rakete 053: This is a general-purpose and anti-personnel variant of the series, and is fitted with the US-3 pre-formed fragmentation warhead. This weighs 3 kg (6.61 lb) and is 326 mm (12.83 in) long for an overall rocket weight of 11.4 kg (25.13 lb) and length of 1145 mm (45.08 in).

USSR

Soviet rockets

The USSR was one of the first protagonists of the unguided rocket for both air-to-air and air-to-surface use, having started development of such weapons in the early 1920s during the Russian Civil War that followed the establishment of the Bolshevik regime in 1917-8. Early rockets were produced in 75 mm (2.95 in), 82 mm (3.23 in) and 132 mm (5.2 in) calibres, the first for air-to-air and the latter primarily for air-to-surface use in the Soviet campaigns against Japan in 1938-9, and during World War 2 against Germany and her allies (1941-5). Development continued after World War 2, the most important of these weapons being the 137 mm (5.4 in) **M-100** 190 mm (7.48 in) **TRS-190** and 212 mm (8.35 in) **ARS-212,** weighing 15.6, 46 and 116 kg (34.4, 101.4 and 255.7 lb) respectively and each fired from single underwing launchers. However, the more recent Soviet air-launched rockets have all been designated in the S series, and produced in prodigious numbers: indeed, it has been estimated that the 55 mm (2.165 in) **S-5** and its variants have been made in numbers exceeding the production of all other rockets combined.

The S-5 is fired from **UB-8-57, UB-16-57, UB-19-57** and **UB-32-57** pods, the first number indicating the number of tubes in each launcher, and the second the actual calibre of the tube, some 2 mm (0.08 in) greater than the folding-fin rocket fired from it. The

S-5 series is produced with nine types of warhead (including hollow-charge anti-tank, HE, fragmentation and chaff), and possesses a burn-out velocity of 620 m (2,034 ft) per second. Other weapons in the S series are the 80 mm (3.15 in) **S-8,** the 130 mm (5.12 in) **S-13,** the 160 mm (6.3 in) **S-16,** the 210 mm (8.27 in) **S-21,** the 210 mm (8.27 mm) **S-21M** with limited command guidance, the 240 mm (9.45 in) **S-24,** the 280 mm (11.02 in) **S-28,** the 320 mm (12.6 in) **S-32,** and the 320 mm (12.6 in) **S-32M** with limited command guidance. The larger rockets are always fired from single launchers under the wings of tactical aircraft, but are rare compared with the S-5 series which is used by virtually all Soviet tactical aircraft and helicopters.

Royal Ordnance 2-in rocket — UK

Type: air-launched rocket (air-to-surface)
Dimensions: diameter 2 in (50.8 mm); length 48 in (1219 mm); fin span 6.7 in (170 mm)
Weight: 9.52 lb (4.32 kg) with AP warhead, or 10.0 lb (4.54 kg) with HE or practice warhead
Warhead: 2.75 lb (1.25 kg) armour-piercing, and 3.25 lb (1.475 kg) HE or practice
Propulsion: one solid-propellant rocket of unrevealed thrust
Performance: data not available

Variant and application

2-in rocket: This is a small-calibre yet powerful weapon derived from a French Matra original. It is suited to a diversity of applications ranging from hovering helicopters to supersonic tactical aircraft, and its advantages are light weight combined with an effective warhead, allowing substantial numbers to be carried even by light tactical aircraft.

BEI Defense Systems Hydra 70 — USA

Type: air-launched rocket (air-to-surface and air-to-air)
Dimensions: diameter 2.75 in (69.85 in); length 46.85 in (1190 mm); fin span 10.0 in (254 mm)
Weight: 18.96 to 26.23 lb (8.6 to 11.9 kg) depending on warhead type
Warhead: see below
Propulsion: one solid-propellant rocket delivering 750 lb (348 kg) thrust for 1.2 to 1.4 seconds (Mk 4 and Mk 40) or 1,500 lb (680 kg) thrust for 2.3 seconds (Mk 66)
Performance: range 550/9,650 yards (505/8825 m)

Variants and applications

Mk 4 2.75-in FFAR: This very widely used tactical rocket with 6.5 in (165 mm) folding tail fins was developed in the USA as the 'Mighty Mouse' series in the air-to-air role with a 6 lb (2.7 kg) warhead, developing in the 1960s to a 10 lb (4.54 kg) warhead and utlimately to a 17 lb (7.7 kg) type with a rocket motor providing 750 lb (340 kg) thrust for between 1.2 and 1.4 seconds. This is sufficient to give the rocket a range between 550 and 6,500 yards (505 and 5945 m), 3,750 yards (3430 m) being the maximum effective range, but its velocity is insufficiently high to free the weapons from the worst effects of gravity drop and cross-wind. The Mk 4 is designed for use with high-performance fixed-wing aircraft.

Mk 40 2.75-in FFAR: This is the Mk 4 variant optimized for use by helicopters. It thus has a scarfed nozzle for greater spin (and thus greater accuracy) when launched from low-speed platforms. Warheads associated with the Mk 4 and Mk 40 versions of the original FFAR are the 10 lb (4.54 kg) M151 HE, white phosphorus or anti-personnel fragmentation, the 10 lb M156 chemical, the 17 lb (7.71 kg) M229 HE white phosphorus or anti-personnel fragmentation, and the Northrop WDU-4A/A flechette warhead. Estimates credit the M151 with the same effect as an 81 mm (3.2 in) mortar bomb, and the M229 with the effect of a 4.2 in (107 mm) mortar bomb. The WDU-4A/A is a result of Project 'Beehive', originally conceived to develop new anti-personnel warheads for 90 and 105 mm (3.54 and 4.13 in) artillery. Picatinny Arsenal developed a special low-cost production technique for flechettes with stabilizing fins and barbed heads, and some 3,000 such flechettes can be packed into the WDU-4A/A warhead.

Mk 66 2.75-in Hydra 70: This is the most important of current weapons, produced in the USA as part of the Hydra 70 system. This model has a higher-impulse rocket motor developed under the auspices of the US Navy, giving a burn-out velocity in the order of 3,280 ft (1000 m) per second. This produces greater maximum range in both the air-to-air and air-to-sufrace roles, and also provides better accuracy in combination with the revised arrangement of four wrap-round fins.

The Mk 66 can be used by fixed- and rotary-wing aircraft, and can carry a useful assortment of warhead types.

The M151 is a high explosive warhead fitted with the M432 or M433 nose-mounted fuses; these are both of the type that can be remotely set from the cockpit to suit altering tactical situations, the former being used to provide airburst capability between 550 and 6,500 yards (505 and 5945 m), and the latter providing ground-burst options varying from super-quick operation to the types of delay needed for capability against targets under the forest canopy, inside buildings or even in bunkers.

The M247 is a hollow-charge anti-tank warhead which, despite its small diameter, has good penetration capabilities when used in conjunction with the M438 base-mounted fuse. Naturally enough, the chances of scoring a decisive hit are improved by the ripple-firing of substantial numbers of rockets.

The M255 is the flechette warhead for use against battlefield targets ranging from other helicopters to exposed personnel; the warhead carries a load of about 2,500 28-grain (0.064-oz/1.8-g) darts and three tracers, and is detonated by a base-mounted M439 remotely-set fuse.

The M261 is a multi-purpose submunition warhead with capabilities against armour, materiel and personnel. The head carries nine M73 hollow-charge fragmentation submunitions, which are released in the air (thereafter descending under spring-opened three-lobe retarders) by the action of the M439 base-mounted fuse; this latter can be remotely set from the cockpit to provide optimum bursting conditions to create an aerial 'wall in space' of submunitions that then fall towards the most vulnerable upper surfaces of a tank or other armoured formation.

The M262 is an illuminating warhead generating 1 million candlepower for 2 minutes, and suitable for the marking of ground targets; the warhead is used with the remotely-set M439 rear-mounted fuse. The M264 is a smokescreen warhead providing up to 5 minutes of screening suitable for obscuration and target marking; the fuse used with this warhead is the M439 base-mounted type with remote setting capability.

The M267 is the practice version of the M261 submunition warhead, fitted with the same fuse but carrying smoke-filled submunitions to mark impact points. Finally there is a chaff warhead as yet unassigned an official designation in the M series. This is the longest warhead in the Hydra 70 series, loaded with 24 circular packs of chaff for jamming and decoy of enemy radars; the type is used with the remotely-set M439 base fuse.

LAU-3A/B: This is a 19-tube launcher for FFARs used by fixed-wing aircraft.

LAU-32/A: This is a multi-tube launcher for FFARs used by fixed-wing aircraft.

LAU-59/A: This is a seven-tube launcher for FFARs used by fixed-wing aircraft.

LAU-61/A: This is a 19-tube launcher for FFARs used by fixed-wing aircraft. The same launcher is used by the US Army for rotary-wing aircraft with the designation **M159A1.**

LAU-68/A: This is a seven-tube launcher for FFARs used by fixed-wing aircraft. The same launcher is used by the US Army for rotary-wing aircraft with the designation **M158A1.**

LAU-69/A: This is a 19-tube launcher for FFARs used by fixed-wing aircraft. The same launcher is used by the US Army for rotary-wing aircraft with the designation **M200A1.**

M157A/B: This is a seven-tube launcher for FFARs used by rotary-wing aircraft.

M227: This is a multi-tube launcher for FFARs used by rotary-wing aircraft.

Hughes M260: This is a seven-tube lightweight launcher developed for the US Army's attack and close-support helicopter force. The launcher is carried on NATO standard 14 in (356 mm) suspension lugs, and its primary dimensional data include a diameter of 10.0 in (254 mm) and a length of 65.0 in (1651 mm). The launcher weighs 34.5 lb (15.65 kg) empty, and can ripple fire its seven FFAR or Hydra 70 rockets at 0.06-second intervals.

Hughes M261: This is a 19-tube lightweight launcher developed for the US Army's attack and close-support helicopter force. The launcher is carried on NATO standard 14 in (356 mm) suspension lugs, and its primary dimensional data include a diameter of 16.0 in (406 mm) and a length of 65.0 in (1651 mm). The launcher weighs 82 lb (37.2 kg)

Though elderly in design, the 5-in (127-mm) Zuni rocket still packs a very powerful punch when delivered with the accuracy inherent in launch platforms such as this AV-8A Harrier of the US Marine Corps, which can carry a variable number of four-round launchers on its underwing hardpoints.

empty, and can ripple fire its 19 FFAR or Hydra 70 rockets at 0.06-second intervals.
BEI Defense Systems BH-7/A: This is the seven-tube launcher developed by BEI specifically for the high-power Hydra 70 series of unguided rockets.
BEI Defense Systems BH-19/A: This is the 19-tube launcher developed by BEI specifically for the high-power Hydra 70 series of unguided rockets.

Yugoslavia

Yugoslav rockets

The Yugoslav **VRZ-57** is very similar to the Soviet S-5, and can carry on HE or anti-tank warhead. The **Lanser L57** pod carries 12 such rockets of 55 mm (2.17 in) actual and 57 mm (2.24 in) nominal calibre. Yugoslavia also uses the 127 mm (5 in) **VRZ-127** rocket based on the US 5 in (127 mm) Zuri Mk 16 rocket.

CRUISE MISSILES

Aérospatiale ASMP

France

Type: air-launched theatre/strategic cruise missile
Dimensions: diameter about 380 mm (14.96 in) and width across inlets about 820 mm (32.28 in); length 5380 mm (211.81 in); span 956 mm (37.6 in)
Weights: total round 900 kg (1,984.1 lb); warhead CEA 100/150-kiloton nuclear, though estimates of 300 kiloton yield have also been quoted
Propulsion: one integral solid-propellant booster rocket delivering unrevealed thrust and one ramjet sustainer delivering unrevealed thrust
Performance: speed Mach 3 at high altitude and Mach 2 at low altitude; range varies from 300 km (186 miles) at high altitude to 80 km (50 miles) at low altitude
CEP (the standard measure of accuracy): not revealed
Guidance: Sagem inertial and terrain-following

Variant and application
ASMP: Designed from 1978 (after an abortive beginning from 1971 as the armament of the proposed Dassault Mirage G and cancelled Dassault Super Mirage), the **Air-Sol Moyenne Portée** (medium-range air-to-surface) is an air-launched cruise missile now optimized for carriage by the Dassault-Breguet Mirage IVP bomber (after conversion

Launch of an ASMP cruise missile from a converted Dassault-Breguet Mirage IV.

Discarding its catapult strop as its lifts over the bows of a French aircraft-carrier, a Dassault-Breguet Super Etendard departs on a training sortie carrying an ASMP operational-level nuclear missile under its starboard wing. The mass of the missile is balanced by a 1200-litre (264-Imp gal) drop tank under the port wing. Note the Aérospatiale Alouette helicopter flying planeguard duty in the background.

from Mirage IVA standard with an armament of one 60/70-kiloton AN-22 free-fall nuclear weapon), the Dassault-Breguet Mirage 2000N low-altitude strike aircraft and the Dassault-Breguet Super Etendard strike fighter.

The ASMP entered service in 1986, and provides the French air forces with a stand-off nuclear capability against large targets such as railway yards, main bridges, and command and communications centres. The type can fly a high-altitude profile ending with a steep supersonic dive, or alternatively a low-altitude terrain-following profile of shorter range.

USSR

AS-15 'Kent'

Type: air-launched theatre/strategic cruise missile
Dimensions: diameter 650 mm (25.6 in); length 6900 mm (271.65 in); span 3450 mm (135.83 in)
Weights: total round 1500 kg (3,306.9 lb); warhead 150/250-kiloton thermonuclear or HE of unknown quantity
Propulsion: one turbojet or turbofan delivering unrevealed thrust
Performance: speed about Mach 0.9; range 2750 or 3000 km (1,710 or 1,865 miles)
CEP (the standard measure of accuracy): estimated 45 m (150 ft)
Guidance: inertial with terrain-following update

Variant and application

AS-15 'Kent': This is the air-launched version of the USSR's standard cruise missile, and began to enter service during the mid-1980s in a number of forms. The AS-15 is currently operated by the Tupolev Tu-95 'Bear-H', and will equip the new Tupolev 'Blackjack' supersonic variable-geometry bomber due in service during 1988. The weapon compares favourably to the US AGM-86 and BGM-109 types, and parallels the American weapon in having SS-C-4 and SS-N-21 land-and sea-launched counterparts.

BL-10 USSR

Type: air-launched strategic cruise missile
Dimensions: diameter data not available; length about 7000 mm (275.59 in); span data not available
Weights: total round about 2000 kg (4,409.2 lb); warhead 200-kiloton thermonuclear
Propulsion: one ramjet or hybrid rocket/ramjet delivering unrevealed thrust
Performance: speed Mach 2+; range about 3250 km (2,020 miles)
CEP (the standard measure of accuracy): data not available
Guidance: data not available

Variant and application
BL-10: Virtually nothing is known of this advanced cruise missile, which is currently under development as the definitive armament of the Tupolev Tu-? 'Blackjack' strategic bomber. There is little reason to doubt, however, that the Soviets have progressed along the same basic design and development path as the Americans, and thus concentrated their efforts on low observability, albeit in a platform possessing markedly higher performance than the US Advanced Cruise Missile. This may mean that while the SL-10 has low electro-magnetic observability, its infra-red emission may be significantly higher than that of the Convair ACM.

Boeing AGM-86B USA

Type: air-launched strategic cruise missile
Dimensions: diameter 27.3 in (693 mm); length 249.0 in (6325 mm); span 144.0 in (3658 mm)
Weights: total round 3,200 lb (1451.5 kg); warhead 270 lb (122.5 kg) 200-kiloton W80-1 nuclear (using oralloy for fission, tritium for fusion and PBX-9502 Improved HE for implosion leading to fission/fusion reaction)
Propulsion: one Williams Research F107-WR-101 turbofan delivering 600 lb (272 lb) thrust
Performance: speed 500 mph (805 km/h); range 1,550 miles (2495 km)
CEP (the standard measure of accuracy range): between 30 and 100 ft (9 and 30.5 m)
Guidance: Litton P-1000 inertial with McDonnell Douglas DPW-23 TERCOM (TERrain COntour Matching) update

Variant and application
AGM-86B: Entering service in 1981, this air-launched cruise missile is one of the US Air Force's most important weapons, some 170 Boeing B-52G bombers being configured

to carry 12 such weapons apiece (six on each of two underwing stations) plus four AGM-69A SRAMs and four free-fall nuclear weapons in the bomb bay. Work is under way on modifying 95 B-52H bombers to the same standard, with provision on both Stratofortress marks for the weapons-bay carriage of the Common Strategic Rotary Launcher for eight AGM-86B and/or AGM-69A SRAM missiles. It is estimated that 85% of the strategic targets in the USSR can be reached by AGM-86Bs launched safely from areas more than 225 miles (360 km) from Soviet airspace, with the SRAMs available for defence suppression should the bombers have to penetrate Soviet airspace.

The new Rockwell B-1B supersonic penetration bomber can carry eight AGM-86Bs on a rotary launcher in the weapons bay, plus another 14 under the wings. Though 4,348 AGM-86Bs were planned, only 1,715 have been built as newer cruise missiles (of the Convair ACM type) are planned with greater capabilities at a time the Soviets are strengthening their capabilities against air-launched cruise missiles. From 1985 in-service rounds have been retrofitted with an ECM package to improve their capabilities against the Soviet defences.

USA

General Dynamics AGM-109H Tomahawk

Type: air-launched airfield attack cruise missile
Dimensions: diameter 21.0 in (533 mm); length 252.0 in (6401 mm); span 100.0 in (2540 mm)
Weights: total round not revealed; warhead dispenser type with BLU-106/B submunitions
Powerplant: one Teledyne Continental J402-CA-401 turbofan delivering 600 lb (272kg) thrust
Performance: speed 500 mph (805 km/h); range 1,555 miles (2502 km)
CEP (the standard measure of accuracy): 65 ft (20 m)
Guidance: Litton inertial plus McDonnell Douglas DPW-23 TERCOM (TERrain Contour Matching) update and Digital Scene-Matching Area Correlation terminal homing

Variants and applications
AGM-109H Tomahawk: The Tomahawk was developed in parallel with the Boeing AGM-86 cruise missile, and was selected for the US Navy and US Air Force in ship/submarine- and ground-launched versions for a number of strategic and theatre applications with nuclear or conventional warheads. These BGM-109A to BGM-109G variants are all powered by the 600 lb (272 kg) thrust Williams Research F107-WR-400 non-afterburning turbofan. The AGM-109H is the proposed medium-range air-launched airfield attack cruise missile for the US Air Force with revised propulsion (the Teledyne Continental J402 turbofan engine) and a conventional warhead with tactical payload of BLU-106/B BKEP (Boosted Kinetic-Energy Penetrator) submunitions.
AGM-109K Tomahawk: This is the proposed medium-range air-launched land attack and sealane control cruise missile for the US Air Force with the J402-CA-401 engine and the WDU-25A/B conventional warhead.
AGM-109L Tomahawk: This is the proposed medium-range air-launched land/ sea attack cruise missile for the US Navy with the J402-CA-401 engine and the WDU-7B conventional warhead.

General Dynamics (Convair) Advanced Cruise Missile

USA

Type: air-launched long-range cruise missile
Dimensions: data not available
Weights: total round probably in the order of 3,300 lb (1496.9 kg); nuclear warhead data not available
Propulsion: one Williams International F112 turbofan delivering unrevealed thrust
Performance: speed about 500 mph (805 km/h); range at least 1,700 miles (2735 km)
CEP (the standard measure of accuracy): not revealed
Guidance: not revealed

Variant and application

Advanced Cruise Missile: Although this important weapon may enter service before 1990, very little is known of its highly classified development. The programme to develop this important successor to the Boeing AGM-86B ALCM was launched in 1977, with the object of improving performance in crucial fields such as low observability and thus increased capability to penetrate into airspace protected by advanced air defences (through the use of 'stealth' technology), improved propulsion for greater range (through installation of greater power with reduced specific fuel consumption), greater survivability (through the incorporation of the most advanced electronic and infra-red countermeasures), and enhanced operational flexibility (through the use of the latest computer hardware and software). That development of the ACM had been awarded to General Dynamics was announced in 1983, and the type is designed for use on Boeing B-52, Rockwell B-1 and Northrop B-2 bombers. Key features are longer range than the AGM-86B combined with a high measure of 'stealth' technology (especially in terms of infra-red emission) to reduce the missile's observability.

AIR-TO-AIR MISSILES

Brazil

Orbita MAA-1 Leo

Type: short-range air-to-air missile
Dimensions: diameter 150 mm (5.9 in); length 2720 mm (107.09 in); span 652 mm (25.7 in)
Weights: total round 86 kg (189.6 lb); warhead 12 kg (26.5 lb) proximity- and impact-fused HE
Propulsion: one Vasconcelos-built Imbel Piquete solid-propellant rocket delivering unrevealed thrust
Performance: speed mach 2+; range 16,000 m (17,500 yards)
Guidance: infra-red homing

Variant and application
MAA-1 Leo: A good example of the considerable strides being made by Brazilian armaments industry, this simple yet capable air-to-air missile entered development in 1976, and was initially a joint-service project known as the Piranha, under the leadership of CTA and later of Vasconcelos. The weapon was twice cancelled for lack of a proper requirement, but with the imminent advent of the EMBRAER-produced version of the Brazilian/Italian AMX attack fighter was then revived as an air force responsibility with final design and manufacture entrusted to Orbita Sistemas Aeroespaciais. By early 1987 final development of the missile was about 40% complete, and the weapon is scheduled to enter service in about 1990. The MAA-1 will probably be retrofitted as replacement for the AIM-9B Sidewinder on in-service Brazilian fighters, but is slated as the primary defensive weapon of the AMX, and will doubtless be offered for export to the growing number of countries dissatisfied with the cost and political constraints associated with weapon purchases from 'more advanced' Western countries.

China

CATIC CAA-2

Type: short-range air-to-air missile
Dimensions: not revealed
Weights: not revealed
Propulsion: one solid-propellant rocket delivering unrevealed thrust
Performance: not revealed
Guidance: infra-red or semi-active radar homing

Variant and application
CAA-2: This is the American designation for an advanced weapon of which virtually nothing is known: the missile resembles the proposed NATO AIM-132A ASRAAM, and is known to have been produced in infra-red and semi-active radar homing versions for use on the Chinese Shenyang J-7 fighter.

CATIC PL-2

China

Type: short-range air-to-air missile
Dimensions: diameter 120 mm (4.72 in); length 2800 mm (110.24 in); span 530 mm (20.87 in)
Weights: total round 70 kg (154.3 lb); warhead 11.3 kg (24.9 lb) infra-red proximity-fused blast fragmentation
Propulsion: one solid-propellant rocket delivering unrevealed thrust
Performance: speed Mach 2; range 5000/7800 m (5,470/8,530 yards)
Guidance: infra-red homing

Variants and application
PL-2: This is essentially the Chinese version of the Soviet AA-2 'Atoll', itself derived from early models of the AIM 9 Sidewinder. Designated **CAA-1** in the standard US terminology for Chinese weapon systems, the PL-2 was developed in the late 1950s and early 1960s, has an uncooled seeker and is thus equivalent to the AIM-9B and restricted to pursuit interceptions at medium and high altitudes.
PL-2A: This is an improved pursuit-course missile introduced in the 1970s with greater discrimination against hot spots as a result of the thermo-electric cooling system for its seeker, allowing operation closer to the ground. The missile is basically similar to the AIM-9E conversion of the AIM-9B Sidewinder.

CATIC PL-5B

China

Type: short-range air-to-air missile
Dimensions: diameter 127 mm (5 in); length 2892 mm (113.86 in); span 530 mm (20.87 in) for the canards and 657 mm (25.87 in) for the fins
Weights: total round 85 kg (187.4 lb); warhead unrevealed quantity infra-red proximity-fused continuous-rod or radio proximity-fused blast fragmentation
Propulsion: one solid-propellant rocket delivering unrevealed thrust
Performance: speed mach 4.5; range 16,000 m (17,500 yards)
Guidance: infra-red homing

Variant and application
PL-5B: With double-delta control surfaces and 'off-boresight' rather than 'all-aspect' engagement capability (equivalent in some respects to the AIM-9L but with a lead-sulphide seeker cooled by compressed air), this is a useful air-to-air missile reflecting China's continued faith in the basic design, and has considerable agility for snap-up and snap-down engagements.

CATIC PL-7

China

Type: short-range dogfighting air-to-air missile
Dimensions: not revealed

Weights: total round not revealed; warhead 12 kg (26.45 lb) proximity- and impact-fused blast fragmentation
Propulsion: one solid-propellant rocket delivering unrevealed thrust
Performance: speed not revealed; range 500/14400 m (545/15,750 yards)
Guidance: infra-red homing

Variant and application
PL-7: This is an 'all-aspect' development of the PL-2/5 series with a continuous-rod warhead and an indium-antimonide seeker cooled by liquid nitrogen. In appearance the missile is similar to the Matra R550 Magic, and is probably designed as a short-range dogfighting missile with engagement capability from anywhere in the target's rear hemisphere.

France

Matra R530

Type: medium-range air-to-air missile
Dimensions: diameter 263 mm (10.35 in); length (semi-active radar version) 3284 mm (129.29 in) or (IR-homing version) 3198 mm (125.90 in); span 1103 mm (43.4 in)
Weights: total round (semi-active radar version) 192 kg (423.3 lb) or (IR-homing version) 193.5 kg (426.6 lb); warhead 27 kg (59.5 lb) proximity and impact delay-fused pre-fragmented or continuous-rod blast fragmentation
Propulsion: (early production models) one Hotchkiss-Brandt/SNPE Antoinette dual-thrust solid-propellant rocket delivering 8500 kg (18,739 lb) thrust for 2.7 seconds and a lower sustain thrust for 6.5 seconds, or (later production models) one SNPE Madeleine dual-thrust solid-propellant rocket delivering unrevealed thrust
Performance: speed Mach 2.7; range 18000 m (19,685 yards)
Guidance: EMD AD26 semi-active radar homing or SAT AD3501 infra-red homing

The Dassault-Breguet Mirage F1C is the cornerstone of the French air-defence force, though currently being supplemented by the more modern Dassault-Breguet Mirage 2000C. This Mirage F1C sports the old/new mix of one Matra R530 medium-range air-to-air missile under the centreline, and two newer Matra R550 Magic short-range dogfighting air-to-air missiles at the wingtips.

Variant and application

R530: Introduced in the early 1960s as armament for the Dassault Mirage III and later used on the Dassault-Breguet Mirage Fl and Vought F-8E(FN) Crusader, the R530 was developed from 1957 as a large weapon with large fixed delta wings and small tail-mounted control surfaces indexed in line with the wings. The basic airframe/powerplant combination was produced with alternative semi-active radar and IR seekers, neither of which has proved particularly successful in combat. The seeker units are interchangeable at squadron level, and the standard operating practice with the R530 is similar to that used with most Soviet air-to-air missiles: fighters generally carry a pair of missiles with different homing heads to provide maximum capability under all electro-magnetic and climatic conditions.

The IR seeker is claimed to be of all-aspect type, but such a facility is limited to engagements under usually advantageous conditions. Despite its limitations, the R530 had little commercial competition, and thus sold to the air forces of 14 countries operating French fighters fitted with the Cyrano Ibis, II and IV radars to which the semi-active radar-homing R530 version was matched. The Aéronavale uses a slightly modified version of the AD26 seeker matched to the Magnavox APQ-94 radar of its Crusader fighters.

Matra Super 530F

France

Type: medium/long-range air-to-air missile
Dimensions: diameter 263 mm (10.35 in); length 3540 mm (139.37 in); span 900 mm (35.43 in)
Weights: total round 250 kg (551.1 lb); warhead 30 kg (66.1 lb) radar proximity-fused Thomson Brandt blast fragmentation
Propulsion: one Thomson Brandt/SNPE Angéle dual-thrust solid-propellant rocket delivering 3875 kg (8,543 lb) thrust for 2 seconds and 2250 kg (5,622 lb) thrust for 4 seconds
Performance: speed Mach 4.6; range 35 km (21.75 miles)
Guidance: EMD Super AD26 semi-active radar homing

Variants and application

Super 530F: Developed from the early 1970s as armament for the Dassault-Breguet Mirage Fl, this high-speed missile is based loosely on the aerodynamics of the R530, revised to resemble those of the American Standard series of surface-to-air missiles with cruciform wings of very low aspect ratio, trailed by a cruciform of slightly greater-span control surfaces indexed in line with the wings and located at the very tail of the missile. Internally the Super 530 is a completely new and advanced weapon matched to the Cyrano IV radar. The missile has twice the range and target-acquisition capabilities of the R530, as well as snap-up capability against aircraft 9000 m (29,530 ft) higher than the launch aircraft: the type also possesses a useful snap-down capability. The Super 530F began to enter service in 1979 after an eight-year development period, and has been produced in substantial numbers for Mirage Fl aircraft in French and export service. The weapon can also be used by Dassault-Breguet Mirage 2000C fighters fitted with the RDM radar.

Super 530D: This is an improved version of the Super 530F so updated in significant features as to virtually a new weapon with digital processing, a new Doppler-mode homing head and increased rocket power. The Super 530D is matched to the RDI and RDM

The Matra Super 530 is the most potent long-range interception missile yet developed outside the USA, its 'snap-up/snap-down' performance providing fighers such as this Dassault-Breguet Mirage F1C with the ability to tackle targets flying much higher or lower than the launch platform.

pulse-Doppler radars of the Dassault-Breguet Mirage 2000 series. This version has a weight of about 265 kg (584.2 lb) and a length of 3800 mm (149.61 in), uses an improved motor for a speed of Mach 4.6+ and a range of up to 60 km (37.3 miles), and has better snap-down capability than the Super 530F. The missile can deal with Mach 3 targets at a height of 24000 m (78,740 ft) and, considerably more importantly, with targets flying fast and low in ground clutter conditions. Thus the Super 530D can be claimed as the only Western missile of non-American origins to match, at least in part, the capabilities of the AIM-54 Phoenix as an all-altitude, all-aspect and all-weather air-to-air missile. Apart from France, the missile has been bought by export customers for the Mirage 2000 such as Egypt, Greece and India.

France

Matra R550 Magic 1

Type: short-range dogfighting air-to-air missile
Dimensions: diameter 157 mm (6.18 in); length 2770 mm (109.06 in); span 668 mm (26.30 in)
Weights: total round 89.8 kg (198.0 lb); warhead 12.5 kg (27.6 lb) proximity and impact delay-fused blast fragmentation
Propulsion: one SNPE Roméo single-stage double-base solid-propellant rocket delivering 2700 kg (5,952 lb) thrust for 1.9 seconds
Performance: speed about Mach 3: range 320/10000 m (350/10,935 yards)
Guidance: SAT AD3601 infra-red homing

Variants and application

R550 Magic 1: Developed from 1968, initially as a private-venture competitor to the US AIM-9 Sidewinder missile but then accepted by the French air ministry, the Magic is designed for 140° rear hemisphere engagement at altitudes up to 18000 m (59,055 ft),

Two of the most important (and commercially successful) weapons produced by the French armaments industry are the Matra R550 Magic air-air missile and the Aérospatiale AM.39 Exocet air-launched anti-ship missile. Both are pictured here under the starboard wing of a Dassault-Breguet Mirage 2000 multi-role fighter. The Exocet has an enviable combat record, and though there remain a number of doubts about the genuine effectiveness of the fusing system, at any but the longest ranges the residual fuel for the SNPE solid-propellant sustainer rocket has proved capable of igniting devastating fires.

the engagement envelope being reduced above this altitude. The Magic's lead sulphide seeker is cooled before launch by liquid nitrogen in a bottle carried inside the launch rail, and the missile can be launched by an aircraft manoeuvring at +6*g* at any speed between 0 and 1300 km/h (808 mph).

Particularly useful features of the R550's design are the ability for snap-shot launch at ranges down to 300 m (985 ft), and the ability to cross safely only 50 m (165 ft) in front of the launch aircraft. The missile's warhead is located in the centre of the body, and the triple set of cruciform aerodynamic surfaces includes freely-rotating fins at the rear, fixed delta canards near the nose and, just aft of the canards, angular control surfaces. The R550 series is generally associated with fighters of French design, but is installationally interchangeable with the Sidewinder.

R550 Magic 2: Entering service in the mid-1980s, this improved version has a genuine all-aspect engagement capability through the adoption of a more sensitive multi-element infra-red seeker, an improved SNPE Richard motor, greater structural strength and better aerodynamic controls for enhanced dogfight manoeuvrability. Just as importantly, the seeker can be slaved to the radar of Dassault-Breguet Mirage 2000 fighters to provide optimum firing conditions and a new electro-magnetic proximity fuse is fitted.

Matra MICA

France

Type: short/medium-range air-to-air missile
Dimensions: diameter 152 mm (6.0 in); length 3100 m (122.05 in); span data not available
Weights: total round 110 kg (242.5 lb); warhead unrevealed Thomson Brandt blast fragmentation
Propulsion: one SNPE solid-propellant rocket delivering unrevealed thrust
Performance: speed Mach 4; range ?/50 km (?/31.1 miles)

Matra has long pursued an independent course as Europe's primary manufacturer of indigenously designed air-to-air missiles, the new and highly impressive MICA (Missile d'Interception et de Combat Aérien, or interception and air combat missile) looking set to continue this trend with substantial sales to the French air arms and to the many air forces using French combat aircraft. The type has a choice between infra-red or active radar homing, offering the pilot a very useful combination to optimize his 'kill' probability under most weather and electro-magnetic conditions. Thrust-vector control is used in powered flight, with aerodynamic surfaces taking over in unpowered flight, and excellent manoeuvrability is claimed for the missile.

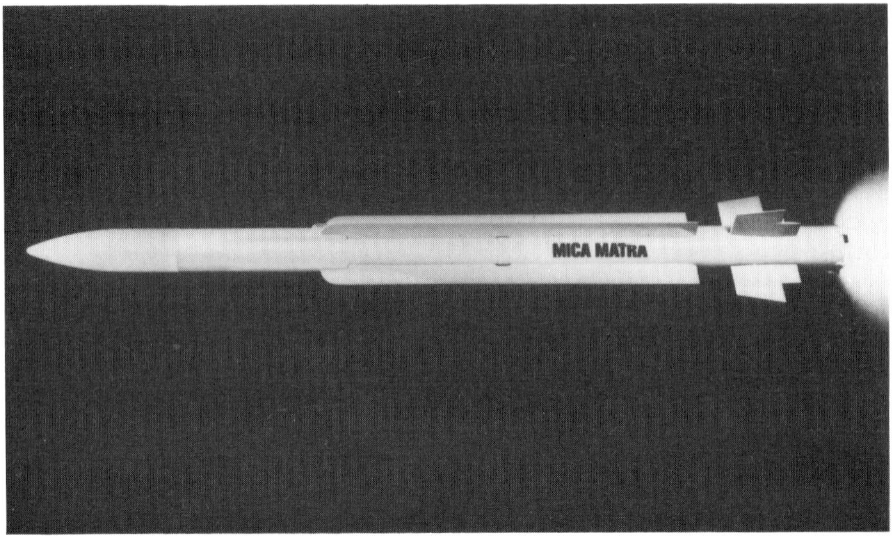

Guidance: strapdown inertial plus infra-red or semi-active homing

Variant and application

MICA: Due to enter service in the 1990s as France's most important short- and medium-range air-to-air missile, the MICA (**Missile d'Interception et de Combat Aérien**, or interception and air combat missile) was first fired in trials during 1982. It is similar in appearance to the US Standard series of surface-to-air missiles, though of course much smaller. The type is slated to arm all current and projected French fighters, and combines the capabilities of other Western air-to-air missiles such as the AIM-120A AMRAAM and AIM-132A ASRAAM.

The MICA is designed for great agility (more than the 50*g* capability of the Magic, and allowing turns immediately after launch to engage targets on any bearing from the launch aircraft other than dead astern), and uses thrust-vector control until motor burn-out, after which the aerodynamic tail controls are used. The MICA is designed for launch on the basis of the parent aircraft's primary sensor, cruising under strapdown inertial guidance and receiving target-data updates before the seeker is turned on. The MICA has a choice of two seekers (the IR type being used for shorter-range air combat and the semi-active radar being used for longer-ranged interception), while its comparatively small size allows aircraft such as the Dassualt-Breguet Mirage 2000 to carry as many as six missiles.

France

Matra Mistral

Type: short-range air-to-air missile

Dimensions: diameter 90 mm (3.54 in); length 1800 mm (70.87 in); span data not available

Weights: total round 17 kg (37.48 lb); warhead 3 kg (6.61 lb) laser proximity-fused Thomson Brandt blast fragmentation

Propulsion: one SNPE dual-thrust solid-propellant rocket delivering unrevealed thrust

Performance: speed Mach 2.6; range 300/6000 m (330/6,560 yards)
Guidance: infra-red homing

Variant and application
Mistral: This is the **Air-Air Tres Courte Portée** (air-to-air very short range) air-launched derivative of the SATCP (Sol-Air Tres Courte Portée, or surface-to-air very short range) missile, and is an extremely capable tube-launched weapon designed for carriage mainly by helicopters, which can carry up to four twin-tube launchers with flick-open eyelid doors. The Mistral can be used for offensive as well as defensive purposes, and targets can be acquired with the helicopter's basic missile sight or, in advanced types, by a crew member's helmet-mounted sight. In both cases the target can be indicated to the missile before launch, as too can the seeker's scanning pattern. After launching the missile the helicopter can turn away as the Mistral covers its maximum 3000 m (3,280 yard) range in a mere 5 seconds against a target manoeuvring at up the 8*g*.

Rafael Armament Authority Python 3

Israel

Type: short-range dogfighting air-to-air missile
Dimensions: diameter 160 mm (6.3 in); length 3000 mm (118.11 in); span 860 mm (33.86 in)
Weights: total round 120 kg (264.6 lb); warhead 11 kg (24.25 lb) proximity and impact delay-fused blast fragmentation
Propulsion: one Rafael Armaments Authority double-base solid-propellant rocket delivering unrevealed thrust
Performance: speed Mach 3; range 500/15000 m (545/16,405 yards)
Guidance: infra-red homing

Variant and application
Python 3: Developed in the late 1970s and early 1980s as the Shafrir 3, the Python is an advanced development of the basic Shafrir with an improved motor offering greater range, a revised seeker with radically improved sensitivity for all-aspect engagement capability, and aerodynamic improvements for enhanced manoeuvrability. The guidance system is claimed to offer the possibility of boresight, uncaged and radar-directed engagement. The close conceptual relationship of the Python (and of its predecessor, the Shafrir) to the Sidewinder is attested by the Israeli missiles' use of Sidewinder-type slipstream-driven 'rollerons' on the rectangular tail surfaces to prevent roll and so ease the task of the guidance system, which works through a cruciform of delta canard surfaces indexed in line with the tail surfaces. The Python 3 is claimed to offer engagement capabilities comparable to those of the AIM-9L variant of the Sidewinder.

Rafael Armament Authority Shafrir 2

Israel

Type: short-range dogfighting air-to-air missile
Dimensions: diameter 160 mm (6.30 in); length 2470 mm (97.24 in); span 520 mm (20.47 in)

The use of rear-fin 'rollerons' marks the Rafael Shafrir as an offspring of the AIM-9 Sidewinder, a fact confirmed by the slim body and canard foreplanes for control.

Weights: total round 93 kg (205.0 lb); warhead 11 kg (24.25 lb) proximity and impact delay-fused blast fragmentation
Propulsion: one Rafael Armaments Authority double-base solid-propellant rocket delivering unrevealed thrust
Performance: speed Mach 2.5; range 500/5000 m (545/5,470 yards)
Guidance: infra-red homing

Variant and application
Shafrir 2: Derived from a preproduction Shafrir 1, the Shafrir 2 began to enter Israeli service in 1969 and is a capable short-range AAM that has performed well in combat with a claimed 65/70 'kill' percentage. The weapon is based loosely on the AIM-9 Sidewinder, but from the start of development in 1961 it was decided to increase the diameter of the body in comparison with that of the Sidewinder, and so simplify the engineering of the weapon as well as making possible the use of a larger warhead containing 4 kg (8.82 lb) of explosive. The missile has the same basic layout as the Python 3, with control effected by the cruciform of canard surfaces which operate in pairs under the influence of pneumatically-operated actuators.

Italy

Selenia Aspide 1

Type: medium/long-range air-to-air missile
Dimensions: diameter 203 mm (8.0 in); length 3700 mm (145.67 in); span 1000 mm (39.37 in)
Weights: total round 220 kg (485.0 lb); warhead 33 kg (72.75 lb) Difesa e Spazio proximity and impact delay-fused blast fragmentation
Propulsion: one SNIA-Viscosa single-stage solid-propellant rocket delivering unrevealed thrust
Performance: speed Mach 4; range 100 km (62.1 miles) from a high-altitude launch dropping to 50 km (31.07 miles) from a low-altitude launch
Guidance: Selenia monopulse semi-active radar homing

Variants and application
Aspide 1: Developed conceptually from the Raytheon AIM-7E Sparrow, the Aspide is like the American missile in being a multi-role type (with important ground-to-air applications in the Spada and Albatros air-defence systems respectively), and entered air-to-air service with the Italian air force in 1979. The type arms the Aeritalia F-104S and Panavia Tornado and, matched to I-band monopulse radar, has greater low-altitude range and snap-down capability than the Sparrow, together with much improved accuracy and resistance to electronic countermeasures. The current production model is the **Aspide**

Based originally on the AIM-7 Sparrow, the Selenia Aspide 1A retains a strong external similarity to its US forebear but is a different weapon internally thanks to its revised propulsion system and Italian electronics. Note the moving mid-body wings and external wiring conduits.

1A with a number of electronic and systems improvements and Selenia is also working on a version with full automatic guidance for 'fire-and-forget' operation.

Armscor V3B

South Africa

Type: short-range dogfighting air-to-air missile
Dimensions: diameter 127 mm (5.0 in); length 2944 mm (116.0); span 533 m (20.86 in)
Weights: total round 73.4 kg (161.8 lb); warhead 10 kg (22.05 lb) proximity and impact delay-fused blast fragmentation
Propulsion: one double-base solid-propellant rocket delivering unrevealed thrust
Performance: speed Mach 3; range 300/5000 m (330/5,470 yards) at high altitude or 300/2000 m (330/2,185 yards) at low altitude
Guidance: infra-red homing

Variants and application
V3A: Designed between 1971 and 1974, this dogfighting missile was introduced to service in 1975 on the French fighters (notably the Dassault-Breguet Mirage III and Mirage F1) operated by the South African air force, and was the world's first AAM used with a helmet-mounted sight for initial (and rapid) designation of the target aircraft. The seeker is cooled by compressed air carried in the missile launcher, and the weapon has an unusual configuration for the 420 mm (16.53 in) span canard control fins, one pair being simple deltas for roll control, and the other pair (indexed at 90° to the plain pair) being double surfaces with swept rectangular surfaces right by the nose trailed by 'double deltas' for

pitch control. At the tail is a cruciform arrangement of rectangular and slightly swept stabilizing fins, which are of low aspect ratio and arranged to rotate freely round the motor nozzle.

V3B: This is an improved model introduced in 1979 with greater seeker and helmet sight acquisition angles, and an upgraded motor. The export version is known as the **V3 Kukri**.

V3C Darter: Due to enter service in 1988, the V3C is a completely upgraded development of the V3B to produce a weapon comparable in most respects to the Magic 2 and AIM-9L Sidewinder. In appearance the new missile is all but identical with the Magic 2, with electronics and surfaces for pitch-and-roll rather than twist-and-steer control. The V3C has a diameter of 6 in (152 mm).

Sweden

Saab Bofors RBS 70 Rayrider

Type: short-range air-to-air missile
Dimensions: diameter 106 mm (4.17 in); length 1320 mm (52.0 in); span 320 mm (12.60 in)
Weights: total round 15 kg (33.07 lb); warhead 1 kg (2.2 lb) laser proximity- and impact-fused blast fragmentation
Propulsion: one Bofors solid-propellant booster rocket delivering unrevealed thrust, and one IMI solid-propellant sustainer rocket delivering unrevealed thrust
Performance: speed supersonic; range 5000 m (5,470 yards)
Guidance: laser beam riding

Variant and application

RBS 70 Rayrider: This Swedish weapon was developed in the mid-1970s as a man-portable short-range surface-to-air missile laser beam-riding guidance for unjammable homing. The type's performance, currently unjammable guidance system and light weight have made it a natural for helicopter use. In this application the launch aircraft can generally carry eight such missiles in two four-round containers, together with the associated Saab Helios roof-mounted sight allowing automatic hand-over of missile guidance to a laser system carried by another helicopter or, conceivably, in the hands of friendly ground forces. The speed and accuracy of the RBS 70 should make the type a formidable offensive and well as defensive weapon in battlefield helicopter operations. The missile is driven from its launch tube by the Bofors booster, which then drops away just before the cruciform wings and tail-mounted control fins deploy. The missile is then accelerated by its powerful sustainer, the onboard control system holding the missile in the centre of the laser beam which the operator keeps fixed on the target. There is also a **Rayrider Mk2** with an improved warhead.

Taiwan

Chung Shan Institute Sky Sword 1

Type: short-range air-to-air missile
Dimensions: not revealed
Weights: not revealed
Propulsion: one solid-propellant rocket delivering unrevealed thrust

Performance: speed supersonic; range between 10000 and 15000 m (10,935 and 16,405 yards)
Guidance: infra-red homing

Variants and application
Sky Sword 1: This is a Taiwanese development of the Sidewinder concept, and bears a very strong similarity to the AIM-9G/H version of the American weapon. The weapon was test-fired for the first time in 1985 or early 1986, but no details of the missile have been divulged. The missile is probably comparable in performance to the AIM-9L Sidewinder (a variant whose export to Taiwan the Americans have not allowed, whereas that of the AIM-9J and AIM-9P has been permitted) but it is conceivable that elements of the Israeli technology developed for the Shafrir and Python have been incorporated. The Sky Sword 1 has all-aspect engagement capability, and is believed to have entered service in 1987.
Sky Sword 2: Due for initial test-firings in 1989, this is an advanced medium-range development of the Sky Sword 1, but no significant details have been released.

AA-2-2C/D 'Advanced Atoll' USSR

Type: short-range air-to-air missile
Dimensions: diameter 120 mm (4.72 in); length (AA-2-2C) 2800 mm (110.24 in) or (AA-2-2D) 3100 mm (122.05 in); span 530 mm (20.87 in)
Weights: total round (AA-2-2C) 70 kg (154.3 lb) or (AA-2-2D) 75 kg (165.3 lb); warhead 6 kg (13.2 lb) impact delay-fused blast fragmentation
Propulsion: one solid-propellant rocket delivering unrevealed thrust
Performance: speed Mach 2.5; range 8000 m (8,750 yards)
Guidance: (AA-2-2C) infra-red homing or (AA-2-2D) semi-active radar homing

Variants and application
AA-2A 'Atoll': Developed with the Soviet design designation **K-13** from the AIM-9B Sidewinder missile (using intelligence information and examples of the missile captured in 1958 and later years), the first-generation AA-2A infra-red homing missile began to enter Soviet service in the mid-1960s with the service designation **RS-3S** and production designation **SB06**. This variant has a range of 5700 m (6,235 yards). The weapon has been produced in vast numbers for use on the whole range of Soviet tactical aircraft, and though the 'Atoll' is being replaced in Soviet service by the AA-8 'Aphid', it is likely to remain in the inventories of the USSR and its clients for many years to come despite its limitations to tailchase pursuit even in optimum conditions.

The missile is believed to parallel the Sidewinder closely in terms of internal arrangement, and uses the same arrangement of slipstream-driven 'rollerons' on the tail-mounted fins for stability. Control is effected via the four canard surfaces, which are linked to work in diametric pairs.
AA-2B 'Atoll': Produced in parallel with the AA-2A, this is the semi-active radar homing version of the first-generation series, and has a range of 8000 m (8,750 yards). 'Atolls' are generally carried in pairs (one or two AA-2As with one or two AA-2Bs) to give fighters the best chance of scoring a hit under all operational and climatic conditions, and this philosophy has been carried over into later Soviet air-to-air missile series.
AA-2-2A 'Advanced Atoll': This is a second-generation version of the infra-red mis-

sile with greater reliability and range, the latter resulting from a longer body containing more propellant.
AA-2-2B 'Advanced Atoll': This is a second generation version of the semi-active radar homing missile with greater reliability and range, the latter resulting from a longer body containing more propellant.

USSR

AA-3A/B 'Anab'

Type: short/medium-range air-to-air missile
Dimensions: diameter 280 mm (11.02 in); length (both versions) about 3600 mm (141.73 in) though earlier estimates put the AA-3A at 4100 mm (161.42 in) and the AA-3B at 4000 mm (157.48 in); span 1300 mm (51.18 in)
Weights: total round 275 kg (606.3 lb); warhead 30 kg (66.1 lb) proximity- and impact-fused blast fragmentation
Propulsion: one solid-propellant rocket delivering unrevealed thrust
Performance: speed Mach 2.5; range (AA-3A) 20 km (12.4 miles) or (AA-3B) 30 km (18.6 miles)
Guidance: (AA-3A) infra-red homing or (AA-3B) semi-active radar homing

Variants and application
AA-3A 'Anab': This was the USSR's second-generation air-to-air missile, and the first all-weather long-range of this type in the Soviet inventory when it began to enter service in 1961. The type is generally associated with Sukhoi Su-11 'Fishpot', Su-15 'Flagon' and Su-21 'Flagon' fighters and with the Yakovlev Yak-28P 'Firebar' interceptor, all fitted with the 'Skip Spin' radar for initial acquisition of the target beyond visual range. Information about the motor is woefully lacking, and it is only assumed that it is of the solid-propellant type, possibly with boost and sustain segments. The design of the missile is characterized by a cruciform of cropped-delta tail surfaces (probably incorporating inset ailerons) and an inline set of small canard surfaces working as diametrically-linked pairs for control in pitch and yaw.
AA-3B 'Anab': Following the Soviet practice of developing air-to-air missiles in infra-red and semi-active radar versions for maximum kill probability under all operational conditions, this is the semi-active radar homing version, generally associated with Sukhoi and Yakovlev fighters. As with the AA-3A, the type is being supplanted by the AA-7 'Apex'.
AA-3-2 'Advanced Anab': The existence of this weapon was first revealed in the West during 1972, but how the weapon qualifies for the epithet 'Advanced' has not been released. If the NATO designation system used for the 'Atoll' has been followed, it can be assumed that this is a longer-ranged variant, and that there are also **AA-3-2A** and **AA-3-2B** variants with infra-red and semi-active radar terminal guidance respectively.

USSR

AA-5A/B 'Ash'

Type: medium-range air-to-air missile
Dimensions: diameter 305 mm (12.0 in); length (AA-5A) 5201 mm (205.12 in) or (AA-

5B) 5500 mm (216.54 in); span 1300 mm (51.18 in)
Weights: total round 390 kg (859.8 lb); warhead about 45 kg (99.2 lb) proximity- and impact-fused blast fragmentation
Propulsion: one solid-propellant rocket delivering unrevealed thrust
Performance: speed about Mach 3; range (AA-5A) 20 km (12.5 miles) or (AA-5B) 65 km (40.4 miles)
Guidance: (AA-5A) infra-red homing or (AA-5B) semi-active radar homing

Variants and application
AA-5A 'Ash': Developed between 1954 and 1959 as the primary armament of the Tupolev Tu-28P 'Fiddler' long-range interceptor (and later of the Mikoyan-Gurevich MiG-25 'Foxbat-A' Mach 3 interceptor), the AA-5 remains in limited service despite its technical obsolescence. AA-5A is the Western designation for the IR-guided version, which first appeared in about 1965 and is generally carried in pairs with two of the AA-5B variant to optimize the launch aircraft's kill capabilities under all weather and electronic conditions. The weapon has a long cylindrical body with a cruciform of long-chord delta wings, and four close-coupled high-aspect-ratio control surfaces indexed in line with the wings.
AA-5B 'Ash': This was the initial production variant of the AA-5, its semi-active radar homing being associated with the 'Big Nose' radar of the Tu-28P. The model entered service in 1961 or 1962 at the same time as the Tu-28P, and Western estimates of the range possessed by the substantial missile may be far short of the real figures.

AA-6A/B 'Acrid' USSR

Type: long-range air-to-air missile
Dimensions: diameter 400 mm (15.75 in); length (AA-6A) 5800 mm (228.35 in) or (AA-6B) 6290 mm (247.64 in); span 2250 mm (88.58 in)
Weights: total round (AA-6A) 750 kg (1,653.4 lb) or (AA-6B) 800 kg (1,763.7 lb); warhead 90 kg (198.4 lb) proximity- and impact-fused blast fragmentation
Propulsion: one dual-thrust solid-propellant rocket delivering unrevealed thrust
Performance: speed Mach 4.5; range (AA-6A) 25 km (15.5 miles) or (AA-6B) 70 km (43.5 miles)
Guidance: (AA-6A) infra-red homing or (AA-6B) semi-active radar homing

Variants and application
AA-6A 'Acrid': Developed in the period between 1959 and 1961 as a long-range bomber-destroying missile for use at medium and high altitudes (for which the missile was designed with large wings), the AA-6 is the world's largest air-to-air missile and entered service in the mid-1960s as the primary armament of the Mikoyan-Gurevich MiG-25 'Foxbat' and Sukhoi Su-21 'Flagon'. Four missiles are generally carried as two pairs each of the IR- and radar-guided versions. The missile has the same basic configuration as the AA-3 'Anab', with four low-aspect-ratio delta wings each carrying a control surface (either ailerons or elevons) and four smaller delta canard control surfaces. The launch aircraft is guided to its operational area by ground-controlled interception techniques, turning on its own radar at a range of some 160 km (100 miles) and then closing to the lock-on range of the missile.
AA-6B 'Acrid': This is the semi-active radar homing version of the AA-6A matched

to the 'Fox Fire' radar of the MiG-25; the Soviet practice is a ripple of two missiles (one of each homing type) to generate maximum kill probability. It is believed that the launch aircraft's extremely powerful radar can lock onto a target at a range of 160 km (100 miles) and then radar-illuminate the target at a range of about 100 km (62 miles), suggesting that Western estimates of the AA-6B's range are well short of reality.

USSR

AA-7A/B 'Apex'

Type: medium-range air-to-air missile
Dimensions: diameter 223 mm (8.78 in); length (AA-7A) 4200 mm (165.35 in) or (AA-7B) 4600 mm (181.1 in); span 1050 mm (41.34 in)
Weights: total round (AA-7A) 300 kg (661.1 lb) or (AA-7B) 320 kg (705.5 lb); warhead 40 kg (88.2 lb) proximity and impact-fused blast fragmentation
Propulsion: one dual-thrust solid-propellant rocket delivering unrevealed thrust
Performance: speed Mach 3.5; range (AA-7A) 20 km (12.5 miles) or (AA-7B) 55 km (34.2 miles)
Guidance: (AA-7A) infra-red homing or (AA-7B) semi-active radar homing

Variants and application
AA-7A 'Apex': Developed in the period from 1971 to 1974, and introduced in the mid to late 1970s, this third-generation Soviet air-to-air missile can be regarded as successor to the AA-3 'Anab' and comparable to the US AIM-7E Sparrow III but with greater manoeuvrability. The basic design is a long cylindrical body mounting a cruciform of middle-sized delta wings in the middle and four control surfaces indexed in line with the wings but located round the motor nozzle. The type is generally associated with the Mikoyan-Gurevich MiG-23MF 'Flogger' and MiG-25E 'Foxbat', and is believed to be optimized for the medium and low-level roles. The Soviet service designation is **R-23T.**
AA-7B 'Apex': This is the semi-active radar homing version of the AA-7A. Near the nose of the missile (in both versions) are four projecting surfaces, at first thought to be extra control surfaces but now considered to be the blister antennae for a semi-active radar homing system working on the interferometry principle. An odd feature of the

Though of poor quality, this photograph shows the comparative size and proportions of two of the air-to-air missile types carried by the Mikoyan-Gurevich MiG-23 'Flogger-B': AA-7 'Apex' missiles under the fixed portions of the wings, and AA-8 'Aphid' missiles under the fuselage.

design is the smaller diameter (0.195 m/7.68 in) of the body's forward section, for which no compelling reason has yet been revealed in the West: it perhaps confirms the lack of an internal dish antenna, the opportunity being taken to area-rule the nose in the region of the projections. The Soviet service designation is **R-23R,** and the variant is associated with the 'High Lark' radar system.

AA-8A/B 'Aphid' USSR

Type: short-range dogfighting air-to-air missile
Dimensions: diameter 120 mm (4.72 in); length (AA-8A) 2150 mm (84.65 in) or (AA-8B) 2350 mm (92.52 in); span 400 mm (15.75 in)
Weights: total round (AA-8A) 55 kg (121.25 lb) or (AA-8B) 60 kg (132.3 lb); warhead 7 to 9 kg (15.4 to 19.8 lb) proximity- and impact-fused blast fragmentation
Propulsion: one dual-thrust solid-propellent rocket delivering unrevealed thrust
Performance: speed Mach 3; range (AA-8A) 10000 m (10,935 yards) or (AA-8B) 15000 m (16,405 yards)
Guidance: (AA-8A) infra-red homing or (AA-8B) semi-active radar homing

Variants and application
AA-8A 'Aphid': Introduced in the mid-1970s as the replacement for the AA-2 'Atoll' series and as the short-range companion to the medium-range AA-7 'Apex', the extremely capable AA-8 is one of the smallest air-launched guided missiles yet developed. The type is used by most modern Soviet fighters, and is notable for its high degree of manoeuvrability, resulting from its small size and advanced configuration with tail-mounted delta wings, canard delta control surfaces and, at the nose just in front of the control surfaces, four blade-like surfaces of rectangular form and very low aspect ratio. The Soviet service designation is **R-60T**.
AA-8B 'Aphid': This is the semi-active radar homing version of the AA-8A. The existence of this weapon, whose Soviet service designation would be **R-60R,** is as yet unconfirmed in the West.

AA-9 'Amos' USSR

Type: medium/long-range air-to-air missile
Dimensions: (estimated) diameter 400 mm (15.75 in); length 4000 mm (157.48 in); span 1225 mm (48.23 in)
Weights: total round about 450 kg (992.1 lb); warhead possibly 80-kg (176.4-lb) proximity- and impact-fused blast fragmentation or (possibly) nuclear
Propulsion: one solid-propellant dual-thrust rocket delivering unrevealed thrust in the boost phase and ?-kg (?-lb) thrust in the sustain phase
Performance: speed Mach 3.5; range at least 130 km (80.8 miles)
Guidance: inertial or semi-active radar (midcourse phase) and active radar (terminal phase)

Variant and application

AA-9 'Amos': Deployed on the Mikoyan-Gurevich MiG-29 'Fulcrum' and MiG-31 'Foxhound' fighters from about 1984, the AA-9 was developed from the early 1970s as a weapon able to deal with low-altitude penetrations of Soviet airspace, particularly by US cruise missiles. Trials have confirmed that the AA-9 can snap-down at least 6000 m (19,685 ft) to engage targets flying at an altitude as low as 50 m (165 ft). US sources credit the missile with a range of 18.5/23.2 km (11.5/14.4 miles) in a head-on low-altitude engagement, or of 8.4/9.25 km (5.2/5.75 miles) in a tail-on low-altitude engagement, or of 39.8/45.4 km (24.75/28.2 miles) in a high-altitude head-on engagement. These figures are almost certainly underestimated as the most likely reason for the missile's unusually large diameter is the need to provide a cross section able to accommodate a large receiver antenna for the long-range semi-active radar homing system, which probably switches to active mode for the terminal phase of the attack. The diameter also suits the missile for the installation of a nuclear warhead for the engagement of strategic targets. The arrangement of the AA-9's flying surfaces is also unusual, the cruciform wings having their swept leading edges aft of the midpoint, and these low-aspect-ratio surfaces being trailed by close-coupled control surfaces which are of greater span and considerably higher aspect ratio than the wings.

USSR

AA-10 'Alamo'

Type: medium-range air-to-air missile
Dimensions: (estimated) diameter (both variants) 185 mm (7.28 in); length (radar variant) 4000 mm (157.48 in) or (IR variant) 3200 mm (125.98 in); span (both variants) 855 mm (33.66 in)
Weights: total round (radar variant) about 200 kg (440.1 lb) or (IR variant) about 155 kg (341.7 lb); warhead unrevealed blast fragmentation
Propulsion: one solid-propellant booster rocket delivering unrevealed thrust and (radar variant only) one solid-propellant sustainer rocket delivering unrevealed thrust
Performance: speed supersonic; range (radar variant) about 40 km (24.85 miles) or (IR variant) about 20 km (12.4 miles)
Guidance: infra-red or semi-active homing

Variant and application

AA-10 'Alamo': Introduced in 1984, this is an all-aspect snap-down missile matched to the radars of new-generation Soviet fighters capable of 'look-down shoot-down' engagements. Few details have been made public in the West, but American official artwork initially gave the impression that the type might be an advanced derivative of the AA-7 'Apex', indicating a similarity of dimensions and weights with those of the earlier weapon. The Mikoyan-Gurevich MiG-29 'Fulcrum' and Sukhoi Su-27 'Flanker' are each thought to be capable of carrying six examples of this missile. However, photographs first released in 1987 show the AA-10 to be totally unlike the AA-7, possessing instead a faint similarity to the obsolescent AA-3 'Anab'.

The most unusual feature of the AA-10 is the use of long-span canard surfaces with pronounced inverse taper: these are fitted to a control ring at the forward end of the very large motor section. The AA-10 is apparently available in two versions: an I/J-band radar-homing model with a long fuselage (presumably to allow the incorporation of a sustainer engine for the greater range appropriate to the launch platform's look-down/shoot-down fire-control system), and an IR-homing model with a short fuselage.

No NATO reporting designations for the variants have been revealed. The IR-homing version is usually carried on underwing hardpoints, and the radar-homing version on underfuselage hardpoints. The latter type has to be launched with the aid of a powerful ejector mechanism to ensure clean separation from the launch aircraft, and this suggests that the missile has a strong airframe, a fact that would also support the suggestion that the AA-10 is highly manoeuvrable.

AA-11 'Archer' — USSR

Type: short-range air-to-air missile
Dimensions: data not available
Weights: data not available
Propulsion: one solid-propellant rocket delivering unrevealed thrust
Performance: speed supersonic; range not known
Guidance: infra-red homing

Variant and application
AA-11 'Archer': This is the latest Soviet short-range AAM, and forms a key part of the inventory for aircraft such as the Mikoyan-Gurevich MiG-29 'Fulcrum' and Sukhoi Su-27 'Flanker'. The type is probably designed as partner or successor to the AA-8 'Aphid', and may possibly have comparable flight performance (in terms of speed and range) coupled with greater manoeuvrability and improved seeker capability.

SA-7 'Grail' — USSR

Type: short-range air-to-air missile
Dimensions: diameter 70 mm (2.76 in); length 1350 mm (53.15 in); span 200 mm (7.87 in)
Weights: total round 9.2 kg (20.3 lb); warhead 2.5 kg (5.51 lb) graze- and impact-fused blast fragmentation
Propulsion: one dual-thrust solid-propellant rocket delivering unrevealed thrust
Performance: speed Mach 1.5; range 800/5000 m (875/5,470 yards)
Guidance: infra-red homing

Variant and application
SA-7 'Grail': This somewhat limited weapon was introduced in the early 1960s as the Soviet army's standard shoulder-fired surface-to-air missile for the protection of infantry formations. The basic missile has the Soviet production designation **9M32** and the service name **Strela 2.** Since 1981 the weapon has been seen in two- and four-round helicopter installations (probably for defensive rather than offensive purposes), most notably on the Soviet Mil Mi-24 'Hind' and Yugoslav Soko-built Aerospatiale SA 341 Gazelle.

Such an installation is greatly facilitated by the SA-7's uncooled seeker, which has the tactical disadvantage of restricting launch to tail-chase engagements. It is sensibly assumed that combat helicopters use their magnifying anti-tank missile sights for the acquisition of targets. It is also likely that a similar installation has been (or will shortly be) made of the SA-7 successor in Soviet service, the **SA-14 'Gremlin'.** This has a considerably improved seeker, and could conceivably be used for offensive as well as defensive purposes on the new generation of Soviet helicopters such as the Kamov Ka-? 'Hokum' battlefield combat helicopter and the Mil Mi-28 'Havoc' gunship and anti-tank helicopter.

UK

BAe Red Top

Type: short-range air-to-air missile
Dimensions: diameter 8.75 in (223 mm); length 130.6 in (3317 mm); span 35.75 in (908 mm)
Weight: total round 330 lb (149.7 kg); warhead 68.3 lb (30.98 kg) infra-red proximity-fused blast fragmentation
Propulsion: one Royal Ordnance Linnet solid-propellant rocket delivering unrevealed thrust
Performance: speed Mach 3.2; range 13,200 yards (12070 m)
Guidance: infra-red homing

Variant and application

Red Top: Introduced in 1964, the Red Top is essentially an upgraded and rationalized Firestreak missile with components arranged in logical order from nose to tail. (The Firestreak's tail controls were operated via pushrods connected to actuators in the nose, the actuators being worked by compressed air piped from a bottle in the tail.) The structure was revised to allow far higher performance and levels of agility, with mid-mounted cruciform wings of large area and tail-mounted delta control surfaces of considerably smaller area. The missile possesses an all-aspect engagement capability plus a somewhat larger blast fragmentation warhead than that of the Firestreak.

UK

BAe Firestreak

Type: short-range air-to-air missile
Dimensions: diameter 8.75 in (223 mm); length 125.5 in (3188 mm); span 29.4 in (747 mm)
Weights: total round 300 lb (136.1 kg) warhead 50 lb (22.7 kg) proximity-fused blast fragmentation
Propulsion: one two-stage solid-propellant rocket delivering unrevealed thrust
Performance: speed Mach 3; range 1,320/7,000 yards (1210/6400 m)
Guidance: infra-red homing

Variant and application

Firestreak: This missile was developed from 1951 as the Blue Jay, but entered service as the Firestreak in 1958. The weapon is now obsolete but still retained in small numbers. The type is limited to pursuit-type engagements, and its seeker head (cooled by nitrogen from the launch aircraft) is slaved to the launch aircraft's radar for lock-on.

The configuration of the weapon is similar to that of the Red Top, but the flying surfaces are smaller (and optimized for lower speeds and smaller lateral accelerations), the tail surfaces are rectangular, and the warhead is an annular type surrounding the tube carrying the exhaust from the mid-mounted rocket to the tail nozzle.

BAe Sky Flash **UK**

Type: medium-range air-to-air missile
Dimensions: diameter 8.0 in (203 mm); length 145.0 in (3683 mm); span 40.0in (1016 mm)
Weights: total round 425 lb (192.8 kg); warhead 66 lb (29.9 kg) Thorn-EMI active radar proximity and impact delay-fused blast fragmentation
Propulsion: one Aerojet or Rockwell Mk 52 Mod 2 solid-propellant rocket delivering unrevealed thrust
Performance: speed Mach 4; range 31 miles (49.9 km) from a high-altitude launch
Guidance: Marconi XJ521 monopulse semi-active radar homing

Variants and application
Sky Flash: This is an advanced British development of the comparatively short-ranged AIM-7E2 Sparrow with an advanced homing system and improved fusing. The weapon entered development in 1969 and retains the classic aerodynamics of the Sparrow, with cruciform of delta control wings mounted on the mid-point of the cylindrical body, and a comparable cruciform of delta stabilizing surfaces mounted at the tail. The weapon

To accommodate a quartet of semi-active radar homing Sky Flash medium-range air-to-air missiles, the Panavia Tornado F.Mk 2/3 series was developed from the GR.Mk 1 baseline variant with a lengthened fuselage to allow semi-recessed carriage of the missiles under the fuselage. Also visible in this superb view are the pair of AIM-9L Sidewinder short-range air-to-air missiles carried on the inboard side of the pylons that also carry 330-Imp gal (1500-litre) long-range tanks.

There can be little doubt about the impressive nature of the Sky Flash's motor ignition. Also visible in this launch from a Panavia Tornado F. Mk 2 are the twin arms of the Frazer-Nash launcher, a pyrotechnically operated system which pushes the missile down through the boundary layer to ensure clean separation of the missile and aircraft.

entered service in 1978, and is an extremely accurate missile with warm-up time reduced from the American original's 15 seconds to a mere 2 seconds. The Sky Flash has excellent capabilities against manoeuvring targets even in conditions of glint and heavy electronic countermeasures, and the type is also used by Sweden with the designation **RB 71.** Existing Swedish stocks of this weapon are being modified (by a consortium of Saab Missiles, Saab Scania Combitech, Ericsson Radio Systems and Volvo Flygmotor) as **RB 71A** missiles with improved motor and guidance systems.

Sky Flash 90: This is the manufacturer's designation for a proposed development based on the Sky Flash Mk 2 cancelled by the British government in 1981, even though its development was nearly complete, in favour of the AIM-120A AMRAAM. The proposed Sky Flash 90 is designed as a lower-cost answer to the requirements of those countries needing a beyond-visual-horizon missile, and the design is based on a control package using a strapdown inertial platform for midcourse guidance and a Marconi active radar for terminal guidance, plus a Royal Ordnance Hoopoe solid-propellant rocket for greater speed and range. These latter factors are aided by thinner wings and a new low-drag rear end to the missile. Sweden has proposed a Mach 4 or 5 development of this active-homing type with a Volvo Flygmotor ramjet as the **RB 73.** Whether using rocket or ramjet propulsion, the Sky Flash 90/RB 73 is likely to prove an important and popular weapon as its active radar guidance removes the need for the continuous-wave illumination of the target by an expensive radar carried by the basic Sky Flash's launch aircraft which has to fly on towards the target until the missile impacts, a course that potentially takes the launch aircraft ever farther into airspace protected by surface-to-air as well as air-to-air weapon systems.

USA

Hughes AIM-4F Falcon

Type: short/medium-range air-to-air missile
Dimensions: diameter 6.63 in (168 mm); length 86.0 in (2184 mm); span 24.0 in (610 mm)
Weights: total round 150 lb (68.0 kg); warhead 28.65 lb (13.0 kg) proximity-fused blast fragmentation
Propulsion: one Thiokol M46 two-stage solid-propellant rocket delivering 4,440 lb (2014 kg) thrust
Performance: speed Mach 4; range 12,350 yards (11295 m)
Guidance: Hughes semi-active radar homing

Variants and application
AIM-4C Falcon: The Falcon was the world's first fully-guided air-to-air missile to enter service, initial operational capability being achieved with the semi-active radar homing GAR-1 (later AIM-4) in 1956 after a development programme launched in 1947. All the early variants were powered by a Thiokol single-stage rocket giving acceleration in the order of 50*g*. The oldest variant still in service is the IR-guided AIM-4C (originally the GAR-2A), which is operational with the Swedish and Swiss air forces under the designations **RB 28** and **HM-58** respectively for use by Saab 35F Draken and Dassault-Breguet Mirage IIIS aircraft fitted with the Hughes Taran radar associated with the RB 27 and HM-55 semi-active radar homing versions of the AIM-26 Super Falcon. The variant has a length of 79.5 in (2019 mm), a diameter of 6.4 in (163 mm), a span of 20.0 in (508 mm), a total weight of 134 lb (60.78 kg), a speed of Mach 3 and a range of 10,560 yards (9655 m).
AIM-4D Falcon: This was the last production version of the Falcon, produced from 1963 in the fighter-combat role. The variant combines the smaller and lighter airframe of the early models with the more powerful motor and the infra-red seeker of the AIM-4G: the result is a capable short-range missile 81.0 in (2057 mm) long, 6.63 in (168 mm) in diameter and 24.0 in (620 mm) in span, and at a launch weight of 145 lb (65.77 kg) capable of Mach 4 over a range of 12,320 yards (11265 m). More than 4,000 AIM-4A and AIM-4C missiles were remanufactured to this standard. The only fighter still equipped with this weapon is the rapidly dwindling number of Convair F-106 Delta Darts.
AIM-4F Falcon: Introduced in 1959, this is a much-improved model matched to the F-106A interceptor and based on a longer body (with greater-span wings) fitted with a two-stage motor.
AIM-4G Falcon: This is a version of the AIM-4F with infra-red guidance, decreasing length by 5 in (127 mm) to 81.0 in (2.06 m) and weight to 145 lb (65.77 kg). Like the rest of the Falcon air-to-air missile family, this can most charitably be described as an obsolete weapon despite its powerful warhead and excellent flight performance.

Raytheon/General Dynamics AIM-7M Sparrow — USA

Type: medium-range air-to-air missile
Dimensions: diameter 8.0 in (203 mm); length 145.0 in (3683 mm); span 40.0 in (1016 m)
Weights: total round 503 lb (228.2 kg); warhead 86 lb (39 kg) proximity and impact delayed-fuse Mk 71 blast fragmentation
Propulsion: one Hercules Mk 58 or Aerojet Mk 65 dual-thrust solid-propellant rocket delivering unrevealed thrust
Performance: speed Mach 4; range 62 miles (100 km) from a high-altitude launch reducing to 27 miles (43.5 km) from a low-altitude launch
Guidance: Raytheon Advanced Monopulse Seeker semi-active radar homing

Variants and application
AIM-7E Sparrow: Introduced in 1962 as the first version of the Sparrow III to enter very large-scale production (the earlier radar beam-riding AIM-7A Sparrow I having totalled 2,000 rounds, the semi-active AIM-7C Sparrow III 2,000 rounds and the liquid-propelled AIM-7D Sparrow III 7,500 rounds), this variant remains in limited service

The second TF-15B (later just F-15B) was used to pioneer many of the features planned for the multi-role F-15E variand of the McDonnell Douglas Eagle, and this fascinating view of the aircraft highlights the conformal FAST (Fuel And Sensor Tactical) pack tucked into the angle of the inlet trunking and the undersurface of the wing, an AIM-7F/M Sparrow carried under the outboard hardpoint, twin AIM-9L/M Sidewinders on lateral projections of the inner hardpoint, and a mass of Mk 82 'slick' GP bombs carried as tandem pairs under the wings and FAST pack tangential hardpoints, and as tandem triplets under the fuselage, for a notional complete load of 28 such 500-lb (227-kg) bombs.

with several important fighter types. The type is powered by a Rocketdyne Mk 38 or Aerojet Mk 52 single-stage motor for a burn-out speed of Mach 3.7, weighs 452 lb (205 kg), has a range of 28 miles (44 km) with its 66 lb (30 kg) warhead (whose casing breaks into some 2,600 fragments on detonation), and is guided by Raytheon continuous-wave semi-active radar guidance matched to a number of radars such as the Rockwell R-21G/H (Aeritalia F-104S), APQ-72, APQ-100, APQ-109, APQ-120 and APG-59 (McDonnell Douglas F-4 Phantom) and AWG-9 (Grumman F-14 Tomcat).

This variant is built in Japan as the **AIM-7J Sparrow.** A limited-production shorter-range and more manoeuvrable version was produced in the USA as the **AIM-7E2 Sparrow** in response to a requirement that became apparent in the Vietnam War, where US fighters were seldom able to use long-range AIM-7E Sparrows because of the political ban on firing missiles against aircraft which had not been visually identified. The AIM-7E2 thus has a reduced minimum range, and the opportunity was taken to provide aerodynamic surfaces that could be plugged in rather than requiring attachment with the aid of special tools. The definitive version still in service is the **AIM-7E3 Sparrow,** a conversion of the AIM-7E2 with improved reliability and greater target-sensing capability.

AIM-7F Sparrow: This is a much redesigned version introduced in 1977 on the McDonnell Douglas F-15 Eagle and now carried also by the McDonnell Douglas F/A-18 Hornet. Whereas the AIM-7E has its forward body (as far aft as the wings) filled with guidance equipment (homing head and autopilot), trailed aft of the wings by a comparatively small warhead and a short motor, the AIM-7F uses the advantages of more compact solid-state electronics to reduce guidance package volume by some 40%, allowing a larger warhead to be incorporated in the space vacated in front of the wings, and a longer motor to be used in the whole of the portion aft of the wings. This variant thus has a much larger engagement envelope through the adoption of solid-state electronics and a Mk 58 or Mk 65 dual-thrust motor. Like the AIM-7E it is 144.0 in (3658 mm) long, but weighs 503 lb (228 kg) and has an 86 lb (39 kg) warhead. The use of continuous-wave and pulse-Doppler guidance with a conical-scan slotted antenna considerably enhances look-down/shoot-down capability, even at the longer ranges possible with this variant. The AIM-7F also possesses superior resistance to electronic countermeasures.

AIM-7M Sparrow: Introduced in 1982 as an interim type pending availability of the AIM-120 AMRAAM, this variant has a new digital monopulse seeker offering performance comparable with that of the British Sky Flash derivative, especially in the longer-range look-down/shoot-down mode. The variant also offers other electronic and engineering improvements to bring down production cost while enhancing reliability and performance at low altitude and in ECM environments. The type also possesses a warhead of greater lethality than those of earlier Sparrows.

AIM-7P Sparrow: Under development in the late 1980s by Raytheon, this is the latest development of the basic Sparrow III air-to-air missile. No details of the programme have been released, but it is likely that the development is centred on electronic rather than aerodynamic or powerplant features.

Ford Aerospace/Raytheon/ Bodensee Gerätetechnik AIM-9L Sidewinder

USA

Type: short-range dogfighting air-to-air missile
Dimensions: diameter 5.0 in (127 mm); length 112.2 in (2850 mm); span 24.8 in (630 m)
Weights: total round 188 lb (85.3 kg); warhead 22.5 lb (10.2 kg) Hughes DSU-21/B active laser-fused blast fragmentation
Propulsion: one Thiokol or Bermite Mk 36 Model 7/8, or reduced-smoke TX-683 solid-propellant rocket delivering unrevealed thrust
Performance: speed Mach 2.5; range 19,360 yards (17700 m)
Guidance: Bodensee Gerätetechnik ALASCA infra-red homing

Variants and application

AIM-9B Sidewinder: Still in the operational inventory of a few nations, this venerable version of the Sidewinder series is 111.4 in (2830 mm) long, spans 22.0 in (559 mm) and entered service in 1956 with a 10 lb (4.54 kg) blast fragmentation warhead and infra-red proximity fuse. From 1982 the fuse has been replaced with the Hughes DSU-21/B active laser fuse in all missiles carrying this warhead, namely the AIM-9B, E, J, N and P variants. Powered by an Aerojet Mk 17 (designed by the Naval Propellant Plant) single-stage motor, the 155 lb (70.3 kg) missile achieves a range of 3,500 yards (3200 m). The **AIM-9B/FWG.2** is an improved European-built model with greater seeker sensitivity. Its length is 114.5 in (2908 mm), but though weight increases to 167 lb (75.8 kg) the range is boosted to 4,000 yards (3660 m). The missile is used by Sweden with the designation **RB 24.**

The **AIM-9E** is a US-built improvement with a Ford thermo-electrically cooled seeker (again for greater sensitivity) plus 'revitalization' of the existing electronics, and the length of the missile is 118.1 in (3000 mm), with weight 164 lb (74.4 kg) and range

The General Dynamics F-16A Fighting Falcon was designed as a lightweight air-combat fighter, but has matured as an extraordinarily versatile multi-role fighter with excellent ground-attack capability. Here an F-16A carrying two tip-mounted AIM-9L/M Sidewinders and an underfuselage drop tank releases a pair of 2000-lb (907-kg) Mk 84 general-purpose bombs.

4,575 yards (4185 M) within a considerably expanded operational envelope. The AIM-9E began to enter service in 1967, and a version with a reduced-smoke motor is designated **AIM-9E2.**

The Japanese **Mitsubishi AAM-1** is virtually a copy of the AIM-9E, though it is slightly smaller (2600 mm/102.36 in long) and lighter (70 kg/154.3 lb) than the American weapon, and has a range of only 7000 m (7,655 yards). Some of the original weapons have been converted with reduced-smoke motors with the designation **AIM-9B2,** and are used for special mission or training.

AIM-9G Sidewinder: Introduced in 1967, this is an improved version of the US Navy's 1965 **AIM-9D Sidewinder** (the real precursor of the latest Sidewinder variants) with an infra-red/radar-proximity fused 22.4 lb (10.2 kg) continuous-rod HE warhead, the Thiokol Mk 36 solid-propellant rocket and the Sidewinder Expanded Acquisition Mode (SEAM) seeker for improved target acquisition and engagement capability, especially at low altitude. This variant is 113.0 in (2870 mm) long and has a fin span of 24.8 in (630 mm), and at a weight of 191 lb (86.6 kg) has a range of 19,360 yards (17705 m).

AIM-9H Sidewinder: Introduced in 1970, this 186 lb (84.4 kg) missile is a further improved AIM-9G with the same warhead, the Mk 36 Mod 6 motor, an improved solid-state seeker providing enhanced tracking speed and all-weather capability, and other modifications to improve manoeuvrability. This type is dimensionally identical with the AIM-9G and possesses the same range, but introduced more powerful actuators to provide greater manoeuvrability.

Multiple engagement capability is all-important in all combat, and this McDonnell Douglas F-15 Eagle is well placed in this respect: apart from its inbuilt M61A1 Vulcan cannon, the aircraft carries four AIM-9L/M short-range and four AIM-7F/M Sparrow medium-range air-to-air missiles.

AIM-9J Sidewinder: Developed from the AIM-9E by converting obsolescent AIM-9Bs (**AIM-9J2**) and making new missiles (**AIM-9J1** and the product-improved **AIM-9J3**), this variant appeared in the mid-1970s and features a Mk 17 solid-propellant motor, a cooled seeker, some solid-state electronics and revised double-delta control surfaces with more powerful actuators to generate greater manoeuvrability. This model weighs 172 lb (78 kg), has a length of 120.9 in (3071 mm) and a span of 22.0 in (559 mm), and can reach a range of 15,850 yards (14495 m), the reduced range of this version being deemed acceptable because of its considerably higher acceleration.
AIM-9L Sidewinder: This is a much improved version introduced in 1976, and combines the latest Sidewinder airframe (featuring an upgraded motor and pointed double-delta foreplanes of yet further refined shape for increased manoeuvrability) with a 25 lb (11.4 kg) annular blast/fragmentation warhead and the AM/FM conical-scan seeker developed by Bodensee Gerätetechnik for the abortive Viper AAM, offering an excellent all-aspect seeker capability combined with a dogfighting airframe. The proximity fusing system incorporates a ring of eight gallium arsenide laser diodes to provide optimum warhead detonation in high-speed manoeuvring combat. The type is used by Sweden with the designation **RB 74,** and is also made under licence in Japan by a consortium headed by Mitsubishi.
AIM-9M Sidewinder: This is the exclusively US-made version of the AIM-9L with a weight of 189.6 lb (86 kg) as a result of using a closed-cycle cooler for improved seeker sensitivity (especially at low altitudes), greater resistance to infra-red countermeasures

and a 'smokeless' Bermite/Hercules Mk 36 Mod 9 motor for reduced visual signature.
AIM-9N Sidewinder: This is a redesignation of AIM-9J1 missiles produced by converting AIM-9B and AIM-9E Sidewinders.
AIM-9P Sidewinder: This is a US-made model (either new-build or upgraded AIM-9B, E and J weapons) to AIM-9L standard with the canards of the AIM-9J series. The variant also offers greater reliability and a weight of 170 lb (77.1 kg). There are three subvariants in the form of the **AIM-9P1** with an active optical proximity fuse, the **AIM-9P2** with 'smokeless' motor, the **AIM-9P3** combining AIM-9P1 and AIM-9P2 features with a new warhead of reduced thermal sensitivity and greater shelf life, and the **AIM-9P4** 'security assistance missile' for the US Air Force with increased target-acquisition and tracking capabilities combined with an advanced fuse.
AIM-9R Sidewinder: This is a 193 lb (87.5 kg) derivative of the AIM-9M (and originally designated **AIM-9M Improved**) under development for production from 1988 with a radically upgraded seeker (of the imaging infra-red variety that does away with the seeker cooling of all other recent Sidewinder variants) for greater acquisition range and better resistance to countermeasures. The type also has an active laser proximity fuse and a range of 21,125 yards (19315 m).

USA

Hughes AIM-26B Super Falcon

Type: medium range air-to-air missile
Dimensions: diameter 11.4 in (290 mm); length 81.5 in (2070 mm); span 24.4 in (620 mm)
Weights: total round 262 lb (118.8 kg); warhead 40 lb (18.1 kg) proximity- fused blast fragmentation
Propulsion: one Thiokol M60 solid-propellant rocket delivering 5,625 lb (2552 kg) thrust
Performance: speed Mach 2; range 10,500 yards (9600 m)
Guidance: Hughes semi-active radar homing

Variant and application
AIM-26B Super Falcon: Introduced in 1960 as the primary weapon of the Convair F-106 Delta Dart (to provide a very high kill probability in head-on engagements with Soviet strategic bombers), the AIM-26A Super Falcon was designed round semi-active radar homing as infra-red homing was thought impractical for collision-course engagements. The comparative inaccuracy of semi-active radar homing dictated the use of a particularly powerful warhead, and the Super Falcon's initial variant was thus fitted with a potent 1.5 kiloton nuclear warhead, this being based on the W25 of the AIR-2 Genie and detonated by a quartet of active radar proximity fuses. Incorporation of this comparatively massive warhead dictated a considerable enlargement of the Falcon's constant-diameter body into a revised contour, tapered forward and aft from bulged centrebody. This also provided volume for the more powerful motor installed to maintain performance with the heavier weapon.

The initial variant was 84.25 in (2140 mm) long and weighed 203 lb (92.1 kg) but is no longer in service, having been succeeded by the conventionally-armed AIM-26B, which still serves with Sweden and Switzerland under the designations **RB 27** and **HM-55** respectively, on the Saab 35F Draken and Dassault-Mirage IIIS fighters fitted with Hughes Taran radar.

Hughes AIM-54A Phoenix

USA

Type: long-range air-to-air missile
Dimensions: diameter 15.0 in (381 mm); length 157.8 in (4008 mm); span 36.4 in (925 mm)
Weights: total round 985 lb (446.8 kg); warhead 132 lb (59.9 kg) Bendix infra-red and Downey Mk 334 radar proximity- and impact delayed-fused continuous-rod blast fragmentation
Propulsion: one Aerojet Mk 60 or Rocketdyne Mk 47 Mod 0 single-stage solid-propellant rocket delivering unrevealed thrust
Performance: speed Mach 4.3+; range 2.4/125+ miles (3.86/201.2+ km)
Guidance: Hughes DSQ-26 system with semi-active homing for the cruise phase and active radar for the terminal phase

Variants and application

AIM-54A Phoenix: Designed for the abortive General Dynamics F-111B swing-wing naval fighter, the Phoenix was brought to fruition (together with the accompanying Hughes AWG-9 pulse-Doppler fire-control system) in the Grumman F-14 Tomcat fleet-defence fighter. The type retains the classic aerodynamics of Hughes's earlier air-to-air missiles, with a substantial body carrying a cruciform of low-aspect-ratio delta wings well aft and trailed by rectangular control surfaces indexed in line with the wings.

The considerable body diameter allows the use of a wide planar-array radar antenna in the nose radome, and also provides volume for the missile's other primary components which are, from front to rear, the electronics (radar avionics, guidance package and Downey Mk 334 radar proximity fuse system), the warhead, the solid-propellant rocket, and (round the nozzle) the autopilot and control-surface hydraulic actuators. After launch from the Tomcat, the missile climbs to a peak altitude of 81,400 ft (24810 m), cruising under control of the onboard autopilot with guidance of the semi-active type using reflections of the radar in the launch aircraft's AWG-9 system. This high-altitude cruise maximizes range by reducing drag and providing the rocket with optimum operating conditions. As the missile dives down to the attack the energy potential of the

The ultimate in long-range tactical weapons is the Hughes AIM-54 Phoenix air-to-air missile used only by the Grumman F-14 Tomcat. This F-14A of VF-32 on USS John F. Kennedy *carries a maximum of six such AIM-54A missiles, four on low-drag pallets under the fuselage and two on hard-points under the outer portions of the fixed wing 'gloves'. Also visible on the lower side of the nose is the port for the muzzle of the Tomcat's 20-mm Vulcan cannon.*

cruise altitude is converted into kinetic energy for greater manoeuvrability in the terminal phase of the flight. The missile's radar switches to the active mode for the final 20,000 yards (18290 m) of the attack, the availability of three fusing modes offering maximum target-destruction capability. The Phoenix began to enter service in 1973. Notable features of the missile and fire-control combination are exceptional range, and the ability to engage six targets simultaneously (using a time-share system so that the AWG-9 can control the missiles, which can lose touch with the radar for 14 seconds before failing to reacquire the designated target) regardless of weather conditions and target aspect. Production continued up to 1980, by which time some 2,566 rounds had been made.

AIM-54B Phoenix: This version entered production in late 1977, and is in essence a product-improved AIM-54A with sheet metal rather than honeycomb aerodynamic surfaces, non-liquid hydraulic and thermal-conditioning systems, and a degree of simplified engineering to ease production.

AIM-54C Phoenix: Developed from 1977 and introduced in 1982, this is a considerably improved missile with digital rather than analog electronics, a new Nortronics strapdown inertial reference unit, solid-state radar and very much more capable electronic counter-countermeasures capability. The weight of this variant is 1,008 lb (457.2 kg), the speed Mach 5 and the ceiling 100,000+ ft (30490+ m). As with the AIM-54A, the length 156.0 in (3962 mm), legend range has been considerably exceeded in service, especially when the F-14 launch aircraft is supported by a Grumman E-2C Hawkeye warning and control system aircraft.

USA

General Dynamics (Pomona) AIM-92A Air-Launched Stinger

Type: short-range air-to-air missile
Dimensions: diameter 2.75 in (69.85 m); length 60.0 in (1524 mm); span 3.6 in (91.4 mm)
Weights: total round 30 lb (13.61 kg); warhead 6.6 lb (2.99 kg) proximity- and impact-fused Picatinny Arsenal blast fragmentation
Propulsion: tandem Atlantic Research solid-propellant rockets delivering unrevealed thrust
Performance: speed Mach 2+; range 5,280 yards (4830 m)
Guidance: infra-red homing

Variant and application

AIM-92A Air-Launched Stinger: This is the air-launched version of the FIM-92A Stinger shoulder-launched surface-to-air missile, and is being developed as a self-protection weapon for helicopters and tactical aircraft as the **Multi-purpose Lightweight Missile System,** the initial application being on the Bell OH-58 Kiowa scout helicopter. This MLMS is a 99 lb (44.9 kg) twin launcher complete with a coolant reservoir for the two missiles' seekers, and the units can be stacked up to four deep to give the helicopter a theoretical complement of 16 short-range missiles (eight on each side of the helicopter). The MLMS can also accommodate the improved Stinger-POST missile, a development of the basic weapon with improved guidance using the latest infra-red counter-countermeasures techniques and the two-colour (infra-red and ultra-violet) Passive Optical Seeker Technique using rosette-scanning for greatly enhanced target-acquisitioning capability.

The Stinger missile is also being adopted for a unique application in a variant of the Lucas helicopter gun turret, normally fitted with a 12.7 mm (0.5 in) FN M3 heavy machine-gun. This adaptation comprises a bank of four Stingers on the basic turret mounting, and has been evolved to meet the German army's requirement for an escort derivative of the VBH, derived from the MBB BO105. With such a turret, aimed by the gunner's Ferranti helmet-mounted sight, the VBH will be able to escort 'convoys' of PAH-1 anti-tank helicopters, using its Stingers for off-boresight engagement of hostile helicopters. A further derivative of the concept is under evaluation for the British army to provide Westland Lynx helicopters with a higher level of off-boresight engagement capability using the new Shorts Starstreak high-velocity missile.

Hughes AIM-120A AMRAAM **USA**

Type: medium-range air-to-air missile
Dimensions: diameter 7.0 in (178 mm); length 143.9 in (3655 mm); span 24.7 in (627 mm)
Weights: total round 335.2 lb (152.05 kg); warhead 45 lb (20.4 kg) proximity delay and impact-fused blast fragmentation
Propulsion: one solid-propellant rocket delivering unrevealed thrust
Performance: speed about Mach 4; range 45 miles (72 km) from a high-altitude launch declining to 34 miles (55 km) from a low-altitude launch
Guidance: Nortronics strapdown inertial for mid-course phase, and Hughes active radar for terminal phase

Variant and application
AIM-120A AMRAAM: Designed in the late 1970s and early 1980s for a service debut in the late 1980s as replacement for the AIM-7F/M variants of the Sparrow, the **Advanced**

Designed to replace the Sparrow in US service, the Hughes AIM-120 Advanced Medium-Range Air-to-Air Missile (AMRAAM) has undergone a protracted development beset by technical, political and economic problems. Four examples of the AMRAAM are carried by this General Dynamics F-16B Fighting Falcon, together with two AIM-9J/N Sidewinders at the wingtips and and ALQ-131 electronic countermeasures pod under the fuselage.

This photo-sequence shows the destruction of a North American/Sperry QF-100 target drone by a Hughes AIM-120A AMRAAM fired from a General Dynamics F-16 Fighting Falcon. The missile was data-linked to the launch aircraft, and activated its active radar seeker only when well into the flight, which terminated in a direct hit on the target.

Medium-Range Air-to-Air Missile is a fire-and-forget BVR (Beyond-Visual-Range) weapon with size and weight little more than those of the AIM-9 Sidewinder but capabilities and performance better than those of the considerably larger Sparrow. The missile is launched from a standard rail, and cruises on a preprogrammed course (with the capability for mid-course update from the launch aircraft's radar) until it reaches the vicinity of the target, where it activates its active radar seeker, which also possesses a 'home-on-jam' mode. The whole programme has been bedevilled by technical problems, cost overruns and political antipathy, and service entry has been delayed to 1989 or 1990.

USA

Douglas AIR-2B Genie

Type: short-range air-to-air rocket
Dimensions: diameter 17.5 in (445 mm); length 116.0 in (2946 mm); span 40.0 in (1016 mm)
Weights: total round 822 lb (372.86 kg); warhead command-detonated W25 nuclear with a yield in the order of 1.5 kilotons (with oralloy as the fissile material and Cyclotrol as the explosive for the implosion leading to the fission reaction)
Propulsion: one Thiokol TU-289 (SR-49-TC-1) solid-propellant rocket delivering 36,000 lb (16602 kg) thrust
Performance: speed Mach 3.3; range 8,800/11,000 yards (9625/10060 m)
Guidance: none

Variant and application

AIR-2B Genie: The Genie is an unguided rocket rather than a missile, and after service entry in the late 1950s as a bomber-destroying weapon is now employed only on a few elderly Convair F-106A interceptors. The rocket is fired on the basis of information from the Hughes MA-1 automated fire-control system, which commands the pilot to arm the weapon, fires it, hauls the aircraft away in a tight turn and then detonates the weapon at the appropriate moment. The missile has a flick-out fins for roll control and reduction of gravity drop, and the warhead's lethal radius is estimated at more than 1,000 ft (305 m).

Vought ASAT

USA

Type: air-launched anti-satellite missile
Dimensions: diameter (first stage) 17.5 in (445 mm) and (second stage) 19.8 in (503 mm); length 212.66 in (5402 mm); span data not available
Weights: total round 2,632 lb (1193.9 kg); warhead (Vought Miniature Vehicle) about 35 lb (15.88 kg)
Propulsion: (first stage) one Lockheed SR75-LP-1 solid-propellant rocket and (second stage) one Altair III solid-propellant rocket delivering unrevealed thrust
Performance: speed hypersonic; range 620 mile (1000 km) orbital altitude
Guidance: infra-red homing

Variant and application
ASAT: One of the world's most advanced weapons, the ASAT was tested during the mid-1980s with a modified McDonnell Douglas F-15 Eagle fighter as the launch platform. The object of the US Air Force is to develop an anti-satellite capability in the early 1990s, with a small quantity of ASAT missiles for launch by F-15 fighters at the apogee of a maximum-power zoom climb, but the programme has run into considerable political antipathy in the US Congress and was cancelled in 1988.

The Vought ASAT is based on the propellant section of the Boeing AGM-69A SRAM missile with a high-velocity second stage carrying the Vought 1005 Miniature Vehicle, which measures 12 by 13 in (305 by 330 mm) and is fired once the vehicle's telescopic infra-red guidance system has locked onto the target satellite. The MV has been designed to shatter the target with the kinetic energy of a 30,680 mph (40375 km/h) impact. In March 1987 it was announced that the USA was so concerned by the threat posed by Soviet reconnaissance satellites that the politically-delayed ASAT programme was being relaunched with investigation into a more capable version, as either an air-launched model with a more powerful booster, or a ground-launched version using the Pershing 2 surface-to-surface missile as its basis.

BBG AIM-132A ASRAAM

West Germany/ UK

Type: short-range air-to-air missile
Dimensions: (estimated) diameter 168 mm (6.6 in); length 2730 mm (107.48 in); span data not available
Weights: total round 100 kg (220.5 lb); warhead 10 kg (22.05 lb) blast fragmentation
Propulsion: one solid-propellant dual-thrust rocket delivering unrevealed thrust
Performance: speed Mach 3+; range 300/15000 m (330/16,405 yards)
Guidance: strapdown inertial plus infra-red homing

Seen on a twin adapter on the starboard outer hardpoint of BAe's Harrier demonstrator, appropriately registered 'G-VTOL', are a pair of AIM-132 ASRAAM (Advanced Short-Range Air-to-Air Missile) dummies. Planned in European partnership to the American AIM-120 AMRAAM, the programme has been tardy in the extreme, and will ultimately result in a missile inferior to the Tail Dog which BAe could have placed in production during the mid-1970s.

Variant and application

AIM-132A ASRAAM: This **advanced short-range air-to-air missile** is under development by the BBG consortium (BAe and Bodenseewerk Gerätetechnik) on the conceptual basis of BAe's **Short-Range Air-to-Air Missile** (originally called Taildog), a wholly successful trials weapon of the 1970s. This was a tube-launched weapon without aerodynamic surfaces and using a helmet-mounted sight system for pre-launch designation of the target. The missile's combination of thrust-vector control and a swivelling motor made for enormous agility, but the SRAAM and its fully-developed millimetre-wave active radar homing were abandoned for political reasons and in face of demands that the new weapon be compatible with existing Sidewinder rail launchers. In its current form ASRAAM is later, less capable and more expensive than the SRAAM, and after an unfortunately protracted development phase is due to enter service in 1992 for use by all NATO countries. The ASRAAM uses lifting-body aerodynamics and has four tail fins for control.

AIR-TO-SURFACE MISSILES

CITEFA ASM-2 Martin Pescador

Argentina

Type: air-to-surface tactical missile
Dimensions: diameter 218.5 mm (8.60 in); length 2940 mm (115.74 in); span 730 mm (28.74 in)
Weights: total round 140 kg (308.6 lb); warhead 40 kg (88.2 lb) impact-fused blast fragmentation
Propulsion: one solid-propellant rocket delivering unrevealed thrust
Performance: speed Mach 2.3; range 2500/9000 m (2,735/9,845 yards) from an aircraft launch, or 2500/4300 m (2,735/4,705 yards) from a helicopter launch
Guidance: radio command

Variant and application
ASM-2 Martin Pescador: This is a simple radio command-guided missile which entered service in 1979 on aircraft such as the FMA IA-58 Pucara. The missile has an impact speed in the order of Mach 1.1 after a maximum flight time of 13 seconds, and CITEFA is currently working on an upgraded version with a heavier warhead and (presumably) a more advanced system of guidance.

Aérospatiale AS.11

France

Type: lightweight air-to-surface tactical missile
Dimensions: diameter 164 mm (6.46 in); length 1201 mm (47.28 in); span 500 mm (19.69 in)
Weights: total round 29.9 kg (65.92 lb); warhead unrevealed Type 140AC hollow-charge anti-tank, Type 140AP02 penetrating blast-fragmentation, Type 140AP59 blast/fragmentation and Type 140CCN anti-ship types
Propulsion: one SNPE Simplet solid-propellant booster rocket delivering unrevealed thrust for 2.85 seconds, and one SNPE Sophie solid-propellant sustainer rocket delivering unrevealed thrust for 31.2 seconds
Performance: speed increases from 360 to 684 km/h (224 to 425 mph) in the cruise phase of the flight; range 3500/7000 m (3,830/7,655 yards)
Guidance: wire command to line of sight

Variants and application
AS.11: This is the helicopter-borne version of the SS.11 surface-launched anti-tank missile, and can be used for a number of tactical applications depending on the type of warhead fitted. The weapon is obsolete as an anti-tank weapon (taking as long as 21 seconds to reach its maximum range), but when used with a gyro-stabilized sight (such as the SFIM APX Bézu 260 carried by the Aérospatiale SA 316/319 Alouette III and the SFIM APX M334 carried by the Aérospatiale SA 341/342 Gazelle) is still effective as a bunker-buster and for attacks on naval targets such as landing craft.

The basic weapon was developed between 1953 and 1955 as the Nord 5210, and was first adopted for helicopter use in 1958. Enormously successful in commercial terms, the SS.11/AS.11 series was produced to the extent of some 180,000 missiles for sale to armed forces in at least thirty countries, many of whom still use the weapon. The Type 140AC warhead can penetrate 140 mm (5.51 in) of armour, the Type 140AP02 detonates 2.6 kg (5.72 lb) of HE after passing through 10 mm (0.4 in) of armour, the Type 140AP59 is an impact-fused fragmentation warhead, and the Type 140CCN can deal with landing craft and attack craft.

AS.11B1: This is an improved version with transistorized guidance and the option of TCA infra-red semi-automatic command to line of sight guidance.

France

Aérospatiale AS.12

Type: air-to-surface tactical missile
Dimensions: diameter 210 mm (8.27 in); length 1870 mm (73.62 in); span 650 mm (25.6 in)
Weights: total round 76 kg (167.5 lb); warhead 28.4 kg (62.6 lb) OP.3C penetrating blast/fragmentation
Propulsion: one SNPE Achille solid-propellant booster rocket delivering unrevealed thrust for 1.15 seconds, and one SNPE Hermione solid-propellant sustainer rocket delivering unrevealed thrust
Performance: speed 338 km/h (210 mph); range 8000 m (8,750 yards)
Guidance: wire command to line of sight

Variant and application

AS.12: This 1955 missile was designed as a scaled-up version of the SS.10 specifically for use against bunkers and warships, and is now of only limited use even in these two roles. Compared with its predecessor, the SS.11/AS.11's key features are double the range and four times the destructive power. The OP.3C warhead can detonate a 28.4 kg

The Aérospatiale AS.12 air-to-surface missile was designed largely for tank-killing. This powerful missile is now obsolete in its original primary role because of its slow speed and simple CLOS (command to line of sight) guidance, but retains useful secondary capabilities in the land role against fortifications and in the naval role against smaller surface vessels. This is a Westland Wasp HAS.Mk 1 with two AS.12s and the missiles' associated roof sight.

(63 lb) charge after penetrating 40 mm (1.57 in) of armour, and in its AS.11 helicopter-carried version is used with the SFIM APX-Bézu 260 or SFIM APX M334 gyro-stabilized sights carried by several types of light tactical helicopter.

Aérospatiale AS.20 France

Type: air-to-surface tactical missile
Dimensions: diameter 250 mm (9.84 in); length 2600 m (102.36 in); span 800 mm (31.5 in)
Weights: total round 143 kg (315.3 lb); warhead 30 kg (66.1 lb) impact-fused blast fragmentation
Propulsion: one SNPE/Aérospatiale Aspic solid-propellant booster rocket delivering unrevealed thrust, and one SNPE/Aérospatiale Icare solid-propellant sustainer rocket delivering unrevealed thrust
Performance: speed Mach 1.7; range 7000 m (7,655 yards)
Guidance: radio command

Variant and application
AS.20: Developed from the radio-commanded AA.20 air-to-air missile, the simple but effective AS.20 was introduced in 1961 and remains in very limited service, mainly as a training round for the AS.30 series.

Aérospatiale AS.30 Laser France

Type: air-to-surface tactical missile
Dimensions: diameter 342 mm (13.46 in); length 3650 mm (143.7 in); span 1000 mm (39.37 in)
Weights: total round 520 kg (1,146.4 lb); warhead 240 kg (529.1 lb) impact- or delay-fused blast fragmentation or semi-armour-piercing HE
Propulsion: one SNPE/Aérospatiale two-stage (composite booster and cast double-base sustainer) solid-propellant rocket delivering unrevealed thrust for 21 seconds
Performance: speed Mach 1.5; range 3000 to 11250 m (3,280 to 12,305 yards); CEP 2 m (6.6 ft)
Guidance: strapdown inertial midcourse guidance, and Thomson-CSF Ariel semi-active laser terminal guidance

Variants and application
AS.30: Introduced in 1960 as the Nord 5401 to provide French Dassault Mirage IIIE fighter-bombers with a moderate stand-off capability, the AS.30 is essentially a scaled-up AS.20. The missile has four delta wings located as a cruciform arrangement mid-way along the body, and its use allows the launch aircraft to come no closer that 3000 m (3,280 yards) to the target, the operator then radio-commanding the missile to the visual line of sight with a CEP of less than 10 m (33 ft). The weapon is 3839 mm (151.14 in) long with the X12 warhead, or 3885 mm (152.95 in) long with the X35 warhead.
AS.30TCA: Introduced in 1964 exclusively for French tactical aircraft, this is a semi-

This French air force Dassault-Breguet Mirage 2000C sports an AS.30 Laser air-to-surface missile on the hardpoint immediately outboard of the landing gear, with an R550 Magic short-range air-to-air missile further out. On the inboard hardpoint is the Thomson-CSF ATLIS 2 laser designator pod used with the AS.30 Laser. Developed in conjunction with Martin Marietta in the USA, the ATLIS 2 pod weighs 160 kg (353 lb), and is 2520 mm (99.2 in) long and 300 mm (11.8 in) in diameter. The pod has a protective shutter over the nose-mounted optics, which comprise a TV camera (with optional wide or narrow fields of vision) and a laser rangefinder/designator. The optics are stabilized and provide automatic two-mode TV tracking, with information presented to the pilot via a TV monitor, head-up display, radar display or helmet-mounted sight.

automatic command to line of sight development, obviating the need for the operator to track both the missile and the target, a severe tactical disadvantage for the AS.30 model. The guidance system for this variant uses an aircraft-carried SAT device to keep track of the flare at the rear of the missile. Another modification is the installation of four rear-mounted flick-open fins, indexed in line with the wings, to replace the control spoilers in the two exhaust nozzles.

AS.30 Laser: Introduced in 1985 after programme launch in 1978 and a first firing in 1980, this much-improved model is designed for use with the Thomson-CSF/Martin Marietta ATLIS 2 (Automatic Tracking Laser Illuminator System Mk 2) pod, allowing the aircraft to launch and break away while the laser pod under the fuselage or wing continues to designate the locked-in target for the missile, whose maximum flight time is 21 seconds. The main carriers of the missile are the Dassault-Breguet Mirage F1 and Mirage 2000, and the SEPECAT Jaguar A. The warhead has a hardened steel casing and, in combination with an impact speed of 1620 km/h (1,007 mph) this provides the missile with the capability to penetrate 2.0 m (6.56 ft) of concrete before detonating.

Launch of an AS.30 Laser from an underwing hardpoint of a French air force SEPECAT Jaguar A attack aircraft. Notable are the twin exhausts from the missile, which half-conceal the ATLIS 2 laser designator pod carried on the aircraft centreline.

Matra BGL 400

France

Type: air-to-surface tactical glide bomb
Dimensions: diameter 403 mm (15.87 in); length 3402 mm (133.94 in); span 789 mm (31.06 in)
Weights: total round 475 kg (1,047.2 lb); warhead 400 kg (882 lb) bomb
Propulsion: none
Performance: speed high subsonic; range 10000 m (10,935 yards)
Guidance: Thomson-CSF TMV 630 EBLIS semi-active laser homing

Variants and application
BGL 250: Currently under development, this is a standard French 250 kg (551 lb) bomb with the same type of laser homing and flight-control system as that used on the in-service BGL 400 series. The diameter of the weapon is 228 mm (8.98 in), length 3329 mm (131.07 in) and span 640 mm (25.2 in).
BGL 400: This useful French weapon is in service with the French air force and several export customers, and was derived in concept from the US 'Paveway' series. It is thus a standard 400 kg (882 lb) HE bomb provided with cruciform wings/fins on a rearward extension of the bomb's body and fitted with a seeker/control section on the nose. The laser seeker is based on the Ariel seeker of the AS.30 Laser missile, and the target can be marked from the ground or by an aircraft using the ATLIS designator system, allowing a lock-on range of between 4000 and 10000 m (4,375 and 10,935 yards). The weapon is designed for release at between 50 and 150 m (165 and 490 ft) by aircraft flying at speeds up to 1100 km/h (684 mph).
BGL 1000: This is a considerably larger version of the same BGL series, but in this instance based on the French standard 1000 kg (2,205 lb) HE bomb, with a diameter of 457 mm (18.0 in), a length of 4212 mm (165.83 in) and a span of 900 m (35.43 in). In 1987 the weapon received the name **Arcole,** and was revealed as a stand-off weapon with a range of 8000 m (8,750 yards) and a specially strengthened nosecap for the destruction of targets as hard as the concrete piers of bridges.

Matra/MBB Apache/CWS

France/West Germany

Type: air-launched stand-off tactical dispenser missile
Dimensions: cross section of the dispenser section includes a width of 630 m (24.8 in) and a height of 480 m (18.9 in); length 4040 mm (159.06 in); span 2530 mm (99.6 in)
Weights: total round (Apache/CWS I) 1000 kg (2,204.6 lb), (Apache/CWS II) 1150 kg (2,535.3 lb), and (Apache/CWS III) 1220 kg (2,689.6 lb); warhead (all three versions) 750 kg (1,653.4 lb) of submunitions
Propulsion: (Apache/CWS I) none, (Apache/CWS II) one solid-propellant rocket, and (Apache/CWS III) one small turbojet or turbofan
Performance: speed (all three versions) up to Mach 0.95 at launch; range at low altitude (Apache/CWS I) 12000 m (13,125 yards), (Apache/CWS II) 25/30 km (15.53/18.64 miles, and (Apache/CWS III) 40/50 km (24.86/31.07 miles)
Guidance: inertial, or (Apache/CWS III) inertial plus an unspecified terminal guidance system

Currently under development by Matra in France and MBB in West Germany, the Apache/CWS is one of the more likely of the current crop of stand-off dispenser weapons. The wings are folded into upper-surface recesses until the vehicle has been dropped by its launch aircraft, then deploy to provide the dispenser with an extended glide capability.

Variants and application

Apache/CWS I: This series of weapons is based on a common airframe, and originated with separate studies by Matra and MBB (the Apache and Container Weapon System respectively) for successors to the Belouga and Durandal. The two concept streams have now been merged into this important family with service entry planned for 1992 on aircraft fitted with the NATO standard 762 mm (30 in) suspension lugs. Designed for launch at high subsonic speed at an altitude between 50 and 70 m (165 and 230 ft), the Apache/CWS then flies at a mission altitude between 50 and 100 m (165 and 330 ft) before descending to release its submunitions at a height below 50 m (165 ft).

The rectangular-section fuselage has flip-out wings, a small tailplane and two ventral fins, and the 2 m (6.56 ft) boxlike central fuselage can be laid out for a number of modular payloads (initially two configurations) ejected sideways from transverse tubes to cover an area 350 m (385 yards) wide by 1000 m (1,095 yards) long. The submunitions carried in the 42 small- or 18 large-diameter tubes (or in a variable mix of the two modular types) are at first to be the same as those used in the MW-1 underfuselage dispenser weapon, namely the 44 mm (1.73 in) KB 44 shaped-charge anti-tank bomblet, the MIFF anti-tank mine, the MUSA fragmentation mine for use against semi-hard targets, the MUSPA fragmentation mine for use against aircraft, the STABO runway-cratering munition and the ASW munition for attacks on hardened aircraft shelters. The dispenser system bears a close resemblance to the Modular Dispenser System (see Free-Fall Weapons section) being developed by MBB for fixed carriage by tactical aircraft such as the McDonnell Douglas F-4F Phantom. The original CWS was planned to carry any of four payloads (unpowered freight container weapon and powered, unpowered napalm or fuel transporter, unpowered weapon container), and such a capability could in due course be restored to the Apache/CWS.

Apache/CWS II: This is the proposed longer-range version with a solid-propellant rocket to provide greater stand-off range for the launch aircraft when faced with the air-defence capabilities of a sophisticated battlefield.

Apache/CWS III: Proposed as the longest-range version, this derivative of the basic vehicle is to have a turbine engine and, almost certainly, a terminal guidance system to ensure that the weapon is accurate in the final stages of its long-range flight.

Israeli Guided Bombs

Israel is believed to use French and US general-purpose bombs, and to have produced some of these in Israel as a means of reducing reliance on strategic imports whose deliveries can be interrupted for political and military reasons during times of crisis. It is not known, however, precisely which models of bombs are produced in Israel though it is reasonable to assume that these include the US Mks 81 to 83 (and possibly the Mk 84).

The **Elbit Opher** uses an infra-red homing/control package comparable to that of the US 'Paveway' laser-homing series in performance terms, but considerably cheaper to produce. The package can be added to many types of free-fall ordnance to create 'semi-smart' weapons: the Opher-fitted bomb is launched either in a dive attack or using the standard continuously-computed impact point system, and homes on the infra-red radiation from fixed or moving targets. At the heart of the Opher system is a simplified high-resolution FLIR with a 7° field of view and providing data to a high-speed digital computer which controls post-launch target lock-on and the actuation of the pneumatically-powered bomb controls. The weapon has been developed primarily as a cost-effective counter to main battle tanks fitted with reactive armour able to 'snuff out' the gas/metal vapour jet of ground-launched hollow-charge projectiles. A typical lock-on range is 1000 m (1,095 yards). The most common application of the Opher kit is on the Mk 82 500 lb (227 kg) bomb to produce a weapon 3430 mm (135 in) long and 325 kg (716.5 lb) in weight.

More precisely comparable to the 'Paveway' series is the **Israel Aircraft Industries Guillotine,** which is a laser-homing weapon, again based on the locally-produced equivalent to the US Mk 82 bomb, which the Israelis have discovered through a mass of practical experience to be the optimum size (balancing destructive effect with the ability to be carried in multiples by most tactical aircraft) for the engagement of the basic range of battlefield targets. The weapon has a cruciform of spring-out delta tail surfaces, and at the nose the normal type of cylindrical section accommodating the seeker (with automatic search and lock onto reflected laser energy) and the powerful delta control surfaces. Israeli sources give the Guillotine a CEP of 2 m (6.56 ft) and a range of 30 km (18.64 miles) from a release altitude of 12000 m (39,370 ft), the Guillotine hitting its target at an angle of 45°.

The **Rafael Pyramid** is an altogether more sophisticated weapon, though again based on the warhead of the Mk 82 bomb. The Pyramid has a very substantial rear-mounted cruciform of delta wings, each trailed at its inboard end by a small control surface, and is a low-cost electro-optical (TV)-guided bomb weighing about 400 kg (882 lb). The tactical rationale of the weapon is the need to provide Israeli aircraft with the ability to hit point targets (the Pyramid has a CEP in the order of 1 m/3.3 ft) from a 30 km (18.64 mile) launch range: this is greater than the range capability of all the Soviet-supplied AA weapons (except the SA-5 'Gammon') deployed by Israel's Arab neighbours.

The Pyramid can be launched from any aircraft able to carry it, for the weapon requires no interface with the launch aircraft as it can be guided by any other aircraft carrying the associated receiver pod and onboard TV monitor/control system. When used against a large target, the Pyramid can be used in a 'launch-and-leave' mode, with the tracker locked onto the target before the weapon is dropped. When used against smaller targets, the weapon is dropped after receiving general directions and is guided towards the target by the pilot, who uses ground features for this coarse guidance phase: the weapon can then be guided right onto the target by the pilot, or allowed to make an autonomous attack after its tracker has been locked onto the target.

Israel

Rafael Armament Development Authority Pop-Eye

Type: air-to-surface missile
Dimensions: data not available
Weights: data not available
Propulsion: one solid-propellant rocket delivering unrevealed thrust
Performance: data not available
Guidance: TV guidance

Variant and application
Pop-Eye: The Pop-Eye air-to-surface missile was developed in Israel to meet indigenous as well as export requirements, and was in late 1987 adopted for the US Air Force. No details of this potentially important weapon have yet been revealed.

Italy

CASMU (Aeritalia/SNIA BPD) Skyshark

Type: air-launched short-range stand-off dispenser missile
Dimensions: body cross section 0.394 m^2 (4.24 sq ft); length 4757 m (187.28 in); span 1500 m (59.06 in)
Weights: total round (glider version) 1050 kg (2,314.8 lb) or (powered versions) 1170 kg (2,579.4 lb); warhead 745 kg (1,642.4 lb) of submunitions
Propulsion: (powered versions) one solid-propellant rocket or (later) one turbofan
Performance: speed about 975 km/h (606 mph); range (initial glider version) 6000/12000 m (6,560/13,125 yards), or (proposed rocket-powered version) 20/25 km (12.43/15.53 miles), or (proposed turbofan-powered version) considerably greater than the rocket-powered version
Guidance: unrevealed guidance and fire-control system

Variants and applications
Skyshark: This interesting vehicle is a product of the company formed by Aeritalia and SNIA BPD especially to develop it, namely the Consorzio Armamenti Spendibili Multi Uso (consortium for multi-role expendable weapons). The Skyshark is designed for stand-off attacks against high-value targets, and is designed for air-launch by most tactical aircraft. In its initial form the vehicle is to be a glider, carrying in a laterally-ejecting module a payload of submunition types as yet unspecified. These can be varied to tackle the specific target envisaged.

The consortium is also planning a rocket-powered medium-range version and, ultimately, a turbofan-powered long-range cruise version. The key to the Italian concept's tactical capabilities is the combination of low radar cross section (resulting from the shape of the vehicle and its construction of radar-absorbent materials), and an extremely advanced onboard guidance and fire-control system. These features should help to ensure that the dispenser reaches its target area safely so that it can then undertake the accurate high-speed delivery of anti-matériel/anti-personnel, anti-runway, anti-tank, area-denial and unitary munitions.

Saab RB 04E

Sweden

Type: air-to-surface tactical missile
Dimensions: diameter 500 mm (19.69 in); length 4450 mm (175.2 in); span 1970 mm (77.56 in)
Weights: total round 616 kg (1,358.0 lb); warhead 300 kg (661.4 lb) blast fragmentation
Propulsion: one IMI Summerfield two-stage cast double-base solid-propellant rocket delivering unrevealed thrust
Performance: speed high subsonic; range varies with launch height to a maximum of 32 km (19.9 miles)
Guidance: Saab two-axis autopilot for midcourse phase, and PEAB (Philips Elektronikindustrier AB) active radar for terminal phase

Variants and application
RB 04C: This was the first European air-to-surface missile with active radar terminal homing for a true all-weather fire-and-forget capability, and entered development in 1949 before the start of one of the world's longest running missile production programmes (1958 to 1978). The type was schemed primarily as an anti-ship weapon, but has alternative capability against land targets. In configuration the RB 04 series is slightly unusual, being fitted at the rear with substantial swept wings tipped by endplate vertical surfaces, and towards the nose with a cruciform arrangement of delta control surfaces indexed in line with the wings. The RB 04C was the initial model, entering service in 1958, and spans 2040 mm (80.31 in). The missile is carried by Saab 35 Draken and Saab 37 Viggen attack aircraft of the Swedish air force.
RB 04D: This is an improved version brought into service during the later 1960s with an upgraded rocket motor and modernized guidance.
RB 04E: This is the final production version with smaller span and modernized guidance as well as an improved structure. Despite the age of the basic concept, this final variant is believed to have very advanced technical features even by current standards.

A Saab RB 05A air-to-surface missile is seen under the port wing of its primary launch platform, the Saab Viggen. Clearly visible here is the mechanical and electronic interface 'shoe' between the hardpoint and the missile.

Sweden

Saab-Bofors RB 05A

Type: air-to-surface tactical missile
Dimensions: diameter 300 mm (11.8 in); length 3600 m (141.73 in); span 800 mm (31.50 in)
Weights: total round 305 kg (672.4 lb); warhead unrevealed proximity-fused Forenarde Fabriksverken blast fragmentation
Propulsion: one Volvo Flygmotor VR-35 dual-thrust prepackaged liquid-propellant rocket delivering 2500 kg (5,511 lb) thrust in the boost phase and 510 kg (1,124 lb) thrust in the sustain phase
Performance: speed Mach 1+; range 9000 m (9,845 yards); CEP less than 10 m (33 ft)
Guidance: radio command

Variant and application
RB 05A: Introduced in the early 1970s, the RB 05A is a fairly large but highly capable air-to-surface missile with genuinely supersonic performance and intended for use against land and sea targets. In configuration the missile is based on a finely streamlined body with low-aspect-ratio delta wings and, just aft of these, slightly swept control surfaces indexed in line with the flying surfaces. The missile is launched at an altitude between 20 and 50 m (65 and 165 ft), rapidly accelerates to a position dead ahead of the launch aircraft and then climbs to some 400 m (1,315 ft) as the operator uses a jam-resistant microwave link to guide it to the target, where the blast fragmentation warhead is detonated by a proximity fuse. The missile can be carried by all Swedish attack aircraft, including the small Saab 105 (SK60).

USSR

AS-7 'Kerry'

Type: air-to-surface tactical missile
Dimensions: diameter 300 mm (11.81 in); length 3500 mm (137.8 in); span 950 mm (37.4 in)
Weights: total round 400 kg (881.8 lb); warhead 100 kg (220.5 lb) blast fragmentation
Propulsion: one solid-propellant rocket delivering unrevealed thrust
Performance: speed Mach 1; range 11100 m (12,140 yards)
Guidance: radio command, or (according to recent Western assessments) beam riding

Variant and application
AS-7: This was apparently the USSR's initial genuine air-to-surface missile even though it first appeared only in the late 1970s, earlier Soviet types having been modified from air-to-air missiles or perhaps the USSR's larger-calibre air-to-surface rockets. The AS-7 is a more capable weapon roughly comparable to the American AGM-12 Bullpup series. Recent thinking about the AS-7 has indicated that the type may have beam-riding rather than radio-command guidance, but has not revealed whether this is of the radar or laser beam-riding type, either of which would make the AS-7 a far more capable weapon than at first thought. It should be noted, however, that either type of beam-riding has clear disadvantages for the launch aircraft over the battlefield.

The missile has the Soviet service name **Grom,** but its service and production designations remain unknown. The weapon can be carried by several types of Soviet tactical aircraft, but the most common launch platforms are believed to be the Mikoyan-Gurevich MiG-27 'Flogger', the Sukhoi Su-17 'Fitter', Su-24 'Fencer' and Su-25 'Frogfoot', and the Yakovlev Yak-38 'Forger'.

AS-9 'Kyle' USSR

Type: air-to-surface anti-radiation tactical missile
Dimensions: (estimated) diameter 500 mm (19.69 in); length 6000 mm (236.22 in); span data not available
Weights: total round about 650 kg (1,433.0 lb), warhead 150 or 200 kg (331.7 or 440.9 lb) blast fragmentation
Propulsion: one solid-propellant rocket delivering unrevealed thrust
Performance: speed Mach 0.8; range 80/90 km (49.7/55.9 miles)
Guidance: passive radar homing

Variants and application
AS-9 'Kyle': Virtually nothing is known of this long-range anti-radar missile designed for use on Soviet tactical aircraft, most notably the Sukhoi Su-24 'Fencer'. Some Western analysts believe that the type is powered by a small turbojet rather than by a rocket and to have supersonic performance to a range of more than 100 km (62.1 miles).
AS-11: The weapon originally thought to have borne this designation is now known as the AS-14, and the AS-11 may thus be a variant of the AS-9 with a different homing system.
AS-12 'Kegler': Virtually nothing is known of this weapon, but some reports suggest that it may be another AS-9 variant with a different homing system. Another possibility is that it is a tactical dispenser weapon.

AS-10 'Karen' USSR

Type: air-to-surface tactical missile
Dimensions: diameter 300 mm (11.81 in); length 3500 mm (137.8 in); span data not available
Weights: total round about 400 kg (881.2 lb); warhead about 100 kg (220.5 lb) blast fragmentation
Propulsion: one solid-propellant rocket delivering unrevealed thrust
Performance: speed Mach 0.9; range 10000 m (10,935 yards)
Guidance: electro-optical or (possibly) semi-active laser homing

Variant and application
AS-10 'Karen': This EO- or laser-homing weapon forms an important part of the inventory for the Mikoyan-Gurevich MiG-27 'Flogger' and Sukhoi Su-17 'Fitter' attack fighters, and for the Sukhoi Su-24 'Fencer' interdictor. As with most Soviet air-to-surface missiles, hard information is woefully lacking, but there are indications that the type may be a derivative of the AS-7 'Kerry' with precision guidance.

USSR

AS-14 'Kedge'

Type: air-to-surface tactical missile
Dimensions: diameter 450 mm (17.72 in); length about 4500 mm (177.17 in); span 1300 mm (51.18 in)
Weights: total round about 600 kg (1,322.75 lb); warhead about 300 kg (661.4 lb) blast fragmentation
Propulsion: one solid-propellant rocket delivering unrevealed thrust
Performance: speed transonic; range 40 km (25 miles)
Guidance: semi-active laser homing

Variant and application
AS-14 'Kedge': Virtually nothing is known of this tactical air-launched missile, which is a precision attack weapon associated with the Mikoyan-Gurevich MiG-27 'Flogger' and other tactical aircraft. The missile has a rear-mounted cruciform of cropped delta wings, and a forward-mounted cruciform of smaller control surfaces indexed in line with the wings. It is believed that the illuminating laser is carried in a pod located under the fuselage or wing.

The missile appears to be an enlarged version of the AS-10, and recent Western analysis suggests that the type may have data-linked mid-course guidance and electro-optical terminal guidance. The missile has a rear-mounted cruciform of cropped delta wings, and a forward-mounted cruciform of smaller control surfaces indexed in line with the wings.

UK

BAe/Marconi ALARM

Type: air-to-surface anti-radiation tactical missile
Dimensions: diameter 8.66 in (220 mm); length 167.0 in (4242 mm); span 28.33 in (720 mm)

The BAe ALARM is currently the world's most advanced and versatile anti-radar missile, and can be carried by a wide assortment of tactical aircraft. The missile is generally associated with advanced combat types such as the Panavia Tornado, but can also be carried by considerably smaller types such as this Westland Lynx HAS.Mk 2/3 naval helicopter, whose Racal MIR-2 'Orange Crop' electronic support measures suite can provide the ALARM with useful pre-launch data on the target emitter's operating frequency and bearing.

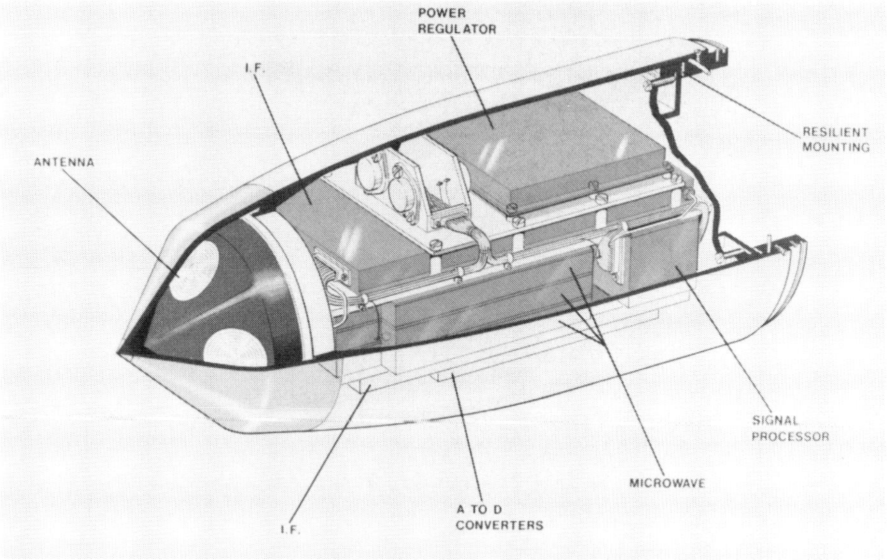

Left *Key to the ALARM anti-radar missile's capability is the sophisticated seeker section provided by Marconi. It has a four-sided antenna section inside the ogival nose, this section carrying four spiral helix broad-band antennae. Signals from the antenna unit are fed to the digital signal processor, which uses modern large-scale integration techniques and inflight-reprogrammable software to cater for different possible targets and assign targeting priorities.*

Below *A manufacturer's illustration shows the ALARM anti-radar missile's loiter operating mode.*

Weights: total round 385 lb (174.6 kg); warhead unrevealed Thorn-EMI laser proximity-fused MBB HE
Propulsion: one Bayern Chemie two-stage solid-propellant rocket delivering unrevealed thrust
Performance: not revealed
Guidance: Marconi passive radar seeking

Variant and application

ALARM: Based aerodynamically on the Sky Flash air-to-air missile, the ALARM (**Air-Launched Anti-Radiation Missile**) is due to enter service in the late 1980s as the primary weapon of British tactical aircraft operating against hostile radars (both land-based and shipborne). The weapon has an extremely advanced and flexible Marconi seeker unit with four spiral helix broad-band receiver antennae (one in each side of the four-sided nose unit), and a computer system that can be pre-programmed with the characteristics of enemy radars likely to be encountered, the target priority list and the operating modes. The seeker system also allows the ALARM to home on radars which have shut down in an effort to break the missile's lock.

The weapon has several operating modes, of which only two have been revealed. In the first the weapon is fired and immediately locks onto a target radar and homes in the standard fashion. In the second the missile is designed to zoom to 40,000 ft (12190 m) after a low-level launch in the vicinity of any possible target, thereupon descending slowly in a nose-down attitude under a small drogue parachute (offering a loiter of several minutes) as the seeker searches for hostile emissions: once a target has been selected, the drogue is discarded and the unpowered missile dives onto the target. In each case the terminal phase of the attack is made 'down the throat' of the hostile radar system, the object being a kinetic and explosive 'kill' of the antenna and other components.

The missile requires no input from the launch aircraft other than an electrical connection for battery fire and motor fire as the weapon is launched. The ALARM is thus a fire-and-forget missile which can be carried by several types of tactical aircraft, the primary British carriers being the BAe Harrier, Panavia Tornado and SEPECAT Jaguar.

Launch of an ALARM anti-radar missile from a Panavia Tornado. At this point the missile chooses between a direct attack on an emitting target, or alternatively a climb to high altitude where a parachute is deployed to give the missile a loiter period so that the terrain below can be searched for a target of opportunity.

The missile was originally to have been powered by a Royal Ordnance Nuthatch rocket, but problems with the development of this single-chamber/dual-grain powerplant in August 1987 caused BAe to revise the design for a simpler dual-chamber rocket, whose integration will inevitably cause delays to the programme.

UK

BAe/GEC/Shorts AALAAW

Type: air-launched tactical anti-armour weapon
Dimensions: cross section 12.6 in (320 mm) square; length about 78.75 in (2000 mm); span about 39.4 in (1001 mm)
Weights: not revealed
Propulsion: none
Performance: speed high subsonic; range 8,800 yards (8045 m)
Guidance: autopilot

The AALAAW (Advanced Air-Launched Anti-Armour Weapon) is still very much at the conceptual stage, being developed by BAe, Shorts and GEC in response to a Ministry of Defence (Air) requirement for a stand-off dispenser to release a useful number of advanced submunitions with millimetre-wave radar terminal guidance. Despite the dispenser's distinctly bluff lines a surprisingly high performance is estimated.

Variant and application

AALAAW: Known so far only as the **Advanced Air-Launched Anti-Armour Weapon** or as Staff Requirement (Air).1238 for use by tactical aircraft, this system has reached only the conceptual stage and the British partners are progressing no further pending the arrival of international collaboration. The present design idea offers a gliding dispenser with a box-like fuselage, finned aft-mounted wings and twin fins at the rear to carry a substantial number of terminally-guided submunitions with millimetre-wavelength seekers and powerful shaped-charge warheads.

BAe/Matra Martel and ARMAT (UK/France)

Type: (AJ.168) air-to-surface tactical missile or (AS.37 and ARMAT) air-to-surface anti-radiation tactical missile
Dimensions: diameter 15.75 in (400 mm); length (AJ.168) 152.4 in (3871 mm) or (AS.37) 4120 mm (162.2 in); span 47.25 in (1200 mm)
Weights: total round (AJ.168) 1,213 lb (550.2 kg) or (AS.37) 530 kg (1,168.4 lb); warhead 330 lb (150 kg) proximity-fused (AS.37) or impact-fused (AJ.168) blast fragmentation
Propulsion: (AS.37) one Hotchkiss-Brandt/SNPE Basile composite solid-propellant booster rocket delivering unrevealed thrust, and one Hotchkiss-Brandt/SNPE Cassandre composite solid-propellant sustainer rocket delivering unrevealed thrust, or (AJ.168) one SNPE composite boost and cast double-base sustainer rocket delivering unrevealed thrust
Performance: speed Mach 2; range 60 km (37.3 miles) from a high-altitude launch declining to 30 km (18.6 miles) from a low-altitude launch
Guidance: (AJ.168) Marconi TV command or (AS.37) ESD AD37 passive radiation seeker

Variants and applications

AJ.168 Martel: The Martel **(MIssile Anti-Radar TELevision)** grew from independent British (Hawker Siddeley Dynamics) and French (Nord) studies in the period 1960-3, and was then developed by France and the UK as one of the first European collabora-

Developed jointly by Matra in France and Hawker Siddeley (now BAe) Dynamics in the UK, the Martel is an underestimated weapon that had, for its time, exceptional range. The French use only the AS.37 with Electronique Serge Dassault passive radiation homing, while the British use both this variant and the AJ.168 with Marconi TV guidance. Seen here *coni TV guidance. Seen here* *under the outer hardpoints of a BAe Buccaneer S.Mk 2B are a pair of the AS.37 variant with, inboard of them, a pair of 430-Imp gal (1955-litre) slipper tanks. The AJ.168 version of the Martel requires a data-link pod, and this generally occupies one of the four underwing hardpoints.*

The detonation of an AS.37 Martel's warhead provides graphic illustration of the destructive potential of this elderly but still potent anti-radar missile, now developed into the ARMAT with a more modern seeker and other improvements.

tive programmes for service introduction in the late 1960s. The British AJ.168 version of the Martel uses command guidance: a small TV camera in the nose of the missile feeds its picture to a screen in front of the operator, who fixes on the target and locks the image into the missile before launching it. The missile then flies under autopilot and barometric altimeter control, being steered by the operator via the data-link pod carried under the launch aircraft's wing. The system provides great accuracy for the decisive placement of the substantial warhead. The primary carrier of the missile is the BAe Buccaneer, which can carry three missiles in addition to the data-link pod.
AS.37 Martel: The French version of the Martel is designed for anti-radiation use with a broad-band passive seeker. This can be used in the air to search through pre-set frequencies for a suitable target (whose operating frequency is then locked into the seeker as the antenna searches through an angle of 90° to localize the target), or locked before take-off if a specific target is to be engaged.
ARMAT: Developed by Matra on the basis of the Martel airframe, this **Anti-Radar MATra** weapon is France's new anti-radiation missile, differing from the AS.37 that it is replacing in having a length of 4150 mm (163.39 in), a span of 1200 mm (47.24 in), a weight of 550 kg (1,212.5 lb) including the 160 kg (352.7 lb) HE warhead, a range of between 15 and 120 km (9.3 and 74.6 miles) depending on launch altitude and flight profile, a higher-impulse rocket motor and a more modern and capable ESD passive radar seeker.

USA

Martin Marietta AGM-12C Bullpup B

Type: air-to-surface tactical missile
Dimensions: diameter 18.0 in (457 mm); length 163.0 in (4140mm); span 48.0 in (1219mm)
Weights: total round 1,785 lb (809.7 kg); warhead 1,1000 lb (454 kg) impact delay-fused blast fragmentation
Propulsion: one Thiokol LR62-RM-2/4 pre-packaged liquid-propellant rocket delivering unrevealed thrust
Performance: speed Mach 1.75; range 17,600 yards (16095 m)

Guidance: radio command to line of sight

Variant and application

AGM-12B Bullpup: The Bullpup was developed to meet a US Navy requirement perceived during the Korean War for a stand-off precision attack weapon to arm tactical aircraft. The weapon was planned round an existing warhead, the 250 lb (113 kg) general-purpose bomb, with an Aerojet General solid-propellant rocket and pneumatically-activated cruciform nose-mounted delta control surfaces (controlled by radio from the launch aircraft) and larger tail-mounted delta wings. The guidance was simple, requiring the operator to keep the missile's tracking flares superimposed over the image of the target.

The original ASM-N-7A was superseded in 1960 by the ASM-N-7B (in 1962 redesignated AGM-12B) with a 12,000 lb (5443 kg) thrust Thiokol LR58 pre-packaged liquid-propellant rocket, 12,000 yard (10975 m) range and a revised warhead. This weapon has a weight of 571 lb (259 kg), a length of 126.0 in (3200 mm) and span of 37.0 in (940 mm), and despite its obsolescence is still in limited service. Though restricted in terms of its guidance system (requiring the aircraft to loiter in the vicinity of the target until weapon impact), it still packs a powerful punch in areas of reduced air-defence capability.

AGM-12C Bullpup B: This is an improved version with 1,000 lb (454 kg) warhead, wider-chord wings and LR62 liquid-propellant rocket.

Texas Instruments AGM-45A Shrike

USA

Type: air-to-surface anti-radiation tactical missile
Dimensions: diameter 8.0 in (203 mm); length 120.0 in (3048 mm); span 36.0 in (914 mm)
Weights: total round 390 lb (176.9 kg); warhead 145 lb (65.8 kg) impact and proximity-fused blast fragmentation
Propulsion: one Rocketdyne Mk 39 or Aerojet Mk 53 solid-propellant rocket delivering unrevealed thrust
Performance: speed Mach 2; range varies with launch altitude and speed from 18 to 25 miles (28.95 to 40.25 km)
Guidance: Texas Instruments passive radar seeking

Variants and application

AGM-45A Shrike: Developed as the ASM-N-10 to meet a US Navy requirement for a simple weapon able to 'take out' enemy land-based radars, the Shrike was introduced in 1963 as the USA's first tactical anti-radiation missile. The weapon was produced in 10 blocks (AGM-45-1 to AGM-45-10), and there are at least 13 different seeker units tailored to individual radar types. This has proved a tactical disadvantage (requiring specific tuning before take-off and thus preventing attacks on targets of opportunity), as does the seeker's lack of inbuilt memory (resulting in a ballistic trajectory if the emitter shuts down). Detection and localization of the radar to be engaged is undertaken by the launch aircraft's own electronic countermeasures or electronic support measures systems, the Shrike's seeker being activated only when the operator knows the weapon

is within the seeker's frequency capabilities and range.
AGM-45B Shrike: This is an improved AGM-45A with the Aerojet Mk 78 solid-propellant rocket, but retaining the standard configuration of the AGM-45A with a mid-mounted cruciform of high-aspect-ratio control surfaces and a tail mounted cruciform of low-aspect-ratio fixed fins. The Shrike series is carried by virtually the whole range of US and Israeli tactical aircraft.

USA

Hughes/Martin Marietta AGM-62A Walleye I

Type: air-to-surface tactical glide bomb
Dimensions: diameter 12.5 in (318 mm); length 136.0 ft (3454 mm); span 45.5 in (1156 mm)
Weights: total round 1,130 lb (512.6 kg); warhead 825 lb (374.2 kg) shaped-charge HE
Propulsion: none
Performance: speed high subsonic; range 1,975/32,400 yards (1805/29625 m) depending on launch altitude
Guidance: electro-optical imaging with nose-mounted TV

Variants and application
AGM-62A Walleye I: Developed in the mid-1960s by the Naval Ordnance Test Station (now the Naval Weapons Center), the Walleye I entered service in 1966 and was made by Hughes and Martin Marietta. The type was designed to provide tactical attack aircraft with the capability for stand-off attacks from heights up to 35,000 ft (10670 m). The system is optimized for use against high-contrast buildings and bridges, whose optical images are easy to pick up and lock into the nose-mounted gyro-stabilized TV camera, allowing the launch aircraft to drop the weapon and depart the area as the missile automatically homes onto its target. The operator in the launch aircraft monitors the progress of the missile and can break into the automatic homing system to redirect the weapon if necessary. The Walleye I's CEP is between 15 and 20 ft (4.6 and 6.1 m).
AGM-62A Walleye I ER/DL: This is the extended-range **Mk 13 Mod 0** variant of the late 1970s with slightly reduced length but fitted with data-link equipment so that the weapon can be dropped in the general direction of the target and have its guidance system locked onto the clear image of the target (by the departing operator) only when the weapon has approached to moderately close range.

USA

Martin Marietta AGM-62A Walleye II

Type: air-to-surface tactical glide bomb
Dimensions: diameter 18.0 in (457 mm); length 159.0 in (4039 mm); span 51.0 in (1295 mm)

Weights: total round 2,340 lb (1061.4 kg); warhead 1,900 lb (861.8 kg) shaped-charge HE
Propulsion: none
Performance: speed high subsonic; range 1,975/49,300 yards (1805/45080 m) depending on launch altitude
Guidance: optical imaging with nose-mounted TV

Variants and application
AGM-62A Walleye II: Developed to overcome the Walleye I's limitations against hard targets, the Walleye II (otherwise **Mk 5 Mod 4**) was introduced in 1974 and has a considerably larger warhead (derived from that of the Mk 84 free-fall bomb) and a revised seeker with a smaller optical gate for increased accuracy. Built by Martin Marietta, the Walleye II requires a target with sharp visual contrasts for good lock-on before launch. The Walleye II's CEP is between 15 and 20 ft (4.6 and 6.1 m).
AGM-62A Walleye II ER/DL: Introduced in 1976, the Extended-Range/Data-Link model was developed so that greater stand-off range can be ensured. The weapon is dropped from high altitude to extend range to 65,500 yards (59895 m), and is locked onto the target late in the flight using the two-way data-link system to launch aircraft, where the operator monitors the TV image in the weapon's guidance system.

Hughes AGM-65 Maverick — USA

Type: air-to-surface multi-role tactical missile
Dimensions: diameter 12.0 in (305 mm); length 98.0 in (2489 mm); span 28.3 in (719 mm)
Weights: total round (AGM-65A) 463 lb (210.0 kg), (AGM-65D) 485 lb (220.0 kg), (AGM-65E/F) 667 lb (302.6 kg) and (AGM-65G) 670 lb (303.9 kg); warhead (AGM-65A/B/D) 125 lb (56.7 kg) Chamberlain shaped-charge HE, (AGM-65C/F) 250 lb (113.4 kg) Mk 19 blast fragmentation, or (AGM-65E/G) 300 lb (136.1 kg) Avco penetrating blast fragmentation
Propulsion: one Thiokol SR109-TC-1 (TX-481) two-stage solid-propellant rocket (or from 1981 one Thiokol TX-633 or Aerojet SR115-AJ-1 reduced-smoke solid-propellant rocket) delivering unrevealed thrust
Performance: speed supersonic; range 985/26,400 yards (900/24120 m)
Guidance: (AGM-65A and AGM-65B) TV imaging, (AGM-65C and AGM-65E) semi-active laser homing, or (AGM-65D, AGM-65F and AGM-65G) infra-red imaging

Variants and application
AGM-65A Maverick: Development of this weapon began in 1965, and it was introduced in 1972 as a US Air Force air-to-surface missile. The Maverick series adheres to the standard configuration of Hughes-developed missiles, with a substantial cylindrical body, a cruciform of swept low-aspect-ratio delta wings, and a cruciform of rectangular control surfaces close-coupled to the rear of the wings. The AGM-65 series is the smallest fully-guided air-to-surface missile family in the US inventory, and certainly the West's most important weapons of the type, largely because in all its versions it is a fire-and-forget type. The AGM-65A is the initial TV imaging version (often known as the **TV Maverick**), and suffers the tactical disadvantage of low magnification for its nose-mounted camera, forcing the pilot of the launch aircraft to fly close to the target to secure lock on before missile launch and subsequent automatic attack. The shaped-charge warhead

contains 83 lb (37.6 kg) of explosive. The CEP of the AGM-65A (and of the AGM-65D) is in the order of 5 ft (1.5 m).

AGM-65B Maverick: This is an improved TV imaging model introduced in 1980 with double the image magnification of the AGM-65A (based on a 2.5° field of view) to overcome the earlier version's tactical disadvantages. The type is often known as the **Scene-Magnification Maverick**.

AGM-65C Maverick: This was a highly capable development designed for the battlefield close support role using Rockwell laser homing compatible with the ground-based ILS-NT200 as well as the aerial 'Pave Knife', 'Pave Penny', 'Pave Spike', 'Pave Tack' and several non-US designation systems. The weapon was known as the **Laser Maverick**, but was cancelled before reaching full service with the US Marine Corps in the close support role against targets generally designated by ground-based laser.

AGM-65D Maverick: This is an imaging infra-red version introduced to USAF service in October 1983, possessing twice the lock-on range of the AGM-65A/B types, and capable of operation in adverse weather conditions and at night. The type is known as the **Imaging Infra-Red Maverick** and uses a Hughes IR seeker designed for deployment in conjunction with FLIR (Forward-Looking Infra-Red) systems or with the LANTIRN (Low-Altitude Navigation and Targeting Infra-Red for Night) pod carried by Fairchild Republic A-10 Thunderbolt II close-support and anti-tank aircraft and General Dynamics F-16 Fighting Falcon fighters in the European theatre, or indeed with the APR-38 radar-warning system of aircraft such as the McDonnell Douglas F-4 Phantom II 'Wild Weasel II' defence-suppression fighter. The AGM-65D is one of the most expen-

The AGM-65 Maverick is generally carried on a special triple installation such as this example carried on the port inner underwing hardpoint of a McDonnell Douglas F-4 Phantom II. Here a US Air Force crewman checks that all is well with the attachment of an AGM-65D Maverick with an imaging infra-red seeker.

sive Maverick variants, but with exceptional range and target-acquisition capabilities is of considerable value for the destruction of high value targets.

AGM-65E Maverick: This **Laser Maverick** model was introduced in 1985 as successor to the cancelled AGM-65C with a less costly Hughes tri-service laser seeker and digital processing, as well as a more powerful 300 lb (136.1 kg) warhead with impact or delay fuses. The CEP of the AGM-65E (and of the AGM-65F) is less than 5 ft (1.5 m), and the main operating service is the US Marine Corps.

AGM-65F Maverick: This is the **Navy Maverick** version for the US Navy combining the airframe and imaging infra-red guidance of the AGM-65D with the 250 lb (113 kg) Mk 19 blast fragmentation warhead and the fuse of the AGM-65E plus software changes to optimize the weapon's capabilities against warship targets. The tactical aircraft most often associated with this variant are the McDonnell Douglas F/A-18 Hornet and Grumman A-6 Intruder.

AGM-65G Maverick: This is the USAF equivalent of the AGM-65F with the 300 lb (136.1 kg) warhead, and is designed for the destruction of 'hard' tactical targets such as bunkers and hardened aircraft shelters, though retaining a capability against tanks. The variant also pioneers a pneumatic actuation system and a new digital autopilot, substantially reducing the cost of this important missile. Other features are a tracking mode that allows the operator to select, before launching the missile, the region of a large target on which the missile should impact, and a lower-altitude trajectory to minimize the possibility of breaking lock in cloud. It has also been suggested that the type will be further developed with a small tactical nuclear (fission) warhead, the need for such a weapon having been expressed by the US Air Force as long ago as 1976.

A US Marine Corps crew chief checks that an AGM-65E Maverick air-to-surface missile is correctly installed on the inboard underwing hardpoint of a BAe AV-8A Harrier close support aircraft. This is a very capable weapon which can be used in conjunction with third-party targeting: a hand-held laser designator provides capability against large or static targets, but a stabilized designator is needed if a point or moving target is to be hit.

Boeing AGM-69A SRAM

Type: air-to-surface operational/strategic missile
Dimensions: diameter 17.5 in (445 mm); length (internal carriage) 168.0 in (4267 mm) or (external carriage) 190.0 in (4826 m); span about 30.0 in (762 mm), given that the three fins are disposed at 120° to each other round the tail with each tip 15.0 in (381 mm) from the centreline
Weights: total round 2,240 lb (1016.1 kg); warhead 300 lb (136 kg) 170/200-kiloton W69 nuclear with plutonium as the fissile material
Propulsion: one Lockheed SR75-LP-1 two-stage solid-propellant rocket delivering unrevealed thrust
Performance: speed Mach 2.8 to 3.2; range 100/137.5 miles (160.9/221.3 km) from a high-altitude launch declining to 35/50 miles (56.3/80.5 km) from a low-altitude launch.
Guidance: Singer-Kearfott KT-76 inertial plus terrain-avoidance radar altimeter.

Variants and application
AGM-69A SRAM: Development of a **Short-Range Attack Missile** to be carried by fighters was initiated in the late 1950s, reflecting the decade's advances in guidance and warhead miniaturization. But the current weapon entered design in 1963 to give larger aircraft a genuine defence-suppression capability of an operational/strategic nature. The type has exceptionally clean lines, without wings as its hypersonic flight performance allows the type to use body lift, but with three low-aspect-ratio tail surfaces indexed at 120° to each other.

The SRAM entered service in 1972 as part of the primary armament for Boeing B-52 Stratofortress heavy bombers (20 SRAMs or eight SRAMs and four free-fall nuclear weapons) and General Dynamics FB-111 medium bombers (six SRAMs or two SRAMs and four free-fall nuclear weapons) as a replacement for the North American AGM-28 Hound Dog stand-off missile. The role at first envisaged for the weapon was stand-off attack against primary defensive installations, but the AGM-69A is now tasked with defence-suppression tasks in association with the launch of Boeing AGM-86B air-launched cruise missiles and the delivery of gravity nuclear weapons.

The SRAM has four basic attack profiles: semi-ballistic from launch to impact; altimeter-supervised terrain-following; ballistic pop-up from screening followed by terrain-following; and combined terrain-following and inertial. Any of these profiles can include preprogrammed 180° course changes, and the warhead can be set for ground- or air-burst. The CEP of the AGM-69A is classified, a figure of 500 yards (457 m) often being quoted, though other sources claim a figure as low as 100 yards (91.4 m). Surviving missiles are being re-engined with a new Thiokol liquid-propellant rocket developed for the upgraded AGM-69B variant that was cancelled in 1977. Since the proposed advent of the SRAM II, the original weapon has frequently been designated **SRAM-A.**
AGM-131A SRAM II: In December 1986 Boeing was selected to develop a new SRAM missile instead of the proposed ASALM (Advanced Strategic Air-Launched Missile). The SRAM II is planned as a very small weapon (only about two-thirds the size of the SRAM-A) using an airframe of largely composite construction as well as an advanced laser ring-gyro inertial navigation system based on the latest VHSIC (Very High Speed Integrated Circuit) technology. The SRAM II will have an advanced two-pulse solid-propellant rocket engine, and the 150/175-kiloton thermonuclear warhead will have capability against hardened defence targets in advance of manned penetration bombers. Like these, the SRAM II is planned as a low-observable or 'stealth' type. The B-1B will be

able to carry 20 SRAM IIs internally on the Common Strategic Rotary Launcher, and the Northrop B-2 will also carry the type. The SRAM II is due to enter service in 1992.

General Dynamics AGM-78B Standard ARM

USA

Type: air-to-surface anti-radiation tactical missile
Dimensions: diameter 13.5 in (343 mm); length 180.0 in (4572 mm); span 43.0 in (1092 mm)
Weights: total round 1,356 lb (615.1 kg); warhead 214.7 lb (97.4 kg) blast fragmentation
Propulsion: one Aerojet Mk 27 Mod 4 dual-thrust solid-propellant rocket delivering unrevealed thrust
Performance: speed Mach 2.5; range varies with launch altitude and speed to a maximum of 56 miles (90 km)
Guidance: Maxson Electronics passive radar seeking

Variants and application

AGM-78A Standard ARM: Derived from the RIM-66 Standard surface-to-air missile (and thus featuring the same configuration with very low-aspect-ratio cruciform wings trailed by larger-span control surfaces indexed in line with the wings), the US Navy's AGM-78A was introduced in 1968 as a longer-range complement to the AGM-45A Shrike. However, like its predecessor, this initial Standard ARM variant uses the same no-memory Texas Instruments seeker with its attendant tactical limitations.

AGM-78B Standard ARM: This is a much improved model with a gimballed wide-band seeker plus memory. Such a seeker requires no pre-launch tuning (making it possible to engage targets of opportunity) and keeps the missile on its committed course even if the emitter shuts down. The missile achieves its best results with naval aircraft carrying the TIAS (Target Identification and Acquisition System) and US Air Force's McDonnell Douglas F-4G 'Wild Weasel' defence-suppression aircraft, whose APR-38 emitter-location system can fulfil the same role as the TIAS in supplying the seeker head with emitter information before launch.

Operating on the 'Wild Weasel' defence-suppression role, this McDonnell Douglas F-4G Phantom II reveals some of the specialist equipment and weapon of this important type. Internal electronics are centred on the APR-38 radar-detection and homing system (with its antennae in the fintop and undernose fairings), external electronics are in this instance the ALQ-131 jammer pod under the fuselage, and the weapons are (from starboard to port) the AGM-45A Shrike, the AGM-78A Standard ARM, the AGM-65 Maverick and the AGM-88A HARM.

AGM-78C Standard ARM: This was a proposed upgraded version of the AGM-78B for the US Air Force and US Navy. Some AGM-78A and AGM-78B missiles have been upgraded to this standard with field modification kits.
AGM-78D Standard ARM: This was a proposed version (together with an AGM-78D2 subvariant) powered by the Mk 69 Mod 0 solid-propellant rocket and marked by further improvement of seeker capabilities.

USA

Texas Instruments AGM-88A HARM

Type: air-to-surface anti-radiation tactical missile
Dimensions: diameter 10.0 in (254 mm); length 164.2 in (4171 mm); span 44.0 in (1118 m)
Weights: total round 796 lb (361.1 kg); warhead 145 lb (65.8 kg) laser proximity-fused blast fragmentation
Propulsion: one Thiokol/Hercules YSR113-TC-1 solid-propellant rocket delivering unrevealed thrust
Performance: speed Mach 3+; range 46+ miles (74+ km)
Guidance: Texas Instruments passive radiation seeking

Variant and application
AGM-88A HARM: The **High-speed Anti-Radiation Missile** began to enter US service in 1983 as successor to the tactically-limited AGM-45 Shrike and the costly (also weighty) AGM-78 Standard ARM. HARM is still a substantial missile, with a large cylindrical body, a cruciform of cropped double-delta control surfaces at the mid-point, and a cruciform of fixed low-aspect-ratio stabilizing fins at the rear, indexed in line with the control surfaces.

The type offers good passive radiation homing with detection from the launch aircraft's sensor suite, in the form of the Itek ALR-45 or McDonnell Douglas APR-38

An AGM-88A HARM (High-speed Anti-Radiation Missile) is seen during trials under a Vought A-7 Corsair II at the US Navy's Point Mugu Test Center on the US Pacific Coast. The HARM is a capable weapon of its type, but both expensive and heavy in comparison with the British ALARM.

radar-warning receivers in US Air Force aircraft, and the AWG-25 equipments of the US Navy's Vought A-7 Corsair II. The missile's high speed also gives the hostile emitter minimum close-down time, and one of the HARM's greatest advantages over the Shrike is the use of a single broad-band seeker rather than a choice of individually tailored seekers, allowing the single seeker unit to be used against land-based and shipborne air-defence radars likely to be encountered by American tactical aircraft.

Two other improvements are a continued homing capability even after the target emitter has been closed down, and signal processing under control of software rather than hardware, allowing the possibility of role expansion to cover the whole range of radars likely to be encountered. The HARM can be used in any of three modes: self-protection mode when the launch aircraft's radar-warning receiver detects a hostile emitter and programmes the missile before launch; pre-briefed mode with the missile fired blind towards a possible target with the seeker searching in flight and commanding a self-destruct if no target is discovered; and target-of-opportunity mode when the seeker of the unfired missile detects and locks onto an emitter.

The aircraft qualified with the weapon are the Grumman A-6E Intruder, Vought A-7 Corsair II, Grumman EA-6B Prowler, and McDonnell Douglas F-4 Phantom and F/A-18 Hornet, and it is likely that the General Electric/Grumman EF-111 will also be adapted for the type, as will the Panavia Tornado ECR.

Rockwell AGM-114A Hellfire USA

Type: air-to-surface tactical missile (optimized for the anti-tank role)
Dimensions: diameter 7.0 in (178 mm); length 64.0 in (1626 m); span 13.0 in (330 mm)
Weights: total round 98.86 lb (44.84 kg); warhead 20 lb (9.1 kg) hollow-charge HE.
Propulsion: one Thiokol TX-657 solid-propellant rocket delivering unrevealed thrust
Performance: speed high subsonic; range 6,500 yards (5945 m)
Guidance: autopilot for midcourse phase, and Martin Marietta semi-active laser homing for the terminal phase

Rapidly superseding the TOW as the US Army's premier helicopter-launched anti-tank missile, the AGM-114A Hellfire is a more versatile weapon without the range limitations imposed by the TOW's wire guidance, and the very distinct tactical advantage of third-party targeting in association with the Hellfire's semi-active laser guidance. This is a Hellfier launch from a McDonnell Douglas AH-64A Apache battlefield helicopter, which can carry up to 16 AGM-114As in four quadruple packages.

Manufacturer's artwork details the key features of the launcher used with the AGM-114A Hellfire laser-homing air-to-surface and anti-tank missile.

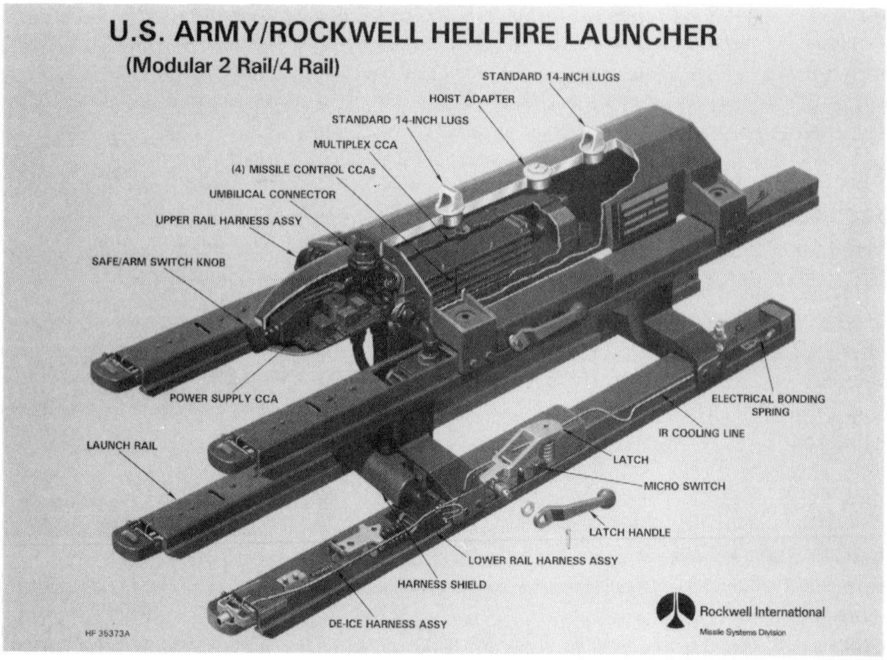

Variants and applications

AGM-114A Hellfire: This is an impressive weapon that opens up the possibility of fire-and-forget helicopter attacks against armoured formations, the missile having proved itself able to home on the reflections of tanks laser-illuminated by either aerial or ground-based designators. The target need only be illuminated once the missile has been launched, thereby reducing the time available for the target to undertake defensive measures and also reducing the launch platform's vulnerability to countermeasures. The configuration of the missile is conditioned largely by the diameter of the warhead, which has to be as large as possible for optimum hollow-charge effect: this results in a portly missile with a cruciform of low-aspect-ratio aft-mounted wings and four trapezoidal control canards indexed in line with the wings.

Rockwell is currently investigating more advanced versions of the missile. In the short term the two most likely developments are the hardening of the seeker section to provide greater resistance to electronic countermeasures, and a pulse rocket propulsion system to boot range performance by as much as 50%. In the longer term Rockwell is working with Marconi Defence Systems of the UK to introduce a Brimfire derivative with millimetre-wave radar guidance. This would be particularly useful on European battlefields as it offers all-weather capability as well as the possibility of fire-and-forget rapid reaction, covert launch capability, seek-and-destroy capability with lock-on only after launch, all-aspect attack capability, and greater capability for multiple target engagements.

AGM-114B Hellfire: This is the version of the AGM-114A for the US Navy and US Marine Corps with a different motor and homing guidance provided by a choice of three different seeker modules.

AGM-114C Hellfire: This is the US Army version of the AGM-114B without the safe arm feature required for US Navy shipboard storage.

RBS 17: Due to enter service by 1988, this is a joint US and Swedish development of the Hellfire for coastal defence purposes with the Swedish army. The missile has been fitted with a Bofors delayed-action blast fragmentation warhead suitable for attacks on landing craft and assault hovercraft, and with revisions to the seeker head weighs a total of 48 kg (106 lb) and has a length of 1626 mm (64.0 in). The missile is transported in a special container which has a loaded weight of 71 kg (156.5 lb), and is fired from a light tripod launcher with the aid of a laser designator.

Motorola AGM-122A Sidearm USA

Type: air-to-surface anti-radiation tactical missile
Dimensions: diameter 5.0 in (127 mm); length 113.0 in (2871 mm); span 24.8 in (630 mm)
Weights: total round 195 lb (88.5 kg); warhead 10 lb (4.54 kg) blast fragmentation
Propulsion: one Rocketdyne Mk 36 Mod 2 solid-propellant rocket delivering unrevealed thrust
Performance: speed Mach 2.3; range 4,000 yards (3660 m)
Guidance: Motorola passive radar seeking.

Variant and application
AGM-122A Sidearm: Currently entering service on US Marine Corps' Bell AH-1 SeaCobra attack helicopters and McDonnell Douglas/BAe AV-8 Harrier STOVL close-support aircraft, the Sidearm is a low-cost self-protection anti-radiation weapon for helicopters and light tactical aircraft. The weapon is a conversion of the semi-active radar-homing AIM-9C Sidewinder air-to-air missile, of which 885 have been held in storage for some years since the US Navy retired the Vought F-8 Crusader which was the only aircraft to operate this semi-active radar homing version of the Sidewinder. The type is light and moderately advanced, but the seeker unit has to be cued before launch by the launch aircraft's radar homing and warning system.

Emerson Defense Systems AGM-123A Skipper 2 USA

Type: air-to-surface boosted tactical glide bomb
Dimensions: diameter 14.0 in (356 mm); length 169.0 in (4293 mm); span 63.0 in (1600 mm)
Weights: total round 1,283 lb (582.0 kg); warhead 1,000 lb (454 kg) Mk 83 bomb
Propulsion: one Rocketdyne Mk 78 solid-propellant rocket delivering unrevealed thrust
Performance: speed transonic; range 34 miles (55 km), limited in practice to 10.25 miles (16.5 km)
Guidance: Texas Instruments semi-active laser homing

Variant and application
AGM-123A Skipper 2: This simple yet highly effective tactical weapon uses off-the-shelf components (the Mk 83 1,000 lb/454 kg 'iron bomb', the rocket motor of the AGM-45B Shrike and a guidance/control system based on that of the Paveway III series guided bomb) to produce a highly accurate stand-off weapon with double the range of the Paveway series. The weapon entered production in 1986, initially for US Navy aircraft such

as the Grumman A-6 Intruder, Vought A-7 Corsair and McDonnell Douglas F/A-18 Hornet, and the manufacturer envisages the conversion of many current Paveway rounds to Skipper 2 standard.

USA

Northrop AGM-136A 'Tacit Rainbow'

Type: air-to-surface anti-radiation tactical missile
Dimensions: diameter 14.0 in (356 mm); length 100.0 in (2540 mm); span 69.4 in (1773 mm)
Weights: total round about 440 lb (199.6 kg); warhead unrevealed impact and proximity-fused FRAG-HE
Propulsion: one Williams International J400-WR-404 turbojet delivering 240 lb (109 kg) thrust
Performance: speed 600 mph (966 km/h); endurance 30 minutes
Guidance: Texas Instruments passive radiation seeking

Variant and application
AGM-136A 'Tacit Rainbow': This is the result of a highly classified programme whose very existence was revealed only in early 1987. The prime contractor is Northrop, which used its BQM-74 Chukar target and reconnaissance drone as the aerodynamic and structural basis of this winged missile. Support is provided by the Boeing Military Aircraft Company, Delco Systems Operations, Singer Kearfott Division, Texas Instruments and Williams International. The last indicates that this advanced anti-radar missile uses miniature jet propulsion, designed to secure longer range and/or extended loiter time, this loiter period being used to search for ground-based emitters, whose locations are then stored in the memory of the missile's advanced guidance and homing package even if they are switched off. The missile can thus select the most important emitter and institute its terminal dive attack with considerable accuracy. The primary role of the 'Tacit Rainbow' missile is to clear the way for manned aircraft. Launched in front of the first wave of attack aircraft, the missile cruises over the intended attack route to select and attack an emitter associated with surface-to air missiles or anti-aircraft artillery. There are also plans to develop a ground-launched version of this important missile.

USA

Rockwell HOBOS/KMU-353A/B

Type: air-to-surface tactical glide bomb
Dimensions: see below
Weights: total round 2,100 lb (952.6 kg); warhead 1,970 lb (893.6 kg) Mk 84 bomb (nominal weight 2,000 lb/907 kg)
Propulsion: none
Performance: speed transonic; range between 1,650 and 26,750 yards (1510 and 24460 m) depending on launch altitude

Variants and application
HOBOS: The **HOming BOmb System** was developed by Rockwell during the 1960s

The planar-wing version of the GBU-15 TV-guided glide bomb is seen in the terminal moments of its flight, in this instance with the warhead of the Mk 84 'iron' bomb against a blockhouse during trials at Eglin AFB in Florida.

Alternatively known as the Cruciform Wing Weapon, the GBU-15(V) is a powerful air-to-surface weapon based on the 2,000-lb (907-kg) general-purpose bomb fitted with aerodynamic surfaces and a guidance package for stand-off accuracy against high-value targets. Seen here under the wing of a McDonnell Douglas F-4E Phantom II also fitted on its port wing with the Northrop ASX-1 TISEO (Target Identification System, Electro-Optical with the image displayed on the rear-seater's radar monitor) is an example of the GBU-15(V)1/B variant with short-chord wings and the TV guidance package. Visible under the aircraft's fuselage is the white pod accommodating the Hughes AXQ-14 two-way data link used to control the GBU-15.

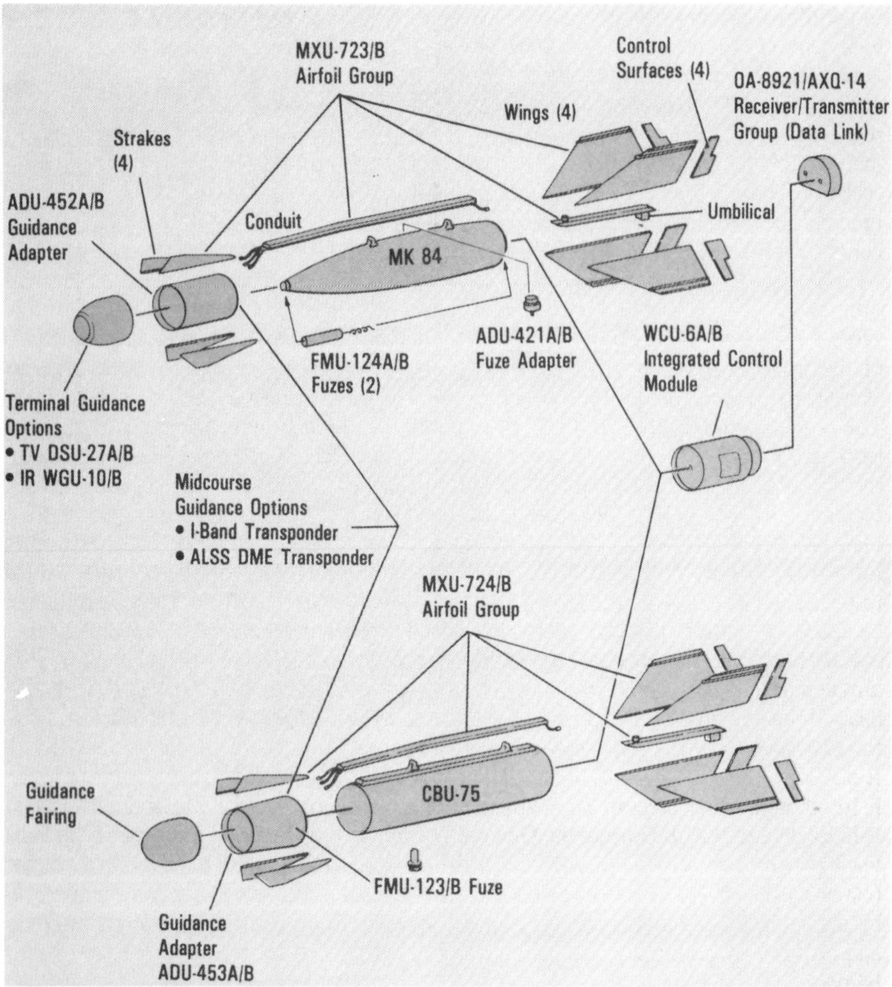

A manufacturer's illustration explains the varying configurations of the GBU-15(V) guided glide bomb and its two core, namely the Mk 84 general-purpose bomb and CBU-75 cluster bomb.

A McDonnell Douglas F-4E Phantom II launches an AGM-130A missile during trials at Eglin AFB in Florida. The AGM-130A is a simple but effective development of the GBU-15(V) glide bomb, the stand-off range of the weapon being extended to a tactically useful degree by the addition of a strap-on rocket motor under the body of the weapon.

in parallel with the Paveway I series within the 'Pave Strike' programme to provide US tactical aircraft to hit pinpoint targets with modified 'iron' bombs. However, in comparison with the Paveway series this type uses electro-optical rather that laser guidance, at first with a TV image-contrast tracker and then with more advanced TV and infra-red seekers for enhanced nocturnal and adverse-weather capability. The three kits for the **GBU-8** production series are the **KMU-353A/B** and **KMU-390/B** image-contrast systems, and the **KMU-359/B** infra-red system, each comprising a nose-mounted guidance section and (connected by strakes along the body of the bomb) a tail-mounted control section with fins. The packages can be added to the 2,000 lb (907 kg) Mk 84 GP bomb (KMU-353A/B and KMU-359/B) and the 3,000 lb (1361 kg M118E1 demolition bomb (KMU-390/B). The CEP of the system is 20 ft (6.1 m).

GBU-15: this is an improved version of the HOBOS concept, developed by Rockwell on the basis of the **Modular Guided Weapon** (otherwise known as the **Cruciform Wing Weapon**) system as a means of extending yet further the range of the HOBOS concept. The core weapon is the 2,000 lb (907 kg) Mk 84 bomb to which is fitted a guidance package, a control package, a two-way data-link package and a cruciform-wing aerodynamic package. The guidance package of the **GBU-15(V)1/B** initial version is the daylight-only DSU-27A/B electro-optical type using the AXQ-14 data-link system, while that of the later **GBU-15(V)2/B** model is the all-weather day/night WGU-10/B imaging infra-red guidance of the AGM-65D Maverick. The guidance system allows the missile to be locked onto its target before or after launch, or alternatively for the weapon to be manually guided by the controller in the launch aircraft. The GBU-15 weighs 2,450 lb (1111.3 kg), and has ranges between 1,650 yards (1,510 m) and 51 miles (82 km) depending on launch altitude, the CEP being in the order of 20 ft (6 m). The weapon has a diameter of 18.0 in (457 m), a length of 154.0 in (3912 mm) and a span of 59.0 in (1499 mm), and the aircraft generally associated with the GBU-15 are the Boeing B-52 Stratofortress, the McDonnell Douglas F-4 Phantom and the General Dynamics F-111, though other aircraft compatible with the weapon are the McDonnell Douglas F-15 Eagle and F/A-18 Hornet, the General Dynamics F-16 Fighting Falcon and the Panavia Tornado. A proposed development is the **GBU-15(V)N/B** with an SUU-54 dispenser (carrying 1,800 BLU-63 and/or BLU-86 anti-tank bomblets) as its payload, and guided by the electro-optical or imaging infra-red nose packages for attacks against airfields and armour, with secondary capability against motor transport and air-defence installations.

AGM-130A: This is a longer-range version of the GBU-15 weighing 2,917 lb (1323.2 kg) with a Mk 84 bomb as payload and powered by a strap-on Hercules solid-propellant rocket weighing 480 lb (217.7 kg) for a range of 15 miles (24 km) from a low-altitude launch. The weapon has been redesigned somewhat, with a more cylindrical body, larger canards and smaller fins with control surfaces. The type has midcourse altitude and heading hold, and the terminal guidance system uses the same type of data-linked electro-optical or imaging infra-red system as that used in the GBU-15. The weapon's principal dimensions include a diameter of 18.0 in (457 mm), a length of 154.5 in (3924 mm) and a span of 59.0 in (1499 mm).

The role of the missile is low-altitude stand-off attack against heavily-defended targets, and the weapon has the same launch qualifications and targeting options as the GBU-15. The direct-attack mode is used when the guidance system can be locked onto the target before the missile is released, allowing a line-of-sight approach to the target, while the indirect-attack mode is used when the target cannot be locked into the guidance system before missile launch. In this latter mode the weapon glides down to a pre-determined altitude of 2,000, 1,000 or 600 ft (610, 305 or 185 m) as a search is made for the target; during the mid-course phase missile altitude can be decreased by the oper-

Seen under the starboard wing of a General Dynamics F-111 attack and interdiction aircraft, these AGM-130A stand-off missiles reveal a clear resemblance to the GBU-15(V) glide bomb in all respects other than the solid-propellant rocket motors attached to their undersides for extended range capability.

ator in steps of 200 ft (60 m), altitude being maintained by a radar altimeter. Some 30 seconds after missile launch the motor is ignited and then jettisoned when it has burned out; the final approach to the target may be made with the seeker locked onto the target, or under manual control of the data-linked operator in the launch aircraft.
AGM-130B: This is the proposed airfield-attack version comparable to the AGM-130A, but carrying as its warhead an SUU-54 (15 BLU-106/B Boosted Kinetic-Energy Penetrator and 75 British-designed Hunting HB876 area-denial submunitions) and having an overall weight of 2,560 lb (1161.2 kg).
AGM-130C: This is the proposed hard-target version comparable to the AGM-130A but carrying as its payload the BLU-109/B Improved 2,000 lb Warhead for decisive effect on hardened point targets.

USA

Brunswick Defense LAD

Type: air-launched low-altitude tactical submunition dispenser
Dimensions: width 26.0 in (660 mm); height 17.33 in (440 mm); length 164.2 in (4.17 m); wing span about 62.0 in (1575 m)
Weights: total dispenser 2,250 lb (1020.6 kg); warhead typically 1,275 lb (578.34 kg) made up of 24 BLU-106/B BKEBP anti-tank and 42 HB876 area-denial submunitions
Propulsion: three-propellant rocket motors delivering unrevealed thrust
Performance: speed 635 mph (1022 km/h); range 24 miles (38.6 km) from a low-altitude launch
Guidance: inertial combined with a moderately sophisticated fire-control package, plus the option for addition of electro-optical, imaging infra-red or millimetre-wavelength terminal guidance

Variant and application
LAD: the **Low Altitude Dispenser** is being developed primarily in the USA by Brunswick Defense (with the collaboration of Garrett Manufacturing in Canada, Agusta and OTO Melara in Italy, and CASA and ENSA in Spain) in response to the **LOCPOD** (LOw-Cost POwered Dispenser) requirement issued by these four NATO countries. The LAD is a multi-role stand-off dispenser designed to provide tactical aircraft of the four countries with a decisive capability and varied submunitions against a number of battlefield targets. The system is being proposed in a number of forms (including a vehicle weighing 3,000 lb/1360.8 kg) with a baseline propulsive unit of three solid-propellant rockets of the type used in the AIM-9 Sidewinder air-to-air missile. With the smaller dispenser this is enough to provide the performance quoted above, together with a cross-range capability of 16 miles (25.75 km) after release at an altitude of 200 ft (61m).

The system is proposed with any of five current ejection systems which have been successfully evaluated (generally side-ejecting horizontal tubes) giving the submunitions launch velocities of up to 325 ft (99 m) per second, with eight successfully evaluated submunition shapes (including the BLU-106/B Boosted Kinetic-Energy Penetrator), and with an inertial navigation system that can be interfaced with advanced avionics on the launch aircraft and, later, with the Global Positioning System as well as terminal sensors. The result will be an exceptionally useful stand-off submunition dispenser with capability against fixed targets (airfields etc) as well as wide-area mobile targets (armoured formations etc). Another option is the return-to-target attack, in which a target of opportunity overflown by the launch aircraft can be engaged by the LAD after it has executed a reversal of course immediately after programming and launch, so removing the need for the launch aircraft to make another pass over a heavily-defended site.

Texas Instruments Paveway I/KMU-351A/B — USA

Type: air-to-surface tactical glide bomb
Dimensions: see below
Weights: total round 2,100 lb (952.6 kg); warhead 1,970 lb (893.6 kg) Mk 84 bomb (nominal weight 2,000 lb/907 kg)
Propulsion: none
Performance: range between 1,650 and 20,000 yards (1510 and 18290 m) depending on launch altitude
Guidance: semi-active laser homing

Variants and application
Paveway I: Introduced in 1967 after the first release of a laser-guided development weapon in 1965, Paveway I is a series of add-on laser-homing kits (marked-target seekers and associated control surfaces developed by Texas Instruments, and weighing about 30 lb/13.6 kg) that can be fitted to standard low-drag bombs for maximum accuracy when used in conjunction with an air or ground-based laser designator. The kits are the **KMU-342/B** (used with the 750 lb/340 kg M117 demolition bomb), the **KMU-351A/B** (used with the 2,000 lb/907 kg Mk 84 general-purpose bomb), the **KMU-370B/B** (used with the 3,000 lb/1361 kg M118E1 demolition bomb), the **KMU-388A/B** (used with the 500 lb/227 kg Mk 82 general-purpose bomb), the **KMU-420/B** (used with the 500 lb/227

lb Rockeye Mk 20 Mod 2 cluster bomb with 358 1.1 lb/0.5 kg anti-tank bomblets) ant the **KMU-421B** (used with the 2,000 lb/907 kg SUU-54B or Pave Storm I cluster bomb with 1,800 1.1 lb/0.5 kg anti-tank bomblets.

The advantage of the system is that the launch aircraft needs no modification and can depart from the target area unless it is the laser-designator aircraft. The system works at night and in poor visibility, and the minimum cloud base for successful operation is 2,500 ft (760 m). The airborne designators most commonly associated with the series are three podded systems, the AVQ-26 'Pave Tack', AVQ-23 'Pave Spike' and LANTIRN (Low-Altitude Navigation and Targeting Infra-Red for Night) carried by many US tactical aircraft. Any aircraft capable of lifting the baseline 'iron' bomb can deliver the Paveway weapon against a target designated by a third party.

Paveway II: An improved series based on the same basic concept but with a simpler (thus cheaper) guidance package and a folding wing aerofoil group (for extra manoeuvrability and additional lateral range) added at the tail. This version entered service in 1980. The weapons within this important basic designation are the **GBU-10E/B** and **GBU-10F/B** (both based on the 2,000 lb/907 kg Mk 84 general-purpose bomb), the **GBU-12D/B** and **GBU-12F/B** (both based on the 500 lb/227 kg Mk 82 general-purpose bomb), the **GBU-16B/B** and **GBU-16C/B** (both based on the 1,000 lb/454 kg Mk 83 general-purpose bomb) and the **Mk 13/18UK** (based on the British 1,000 lb/454 kg Mk 13/18 general-purpose bomb).

The GBU-10 type has a diameter of 18.0 in (457 mm) and an overall length of 169.9 in (4315 m). The GBU-12 type has a diameter of 10.75 in (273 mm) and an overall length of 131.15 in (3331 mm). In each case the weapon functions in a manner basically similar to that of the Paveway I series: the 3.25 in (82.5 mm) diameter freely-gimballed laser seeker at the extreme nose is aligned by its 6.25 in (159 mm) ring tail with the velocity vector of the falling bomb, and detects laser radiation (reflected from the 'illuminated' target) at different strengths on each of its four internal quadrants; this strength differential is used to generate error signals passed to the computer in the 8 in (203 mm) section just forward of the canard control fins; the computer turns these inputs

A pair of GBU-24 Paveway III laser-guided bombs is seen on the swivelling underwing hardpoints of a General Dynamics F-111 attack fighter, one of the most significant aircraft qualified to use this important weapon system. The excellent trajectory-shaping capabilities of the Paveway III can be deduced in part from the large size of the control surfaces just behind the seeker head.

into commands to the four solenoids operating the cruciform of control fins (spanning 31.25 in/794 mm in the GBU-10 and 17.65 in/448 mm in the GBU-12), steadily reducing the angular difference between the seeker's line of sight and the velocity vector of the bomb to zero, at which point an equal strength of laser radiation is being received by each of the seeker's quadrants as the weapon is falling directly towards the target.
Paveway III: Entering service in 1987, this is an improved version of the Paveway II with microprocessor controls and a digital autopilot (with BAe Dart precision gyro) intended mainly for use in the degraded weather and high-threat scenario of European operations. The Paveway III is also designated the **Low-Level Laser-Guided Bomb,** and was designed for release at low level (in either level flight of a zoom climb) and thus fitted with high-lift folding wings, but can also be released at higher altitudes in dives as steep as 60°. The result is a weapon of far greater operational flexibility than its predecessors.
The variant fitted with the 2,000 lb (907 kg) Mk 84 general-purpose bomb as its warhead has the service designation **GBU-24/B,** and also features an improved seeker and more advanced microprocessor technology, allowing low-level and off-axis stand-off delivery in conditions of low visibility, with the mid-course guidance providing the possibility of delayed laser-designating and trajectory shaping with all their tactical advantages of a considerably expanded launch envelope. The primary data for the GBU-24, which is the only version being procured (though a 500 lb (227 kg) variant was planned) include a weight of 2,315 lb (1050.1 kg), a length of 172.76 in (4388 mm) and spans of 36 in (914 mm) folded and 81.6 in (2073 mm) open.

Dornier/Aérospatiale SR-SOM

West Germany/France

Type: air-launched short-range tactical dispenser missile
Dimensions: cross section of the dispenser section and height data not revealed; length (SR-SOM 1) 3400 mm (133.86 in or (SR-SOM 2) 4300 mm (169.3 in); span 2600 mm (102.36 in)
Weights: total round (SR-SOM 1) 720 kg (1,587.3 lb), or (SR-SOM 2) 1400 kg (3,086.4 lb); warhead (SR-SOM 1) 350 kg (771.6 lb) or (SR-SOM 2) 900 kg (1,984.1 lb) of submunitions
Propulsion: one solid-propellant rocket delivering unrevealed thrust
Performance: speed Mach 0.8 at low altitude; range (SR-SOM 1) 20 km (12.43 miles) or (SR-SOM 2) 40+ km (24.86+ miles) at low altitude
Guidance: 'intelligent sensor' midcourse guidance and high-precision terminal guidance of types yet to be determined

Variants and application

SR-SOM 1: Under the leadership of Dornier, a Franco-West German team is developing the all-weather fire-and-forget SR-SOM (**Short-Range Stand-Off Missile**) in two forms. The SR-SOM 1 is planned for lighter tactical aircraft with NATO standard 356 mm (14 in) suspension lugs, and the core of the present configuration is the MoBiDiC (Modular Bird with Dispenser Container), a modular stand-off weapon schemed on the basis of Dornier's SR-SOM and Aérospatiale's Pégase II. Development of the payload section and submunitions has been entrusted to Diehl in West Germany and to Thomson Brandt in France, the proposed submunition type being designed to give the weapon autonomous capability against hardened shelters and moving armour.

Above *Carry trials of a possible SR-SOM configuration under the port wing of a West German air force Dassault-Breguet/Dornier Alpha Jet A light attack aircraft. Also visible in this photograph is the underfuselage installation of one Mauser BK 27 cannon carried by this Alpha Jet variant.*

Right *An ejection trial at the Dornier proving grounds shows the manner in which the submunitions are dispensed laterally from the proposed SR-SOM dispenser weapon, using small pyrotechnic charges.*

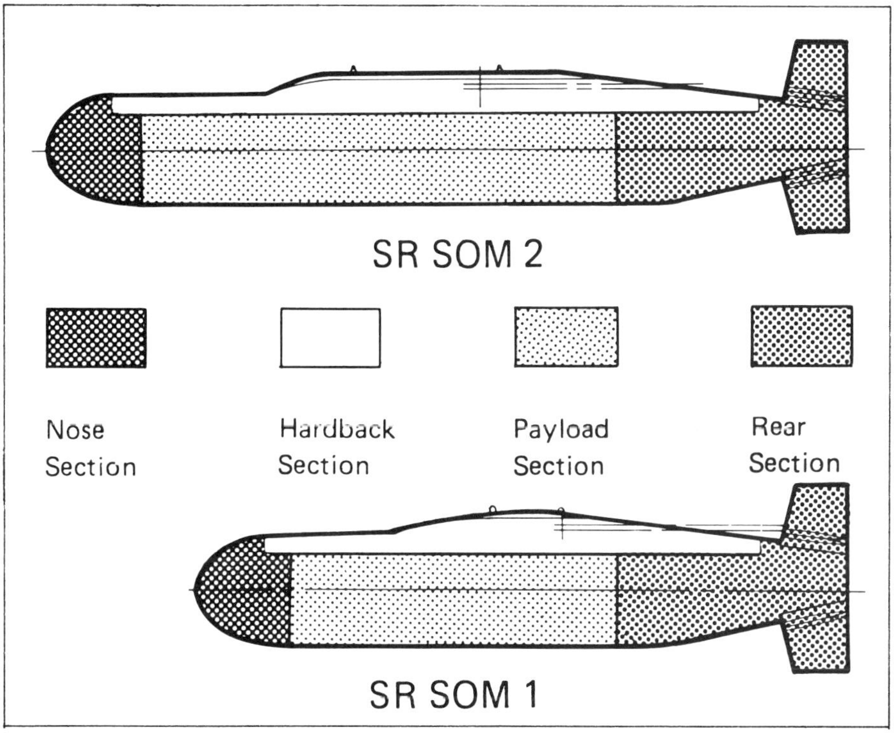

The SR-SOM modular dispenser weapon is planned for use by aircraft with NATO standard 356-mm (14-in) and 762-mm (30-in) suspension lugs: the SR-SOM 1 for the former and SR-SOM 2 for the latter, each having the same nose and tail sections, with differing payload sections built up from the same types of module, and slightly different hardback sections.

Thomson Brandt had previously studied submunitions for the Pêgase II, these including fuel/air explosive, TACED sensor-aided anti-armour, BAP 100 anti-runway, BAT 120 anti-vehicle, AMV 120 anti-tank, ABL anti-light armour, and area-denial types, which may form the basis of some of the submunitions currently under development. The MoBiDiC vehicle is based on a beam supporting the aircraft interface lugs, rectangular opening wings and tail section (the last incorporating the rocket motor and opening cruciform tail surfaces indexed at 45° to the wings). Under the basic structural beam is the modular dispenser section with provision for sideways launching of submunitions, and at the nose is an aerodynamic fairing.

SR-SOM 2: This is the larger version of the basic SR-SOM, designed for higher-performance aircraft with NATO standard 762 mm (30 in) suspension lugs and possessing greater range. The design team is also evaluating the possibility of a turbine-engined version for considerably better range capability.

ANTI-TANK MISSILES

Argentina

CITEFA Mathogo

Type: helicopter-launched short/medium-range tactical anti-tank missile
Dimensions: diameter 102 mm (4.02 in); length 998mm (39.3 in); span data not available
Weights: total round 11.3 kg (24.9 lb); warhead 2.8 kg (6.17 lb), hollow-charge HE
Propulsion: one dual-thrust solid-propellant rocket delivering unrevealed thrust
Performance: speed 325 km/h (202 mph); range 350/2100 m (385/2,295 yards)
Guidance: wire command to line of sight

Variant and application
Mathogo: This missile entered production in the late 1970s, and is by Western standards an obsolescent weapon resembling the Swedish Bantam. The type was developed as a surface-launched weapon, but has now been fully qualified as a helicopter-launched missile. The Mathogos advantages are light weight, compact dimensions and a warhead capable of penetrating 400 mm (15.75 in) of armour, while its disadvantages are low speed and a comparatively poor maximum range. These latter would force the launch helicopter to approach its target so closely that it would come under severe attack during the missile's flight time of 23 seconds to maximum range. In 1986 it was revealed that there are two subvariants of the Mathogo, one with the standard range quoted above, and the other with the usefully increased maximum range of 3000 m (3,280 yards).

China

NORINCO Red Arrow 8

Type: helicopter-launched medium/long-range tactical anti-tank missile
Dimensions: diameter 120 mm (4.72 in); length 875 mm (34.45 in); span 320 mm (12.6 in)
Weights: total round 11.2 kg (24.69 lb); warhead of unrevealed weight, hollow-charge HE
Propulsion: one dual-thrust solid-propellant rocket delivering unrevealed thrust
Performance: speed 865 km/h (537 mph); range 100/3000 m (110/3,280 yards)
Guidance: wire command to line of sight

Variant and application
Red Arrow 8: This is China's most advanced anti-tank missile and, though the weapon has so far been seen only in the surface-launched role, it is likely that the type is used for helicopter applications. The weapon has the Chinese designation **Hong Jian 8**, and is an optically-tracked wire-guided missile with semi-automatic command to line of sight guidance. The likely armour penetration of the warhead is 800 mm (31.5 in), and in many respects the missile bears a striking similarity to the Franco/West German Euromissile Milan.

NORINCO Red Arrow 73

China

Type: helicopter-launched medium/long-range tactical anti-tank missile
Dimensions: diameter 120 mm (4.72 in); length 868 mm (34.17 in);span 393 mm (15.47 in)
Weights: total round 11.3 kg (24.91 lb); warhead 3 kg (6.6 lb) hollow-charge HE
Propulsion: one dual-thrust solid-propellant rocket delivering unrevealed thrust
Performance: speed 430 km/h (267 mph); range 500/3000 m (545/3,280 yards)
Guidance: wire command to line of sight

Variant and application
Red Arrow 73: This is essentially a Chinese development of the Soviet AT-3 'Sagger', to which it is basically similar in all respects other than its smaller wings. The weapon is known to the Chincsc as thc **Hong Jian 73**, and its warhead can penetrate 500 mm (19.7 in) of armour at 0° or 150 mm (5.9 in) at 65°.

Aérospatiale Eryx

France

Type: short-range light tactical anti-tank missile
Dimensions: diameter 160 mm (6.3 in); length 925 mm (36.42 in); span data not available
Weights: total round 14.4 kg (31.75 lb); warhead 3.6 kg (7.94 lb) hollow-charge HE
Propulsion: one SNPE solid-propellant rocket delivering unrevealed thrust
Performance: speed 72 km/h (44.7 mph) rising to 303 km/h (188 mph); range 25/600 m (27/655 yards).
Guidance: wire semi-automatic command to line of sight.

Variant and application
Eryx: Originally known as the **ACCP** (**Anti-Char Courte Portée**, or short-range anti-tank) missile, this is a lightweight anti-tank weapon selected for the French army in 1986. The type has an extremely small motor resulting in very low launch speed from the Kevlar launch tube, but this tactical limitation is offset by the type's minimal back-blast and (in its ground-launched version) its ability to be fired from inside a building. The warhead can penetrate 900 mm (35.4 in) of armour, and after introduction as a ground-based weapon the Eryx is being developed as a helicopter-launched system for French and export orders. Though the light weight of the missile is an obvious advantage, allowing even the lightest helicopters to carry a multiple installation, the short range of the missile is a tactical disadvantage as launch helicopters will have to approach very close to their prospective targets.

Euromissile HOT

France/West Germany

Type: helicopter-launched long-range heavyweight tactical anti-tank missile
Dimensions: diameter 165 mm (6.5 in); length 1275 mm (50.20 in); span 312 mm (12.28 in)

The air-launched anti-tank missile of many European air arms is the Euromissile HOT, this is a quadruple installation on a French army Aérospatiale SA 342M Gazelle. The SFIM APX M334 stabilized sight associated with the HOT installation can be seen above the port side of the cockpit.

Weights: total round 23.5 kg (51.8 lb); warhead 6 kg (13.2 lb) hollow-charge HE
Propulsion: one SNPE Bugeat solid-propellant booster rocket delivering unrevealed thrust for 0.9 second, and one SNPE Infra solid-propellant sustainer rocket delivering 24 kg (53 lb) thrust for 17.4 seconds
Performance: speed 865 km/h (537 mph); range 75/4250 m (82/4,650 yards)
Guidance: SAT/Eltro wire semi-automatic command to line of sight

Variants and applications
HOT: Entering service in the early 1970s, the HOT (**Haute subsonique Optiquement téléguide tiré d'un Tube**, or high-subsonic optically-guided tube-launched) missile is a powerful anti-tank weapon designed for static ground use, or for installation in armoured fighting vehicles and helicopters. The missile can be used in single or multiple tube launchers, and can penetrate more than 800 mm (31.5 in) of armour under optimum conditions, though a more realistic figure is 280 mm (11 in) at 65° impact angle. The type's primary disadvantage is a modest speed, which means that the missile takes some 17 seconds to reach its maximum range even when launched from a helicopter, which thus has to remain exposed for this time: intermediate times are 8.7 seconds to 2000 m (2185 yards), 12.5 seconds to 3000 m (3,280 yards) and 16.3 seconds to 4000 m (4,375 yards). The stabilized sight generally associated with the HOT in the Aérospatiale SA 342 Gazelle and MBB BO 105P is the SFIM APX M397 unit mounted in the roof of the launch helicopter.
HOT 2: This is an improved version of the basic weapon featuring much-enhanced armour penetration as a result of its larger 150 mm (5.9 in) diameter warhead contain-

ing 4.1 kg (9.04 lb) rather than 3 kg (6.61 lb) of HE carried in the HOT's 136 mm (5.35 in) diameter warhead. The HOT 2 is also faster than its predecessor, with all the consequent tactical advantages of this fact. Further improvement in capability will be provided by adoption of the stabilized Viviane day/night sight currently under development on the basis of the Victor thermal camera pioneered in the Venus system used for the Aérospatiale SA 365M Panther light battlefield helicopter: this provides a genuine nocturnal capability, and the West Germans are developing a similar capability with the integration of a licence-built Texas Instruments FLIR sight with the HOT system.

Israel Military Industries Mapats — Israel

Type: helicopter-launched medium/long-range heavy anti tank missile
Dimensions: diameter 148 mm (5.83 in); length 1450 mm (57.09 in) with probe extended; span data not available
Weights: total round 18.5 kg (40.78 lb); warhead 3.6-kg (7.94-lb) hollow-charge HE
Propulsion: one solid-propellant booster rocket delivering unrevealed thrust, and one solid propellant sustainer rocket delivering unrevealed thrust
Performance: speed 1135 km/h (704 mph); range 65/4,500 m (70/4,920 yards)
Guidance: laser beam-riding

Variant and application
Mapats: Sometimes known colloquially as the 'Togger' because it is based on the US BGM-71 TOW but has features of the Soviet AT-3 'Sagger' anti-tank missile, the Mapats is a highly capable weapon with advanced unjammable guidance and capable of penetrating 800 mm (31.5 in) or, according to some sources, 1000 mm (39.37 in) of multi-layer armour. The weapon was developed from the late 1970s, and was initially deployed in 1983 as a ground- and vehicle-launched missile. The missile is supplied as a certified round in a fibreglass container which serves as the launch tube, the complete missile/container assembly weighing 29 kg (63.9 lb) and measuring 1500 mm (59.06 in) in length.

After launch the missile's booster drops away, and the sustainer ignites to power the missile to the maximum range, which is attained in 23.5 seconds. The vehicle-carried laser has a modulated beam that is used to track the target, sensors on the missile's rear detecting angular deviation from the laser's line of sight to allow the onboard digital generation of steering commands designed to bring the missile back onto the laser's line of sight. The sensitivity of the system is such that tracking rates (in both azimuth and elevation) of 8° per second are possible. From 1987 the Mapats has been used in an air-launched role, the missile's agility giving it a secondary capability against helicopters. The laser guidance system has ×13 magnification optics for the operator and, in its ground form, the capability to traverse through 360° and to elevate from −20° to +30°: it is unlikely that traverse in the helicopter version is so great, though the elevation angle may be retained.

Kawasaki Type 64 — Japan

Type: helicopter-launched short/medium-range tactical anti-tank missile

Dimensions: diameter 120 mm (4.72 in); length 1000 mm (39.37 in); span 600 mm (23.6 in)
Weights: total round 15.7 kg (34.6 lb); warhead unrevealed, hollow-charge HE
Propulsion: one Daicel/Nippon Oil two-stage solid-propellant rocket delivering unrevealed thrust
Performance: speed 305 km/h (190 mph); range 350/1800 m (385/1,970 yards)
Guidance: Nihon Electric wire command to line of sight

Variant and application
Type 64: Introduced in 1965 and otherwise known as the **KAM-3D,** this is a simple, slow yet effective missile now being replaced in the ground role by the Type 79 weapon. In the ground-launched application the system requires a three-man crew, and the operator can control one or more laterally-separated missiles from a central firing position. The missile is launched from a small ground support/baseplate, and can also be fitted on helicopters.

Sweden

Bofors RB 53 Bantam

Type: helicopter-launched short/medium-range tactical anti-tank missile
Dimensions: diameter 110 mm (4.33 in); length 850 mm (33.46 in); span 400 mm (15.75 in)
Weights: total round 7.6 kg (16.75 lb); warhead 1.9 kg (4.2 lb) hollow-charge HE
Propulsion: one Bofors dual-thrust solid-propellant rocket delivering unrevealed thrust
Performance: speed 305 km/h (190 mph); range 300/2000 m (330/2,185 yards)
Guidance: wire command to line of sight

Variant and application
RB 53 Bantam: This is a first-generation anti-tank missile (in its basic ground application carried in and launched from a small container) with simple guidance and low cruising speed, the operator being afforded a measure of protection by a separation of up to 120 m (130 yards) between the launcher and the control position. The weapon was adopted in 1963, and its warhead can penetrate 500 mm (19.68 in) of armour. The type can also be launched from light helicopters, this application of the missile being aided by the Bantam's light weight and compact dimensions.

Sweden

Bofors RBS 56 Bill

Type: helicopter-launched medium-range tactical anti-tank missile
Dimensions: diameter 150 mm (5.91 in); length 900 mm (35.43 in); span 410 mm (16.14 in)
Weights: total round 10.7 kg (23.59 lb); warhead unrevealed, hollow-charge HE
Propulsion: one Royal Ordnance dual-thrust solid-propellant rocket delivering unrevealed thrust
Performance: speed 720 km/h (447 mph); range 150/2000 m (165/2,185 yards)
Guidance: wire semi-automatic command of sight

Variant and application

RBS 56 Bill: Entering service in the later 1980s, the Bill (pick) is a highly advanced anti-tank missile fired (in its baseline ground-launched form) from a container tube fitted with a sight unit and mounted on a tripod; in the air-launched application the missile container (delivered complete with the missile as a certified round of ammunition and weighing 20.5 kg/45.19 lb) can be fitted to the outside of a helicopter in the standard fashion for such weapons. The keys to the Bill's capabilities are its guidance system, which causes the missile to fly 1 m (39.4 in) above the operator's line of sight, and its combination of proximity fuse and shaped-charge warhead, which is angled to fire 30° downwards from the missile centreline as the weapon overflies the target's most vulnerable upper surfaces. This allows the jet of gas and vaporized metal formed by the warhead's detonation to burn through the target tank's thinnest armour.

The missile's flight times are 6 seconds to 1200 m (1,315 yards) and 11 seconds to 2000 m (2,185 yards). In common with other anti-tank missiles the RBS 56 had greater capability against stationary targets, but the tactical realities of the modern battlefield were not ignored in the design and the RBS 56 can thus engage crossing targets, moving at a maximum speed of 36 km/h (22.4 mph), at range of between 300 and 2000 m (330 and 2,185 yards).

AT-2 'Swatter-C' USSR

Type: helicopter-launched medium/long-range heavyweight tactical anti-tank missile
Dimensions: diameter 132 mm (5.2 in); length 1160 mm (45.67 in); span 660 mm (26.0 in)
Weights: total round 32.5 kg (71.65 lb); warhead of unrevealed weight, hollow-charge HE
Propulsion: one solid-propellant rocket delivering unrevealed thrust
Performance: speed 540 km/h (336 mph); range 500/4000 m (545/4,375 yards)
Guidance: radio semi-automatic command to line of sight and infra-red terminal homing

Variant and application

AT-2 'Swatter-C': the 'Swatter' series was introduced in the mid-1960s as the 'Swatter-A' ground-launched anti-tank missile with the Soviet designation **PUR-62 Falanga,** simple radio-command guidance (with all its vulnerability to contermeasures) and a warhead able to penetrate 480 mm (18.9 in) of armour under optimum conditions. The 'Swatter-B' is an improved version with greater range and a warhead able to pierce 510 mm (20.18 in) of armour. The 'Swatter-C' is the dedicated helicopter-launched version of the basic weapon for the Mil Mi-8 'Hip' and Mi-24 'Hind' battlefield helicopters, and possesses semi-automatic command to line of sight guidance and yet greater range.

AT-3 'Sagger' USSR

Type: helicopter-launched medium/long-range tactical anti-tank missile
Dimensions: diameter 119 mm (4.69 in); length 860 mm (33.86 in); span 460 mm (18.1 in)
Weights: total round 11.29 kg (24.9 lb); warhead 3 kg (6.6 lb) hollow-charge HE

Propulsion: one solid-propellant booster rocket and one solid-propellant rocket delivering unrevealed thrust
Performance: speed 430 km/h (267 mph); range 300/3000 m (330/3,280 yards)
Guidance: wire command to line of sight

Variants and application
AT-3 'Sagger-A': Introduced in about 1970, the AT-3 is a more capable and versatile weapon that the AT-2, with a countermeasures-proof wire guidance system and more powerful sustainer for greater (but still comparatively slow) speed. The type is designated **PUR-64 Malyutka** in the USSR, and its warhead can penetrate 410 mm (16.14 in) of armour.
AT-3 'Sagger-B': Introduced in 1973, this upgraded model has an improved motor for greater speed and reduced flight time.
AT-3 'Sagger-C': Introduced in the late 1970s, this is a much developed version with semi-automatic command to line of sight guidance.

USSR

AT-5 'Spandrel'

Type: helicopter-launched long-range heavyweight tactical anti-tank missile
Dimensions: (estimated) diameter 170 mm (6.7 in); length 1300 mm (51.18 in); span data not available
Weights: total round 17 kg (37.5 lb); warhead unrevealed, hollow-charge HE
Propulsion: one solid-propellant rocket delivering unrevealed thrust
Performance: speed 720 km/h (447 mph); range 100/4000 m (110/4,375 yards)
Guidance: wire semi-automatic command to line of sight

Variant and application
AT-5 'Spandrel': Introduced in about 1980, the AT-5 is a powerful second-generation vehicle-launched weapon using technology 'gleaned' from such Western types as the Euromissile HOT. The HEAT warhead can penetrate 750 mm (29.53 in) of armour and, though there is no concrete evidence of the fact, it is likely that this useful weapon can be launched from Soviet battlefield helicopters.

USSR

AT-6 'Spiral'

Type: helicopter-launched long-range heavyweight tactical anti-tank missile
Dimensions: (estimated) diameter 140 mm (5.51 in); length 1800 m (70.87 in); span data not available
Weights: (estimated) 32 kg (70.55 lb); warhead unrevealed hollow-charge HE
Propulsion: one dual-thrust solid-propellant rocket delivering unrevealed thrust
Performance: (estimated) speed 1000 km/h (621 mph); range 100/7000 m (110/7,655 yards)
Guidance: radio command to line of sight (or possibly laser beam-riding)

Variants and applications
AT-6 'Spiral': So far seen only on the Mil Mi-24 'Hind' attack helicopter, the AT-6 is the first Soviet third-generation anti-tank missile, its good range being matched by the excellent armour penetration figure of about 800 mm (31.5 in).

Hughes BGM-71A TOW

USA

Type: helicopter-launched medium/long-range heavyweight tactical anti-tank missile
Dimensions: diameter 6.0 in (152 mm); length 46.25 in (1174 mm); span 13.5 in (343 mm)
Weights: total round 49.6 lb (22.5 kg); warhead 8.6 lb (3.9 kg) Picatinny Arsenal hollow-charge HE
Propulsion: one Hercules K41 two-stage solid-propellant rocket delivering unrevealed thrust
Performance: 700 mph (1127 km/h); range (pre-1976 models) 70/3,280 yards (65/3000 m) or (post-1976 models) 70/4,100 yards (65/3750 m)
Guidance: Emerson Electric wire semi-automatic command to line of sight

Variants and applications
BGM-71A TOW: Entering service in 1970, the TOW (**Tube-launched Optically-tracked Wire-guided**) missile is the West's most important anti-tank weapon, and is a heavyweight type designed for vehicle- or helicopter-borne launchers. It has proved its capabilities in several conflicts. The warhead contains 5.3 lb (2.4 kg) of explosive, and has shown itself able to penetrate 600 mm (23.62 in) of armour. The range of this initial model is limited to 3,280 yards (1000 m) by the length of its guidance wires.
BGM-71B Extended-Range TOW: This is the semi-official designation of the basic weapon produced from 1976 with greater range than the original model as the result of the lengthening of the guidance wires to 4,100 yards (3750 m).
BGM-71C Improved TOW: Introduced in the early 1980s in response to the development of Soviet tanks with improved armour, this interim model features a larger-diameter 5 in (127 mm) warhead with LX-14 explosive and a telescoping 15 in (381 mm) nose probe that extends in flight to ensure a perfect stand-off distance for the denotation of the shaped-charge warhead. This model has a length of 70 in (1778 mm) with the probe extended, and weighs 56.65 lb (25.7 kg). Armour penetration with the new warhead increases to 700 mm (27.56 in).

One of the most important armament fits for Western battlefield helicopters comprises eight BGM-71 TOW anti-tank missiles. These are most frequently encountered as two quadruple installations on the outer hardpoints of helicopters such as this Agusta A 129 Mangusta, seen here without the drag-reducing fairings now common with the installation. The whole M65 suite comprises the two launcher units, the FACTS (FLIR-Augmented Cobra TOW Sight) flexible mounting in the underside of the nose and surmounted by the small sensor for the pilot's night vision system, and the fuselage-located electronics. Also visible on this A 129's nose are the two cheek-mounted passive radar-warning receivers covering the helicopter's forward hemisphere.

Fitted on a Swedish MBB BO 105C helicopter, the Emerson/Saab HeliTOW system combines two two-round launchers for the BGM-71 TOW anti-tank missile and a Helios stabilized sight mounted above the cabin roof for use by the co-pilot/gunner.

BGM-71D TOW 2: Introduced in 1983 as the standard US weapon for tackling the latest Soviet tanks at long range, this model has an improved double-base motor (offering about 30% greater impulse) and a 13.2 lb (6 kg) warhead increased in diameter to 6 in (152 mm) and provided with a 21.25 in (540 mm) nose probe ensuring optimum stand-off distance for penetration of 800 mm (31.5 in) of armour. This significant variant weighs 61.95 lb (28.1 kg) and, though it can be fired from the original analog-electronics launcher, it is designed for use with an improved digital-electronics launcher fitted with thermal as well as optical sights, and with a more advanced guidance package.

Hughes is also proposing four developments of this powerful weapon with the provisional designations **TOW 2A** (a variant with an improved direct-attack warhead), **TOW 2B** (a top-attack version with a millimetre-wave radar or electro-optical fuse), **TOW 2D** (of which no details have been released), and **TOW 2N** (with wire-less command system of unspecified type, supersonic speed and a range of 10,000 yards/9145 m). Tested in 1987 the TOW 2A variant has a small explosive charge on its nose probe designed to trigger the detonation of the target's reactive armour, so opening the way for the gas/metal vapour jet of the missile's main charge. Though the system works in its basic elements, its operational utility is doubtful because operators have found it very difficult to control the nose-heavy missile.

Hughes Aircraft M65: This is the aerial application the TOW missile, and was originally developed for the Bell AH-1 HueyCobra series of close support helicopters, resulting in the AH-1J, AH-1S, AH-1T and AH-1W variants. The system has also been adapted for other helicopter types, most notably the McDonnell Douglas Helicopters MD-500/530 Defender (originally designed as the Hughes Model 500/530 Defender), the Bell Model 206L-1, the MBB BO105C, the Agusta A 109 and A 129 Mangusta, and the Westland Lynx. Depending on the configuration of the helicopter and the preference of the operator service, the stabilized sight (for target-acquisition and missile-guidance functions) can be fitted as a chin-, roof- or mast-mounted installation. The basic sight can also be upgraded considerably with FLIR (forward-looking infra-red) augmentation for night and adverse-weather operations as the **FACTS (FLIR-Augmented Cobra TOW Sight)**, or with laser augmentation for accurate rangefinding as the **M65L**, otherwise known as the **LAAT (Laser-Augmented Airborne TOW)** sight. In 1987 Hughes began production of the **Cobra-NITE** sight to upgrade the both M65 and M65L sights used in the AH-1S HueyCobra with a night and adverse-weather capability.

Emerson Electric/Saab-Scania HeliTOW: This is an advanced missile-guidance package developed to provide light helicopters in the field with heavyweight anti-tank capability. Excluding its four missiles, the system weighs 310 lb (140.6 kg) complete with the Helios sight augmented by the optional Honeywell MiniFLIR or Ericsson FLIR, and can be fitted in thirty minutes.

ANTI-SHIP MISSILES

CPMIEC C-601 — China

Type: air-launched medium-range tactical anti-ship missile
Dimensions: diameter 900 mm (35.43 in); length 7380 mm (290.55 in); span 2400 mm (94.49 in)
Weights: total round 2440 kg (5,379.2 lb); warhead 225 kg (496.0 lb) blast fragmentation
Propulsion: one solid-propellant booster rocket delivering unrevealed thrust, and one sustainer turbojet delivering unrevealed thrust
Performance: speed 1100 km/h (684 mph); range 100 km (62 miles)
Guidance: strapdown inertial for midcourse phase, and monopulse active radar for terminal phase

Variant and application
C-601: Little of a definite nature is known of this useful anti-ship missile, which began to enter service in 1982 as the primary armament of the H-6D anti-shipping version of the Chinese-built version of the Soviet Tupolev Tu-16 'Badger' bomber. The H-6D can carry two of the type on underwing hardpoints for launch after a suitable target has been acquired by the aircraft's substantial undernose search radar. The C-601 has large delta wings and triple tail surfaces set at an angle of 120° to each other; the weapon is clearly derived ultimately from Soviet SS-N-2 'Styx' weapon with features of the AS-5 'Kelt', and in the US standard terminology is known as the **CAS-N-1.**

CPMIEC C-801 — China

Type: air-launched medium-range tactical anti-ship missile
Dimensions: diameter 360 mm (14.17 in); length 5814 mm (228.9 in); span 1118 mm (44.02 in)
Weights: total round 815 kg (1,796.7 lb); warhead 165 kg (363.75 lb) semi-armour-piercing HE
Propulsion: one solid-propellant booster rocket and one solid-propellant sustainer rocket delivering unrevealed thrust
Performance: speed 1100 km/h (684 mph); range 50 km (31.1 miles)
Guidance: strapdown inertial for midcourse phase, and monopulse active radar for terminal phase

Variant and application
C-801: Introduced to Chinese service in 1983, the C-801 is an advanced anti-ship missile of which little is known with certainty. In its surface-launched applications the type is said to be compatible with the fire-control system used with the Chinese HY-2 and FL-1 derivatives of the Soviet SS-N-2 'Styx', and the missile can also be launched from aircraft, shore batteries, ships and surfaced submarines. The weapon cruises at a preset

altitude of 20 or 30m (66 or 98 ft) before descending to a height of between 5 and 7 m (16.4 and 23 ft) for the radar-homing phase and final diving attack. The weapon is known in Chinese service as the **Ying Ji-6,** YJ standing for eagle strike.

China

CPMIEC HY-4

Type: air-launched medium-range tactical anti-ship missile
Dimensions: diameter 760 mm (29.92 in); length 7360 mm (289.76 in); span 2750 mm (108.27 in)
Weights: total round 1740 kg (3,836 lb); warhead 500 kg (1,102.3 lb) semi-armour-piercing HE
Propulsion: one turbojet delivering unrevealed thrust
Performance: speed between 980 and 1040 km/h (609 and 646 mph); range 150 km (93.2 miles)
Guidance: strapdown inertial for midcourse phase, and monopulse active radar for terminal phase

Variant and application
HY-4: This is a useful development of the HY-2, the Chinese version of the Soviet SS-N-2 'Styx' surface-launched anti-ship missile, with a small turbojet (fed via the same type of ventral inlet as many other air-breathing anti-ship missiles) replacing the earlier missile's rocket propulsion for considerably enhanced range. The HY-4 is designed for launch by the H-6D anti-ship variant of the H-6 bomber (Chinese-built version of the Soviet Tupolev Tu-16 'Badger'), which can carry two such missiles under its wings. The missile cruises at an altitude between 70 and 200 m (230 and 655 ft), and its effective range is from 35 to 135 km (21.75 to 83.9 miles). The type entered service in 1971, and there is also a shore-launched version with a jettisonable booster rocket that increases overall weight to 2000 kg (4,409.2 lb). The full Chinese designation is **Hai Ying-4**, HY standing for sea eagle.

France

Aérospatiale AS.15TT

Type: helicopter-launched short-range tactical anti-ship missile
Dimensions: diameter 188 mm (7.40 in); length 2300 mm (90.55 in); span 564 mm (22.2 in)
Weights: total round 103 kg (227.1 lb); warhead 29.7 kg (65.5 lb) impact-fused blast fragmentation
Propulsion: one SNPE/Aérospatiale Anubis solid-propellant rocket delivering unrevealed thrust for 45.2 seconds
Performance: speed 1000 km/h (621 mph); range 16000 m (17,500 yards)
Guidance: (height) Thomson-SCF preprogrammed autopilot designed to produce a reducing cruise altitude to a height determined by TRT radar altimeter, and (bearing) radar command

Variant and application
AS.15TT: Entering service in the mid-1980s, the AS.15TT (**Air-Sol.15 Tous Temps**, or

An Aérospatiale SA 365F Dauphin 2 light helicopter fires one of its maximum four AS.15TT lightweight anti-ship missiles. The missile flies under autopilot control down to sea-skimming height under control of a radar altimeter, and is guided in bearing by radio command using the helicopter's Thomson-CSF Agrion 15 radar, whose antenna is visible under the helicopter's nose. The receiver pods are mounted at the tips of the missile's four wings.

all-weather air-to-surface type 15) was largely financed by Saudi Arabia as part of the enormous 'Sawari' defence deal with France. The missile was designed as successor to the AS.12 as the primary weapon of ship- and shore-based light helicopters tasked with countering fast attack craft and other comparatively small naval units. The Aérospatiale SA 365 Dauphin 2 and Aérospatiale AS 332 Super Puma helicopters are the two major types currently qualified to operate the AS.15TT. The missile is controlled in bearing by the launch helicopter's Thomson-CSF Agrion-15 radar, and in height by a programme that lets the missile down to sea-skimming height for the approach and to wavetop height for the final 300 m (985 ft) of the attack. Aérospatiale has also developed an MM.15 surface-launched version for use by fast attack craft.

Aérospatiale AM.39 Exocet

France

Type: air-launched medium-range tactical anti-ship missile
Dimensions: diameter 350 mm (13.75 in); length 4690 mm (184.65 in); span 1100 mm (43.31 in)
Weights: total round 655 kg (1,444.0 lb); warhead 165 kg (363.75 lb) GP1 proximity and impact delay-fused blast fragmentation
Propulsion: one SNPE Condor solid-propellant booster rocket delivering unrevealed thrust for 2 seconds, and one SNPE Hélios solid-propellant sustainer rocket delivering unrevealed thrust for 150 seconds
Performance: speed 1140 km/h (708 mph) or Mach 0.93; range between 50 and 70 km (31.1 and 43.5 miles) depending on launch altitude
Guidance: inertial plus TRT RAM.01 radar altimeter for midcourse phase, and EMD ADAC monopulse active radar for terminal phase

Variants and application

AM.38 Exocet: The most widely produced anti-ship missile of Western origins, the Exocet was originally designed in the late 1960s as the MM.38 ship-launched missile and began to enter service in 1974. Since that date the missile has been developed in several forms and seen extensive operational use, notably in the Anglo-Argentine Falklands War of 1982 and in the continuing Iraqi-Iranian Gulf War. Development of the Exocet's air-launched version began in 1975, and the first round was test fired in December 1976 to allow service deliveries from July 1977. This initial AM.38 was a limited-production helicopter-launched version of the MM.38 using SNPE Epervier booster and SNPE Eole V sustainer rocket motors with concentric nozzles. A 1-second ignition delay was built into the booster to avoid damage to the launch platform.

AM.39 Exocet: This is the full-production aircraft-launched version of the MM.38 with a revised propulsion arrangement in a shorter body, reducing weight but increasing range. The missile is launched towards the target on range and bearing data provided by the launch aircraft's sensors and fire-control system, and cruises at low altitude until some 10000 m (10,935 yards) from the anticipated target position, when the I/J-band monopulse active seeker head is turned on, the target acquired and the terminal phase initiated at the one of three heights preselected at launch on the basis of sea state. Residual fuel adds considerably to the effects of the warhead detonation, which has in itself proved somewhat troublesome and unreliable. The French launch platforms most commonly associated with the missile are the Aérospatiale SA 321 Super Frelon helicopter, the Dassault-Breguet Atlantique 2 maritime patrol aircraft, the Dassault-Breguet Mirage FlE multi-role fighter and the Dassault-Breguet Super Etendard carrierborne attack fighter. Late-production rounds have the Super ADAC homing radar, which offers the considerable tactical advantage of improved resistance to electronic countermeasures. There are also an SM.39 submarine version (in a capsule for tube launch) and a much improved MM.40 ship-launched version of the Exocet.

France/West Germany

Aérospatiale/MBB ANS

Type: air launched short/medium-range tactical anti-ship missile

Dimensions: diameter 350 mm (13.78 in); length 5700 m (224.4 in); span 1100 mm (43.31 in)

Weights: total round 950 kg (2,094.4 lb); warhead 180 kg (396.8 lb) penetrating blast fragmentation

Propulsion: one MBB integral rocket-ramjet serving first as a solid-propellant booster rocket delivering unrevealed thrust, and then as a rocket-ramjet sustainer delivering unrevealed thrust

Performance: speed Mach 2 at sea level and Mach 3 at altitude; range 6/180+ km (3.7/112+ miles)

Guidance: strapdown inertial for midcourse phase, and ESD Super ADAC active radar for terminal phase

Variant and application

ANS: The **Anti-Navire Supersonique** (supersonic anti-ship) missile is currently under development by Aérospatiale and MBB for service in the early 1990s as successor to the Exocet and Kormoran air-launched missiles in French and West German service respectively, though selected export customers are likely to be offered the weapon. The

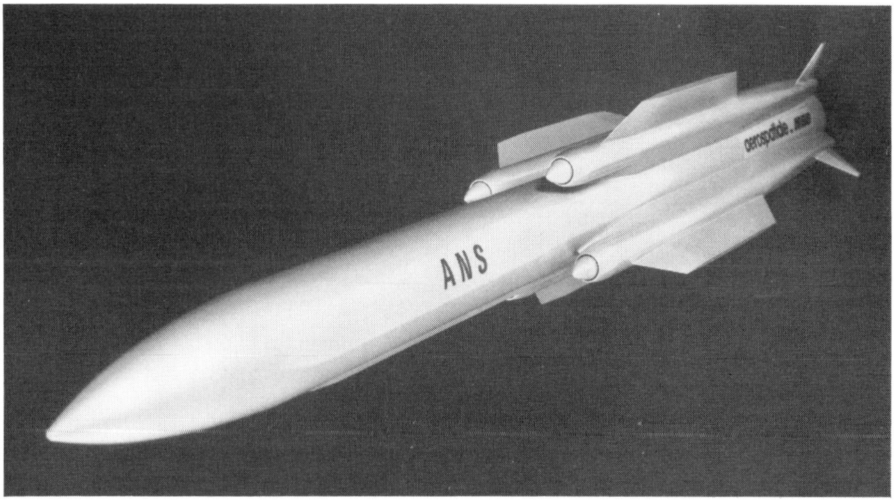

The ANS being developed jointly by Aérospatiale in France and MBB in West Germany (to replace the AM.39 Exocet and Kormoran respectively) is believed to be conceptually the world's most advanced anti-ship missile. The key feature of the design is the hybrid rocket/ramjet propulsion system, which offers high supersonic performance as a means of reducing the advance warning offered to the target by the approach of the missile. This means that although the active radar homing system is not that much better than the type carried by the Exocet, the target will have virtually no time in which to decoy the missile or alternatively destroy it with the ship's point defence gun and/or missile system.

ANS features an advanced propulsion system, with an integral rocket for acceleration to Mach 2 cruising speed, at which it becomes a hybrid rocket ramjet supplying fuel-rich gas to be mixed with liquid fuel for burning in the ramjet combustion chamber. This all makes for very high attack speed (reducing the time available to the target for the implementation of countermeasures), and the terminal guidance will probably incorporate a home-on-jam mode. The type is also planned with exceptional counter-countermeasures capability.

IMI Gabriel Mk IIIA/S

Israel

Type: air-launched medium-range tactical anti-ship missile
Dimensions: diameter 340 mm (13.4 in); length 3840 mm (151.18 in); span 1100 mm (43.31 in)
Weights: total round 558 kg (1,230.2 lb); warhead 150 kg (330.7 lb) impact and proximity delay-fused blast fragmentation
Propulsion: one solid-propellant rocket delivering 3600 kg (7,937 lb) thrust
Performance: speed transonic; range 40 km (25 miles)
Guidance: inertial plus radar altimeter for midcourse phase, and active radar for terminal phase

Variants and application
Gabriel Mk IIIA/S: This is the air-launched version of the ship-launched Gabriel Mk III, itself derived from the earlier Gabriel Mks I and II anti-ship missiles. The Gabriel Mk I was developed in the 1960s as a 21 km (13.05 mile) range missile for deployment on fast attack craft, while the Gabriel Mk II introduced a longer body for more fuel as a means of boosting range to 36 km (22.4 miles). The ship-launched Gabriel Mk III introduced a frequency-agile radar seeker, though the optical and semi-active radar homing systems of the Gabriel Mks I and II can also be used to provide greater tactical flexibility and continued operability when the enemy is using electronic countermeasures. The weapon can thus have three guidance modes, namely fire-and-forget, fire-

and-update via a data-link from a targeting helicopter, and fire-and-command using the launch vessel's radar for better targeting data. The missile cruises at 100 m (330 ft) and then descends to 20 m (66 ft) for the approach to the target, the attack phase being flown at a height pre-set at 1.5, 2.5 or 4 m (4.9, 8.25 or 13.1 ft) depending on the sea state.

The Gabriel Mk IIIA/S is the air-launched version of the ship-launched missile with a slightly longer body, reduced-span wings and lighter weight, which all contribute to higher speed but shorter range. The missile has three operating modes: range-and-bearing launch (using radar-derived data), range-and-bearing launch (using manually-entered data), and bearing-only launch.

Gabriel Mk IIIA/S ER: This is an extended-range version of the Gabriel Mk IIIA/S with a longer sustainer rocket that increases weight to 600 kg (1,322.75 lb) and range to 60 km (37.3 miles), which is about that of the ship-launched Gabriel Mk III variant. Both the Gabriel Mk IIIA/S and Mk IIIA/S ER have the same options as the basic Gabriel Mk III as detailed for the Gabriel Mk IIIA/S.

Gabriel Mk IV: This is an updated version planned for development in the late 1980s and early 1990s with a length of 4700 m (185.04 in) and powered by a small turbojet for a range of 200 km (124 miles). There will doubtless be provision for mid-course update of the inertial navigation system of this variant, whose development is anticipated in ship- and air-launched versions.

Italy

Sistel Sea Killer Mk 2

Type: helicopter-launched short/medium-range tactical anti-ship missile
Dimensions: diameter 206 mm (8.11 in) for body and 316 mm (12.44 in) for warhead; length 4832 mm (190.24 in); span 978 mm (38.5 in)
Weights: total round 340 kg (749.6 lb); warhead 310 kg (463 lb) proximity- and impact-fused semi-armour-piercing blast fragmentation
Propulsion: one SEP 299 double-base solid-propellant booster rocket delivering 4400 kg (9,700 lb) thrust for 1.6 seconds, and one SEP 300 composite solid-propellant sustainer rocket delivering 100 kg (220 lb) thrust for 73 seconds
Performance: speed 1080 km/h (671 mph); range 25 km (15.5 miles)
Guidance: combination of autopilot (azimuth) and radar altimeter (height) for midcourse phase to sea-skimming height, and SMA active radar for terminal phase

Variants and applications

Sea Killer Mk 2: The Sea Killer is a lightweight anti-ship weapon, originally designed as the Sea Killer Mk 1 for use from fast attack craft and possessing a range of only 10000 m (10,935 yards) with its 170 kg (374.8 lb) warhead as it had only a single-stage propulsion system. The Sea Killer Mk 1 used the Contraves Sea Hunter Mk 2 radar for guidance and was a sea-skimming missile now no longer in service. The Sea Killer Mk 2 introduced a tandem-propulsion system for greater range with a larger warhead, and in its ship-launched application uses the Contraves Sea Hunter Mk 4 radar for guidance. The specifically air-launched version of the Sea Killer Mk 2 is part of the **Marte** system, in which the missile is fired from a helicopter fitted with the SMA APQ-706 radar: smaller helicopters such as the Agusta (Bell) AB.212ASW can carry one missile and the appropriate guidance equipment, but larger machines such as the Agusta (Sikorsky) ASH-3 can carry two missiles, the weight of the complete installation with two missiles being 1165 kg (2,568.3 lb).

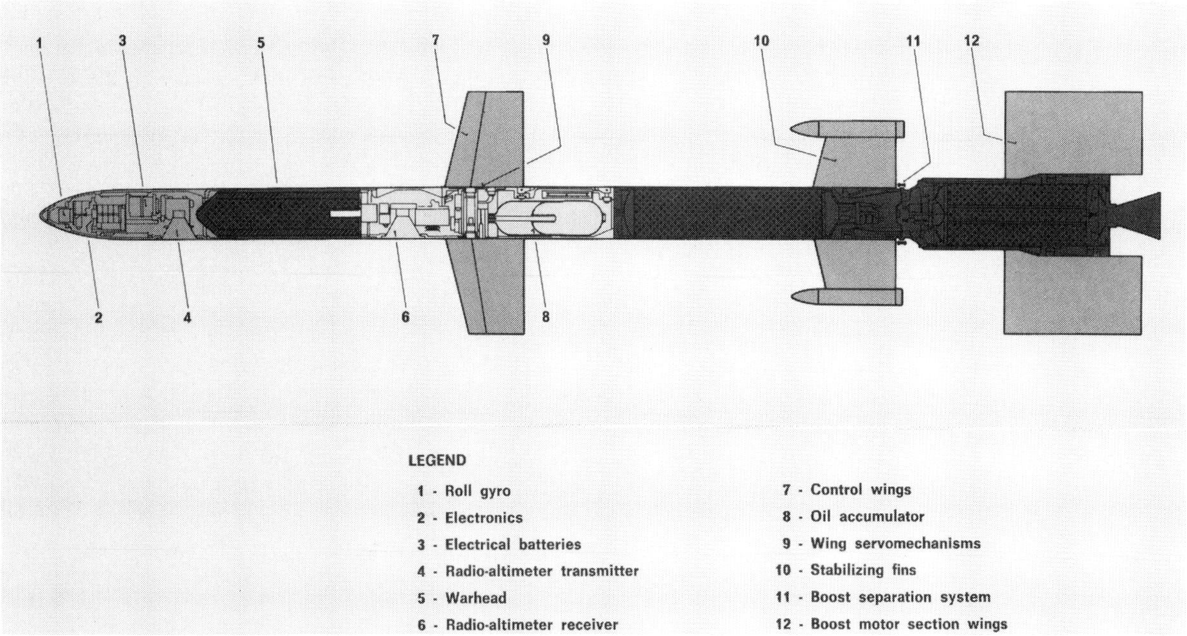

Manufacturer's artwork reveals the salient features of the early version of the Sea Killer Mk 2. The full-standard missile has a revised homing system and a larger warhead.

The original **Marte Mk 1** system uses a simple development of the Sea Killer Mk 2 missile with the original parallel-section warhead weighing 170 kg (374.8 lb) within an overall missile weight of 300 kg (661.4 lb) and length of 4700 mm (185.04 in) complete with the 1060 mm (41.73 in) booster stage; guidance over the maximum range of 20 km (12.4 miles) is entrusted to a mixture of autopilot and radar altimeter for the midcourse phase and Sistel radar beam-riding and/or optical guidance for the terminal phase. The **Marte Mk 2 system** uses the improved Sea Killer Mk 2 described in the specification above: this has a revised warhead and guidance section based on that of the Otomat surface-launched anti-ship missile. The warhead weighs 210 kg (463.0 lb) including 60 kg (132.3 kg) of explosive, and includes active radar homing to provide the missile with true 'fire-and forget' capability. This improved Sea Killer Mk 2 is also planned for the Aeritalia/Aermacchi/EMBRAER AMX and Aermacchi M.B.339 fixed-wing light attack aircraft, though the higher speeds of these launch platforms allows the missile to be used without its 1087 mm (42.8 in) booster stage.

Mitsubishi Type 80 (ASM-1) — Japan

Type: air-launched medium-range tactical anti-ship missile
Dimensions: diameter 350 mm (13.78 in); length 4000 mm (157.48 in); span 1200 mm (47.25 in)
Weights: total round 610 kg (1,344.8 lb); warhead 200 kg (440.9 lb) blast fragmentation
Propulsion: one Nissan Motors single-stage solid-propellant rocket delivering unrevealed thrust
Performance: speed 1100 km/h (684 mph); range 80 km (49.7 miles)
Guidance: Japan Aviation Electronics strapdown inertial plus TRT radar altimeter for midcourse phase, and Mitsubishi Electronics active radar for terminal phase

Variant and application

Type 80: Developed between 1973 and 1980, and introduced in 1981 as the primary anti-ship armament of the Mitsubishi F-1 attack fighter, the **ASM-1** is a simple yet effective anti-ship missile of the fire-and-forget type. The weapon is notable for its comparatively high speed. The type has also been developed into the SSM-1 coast-defence missile with a jettisonable solid-propellant booster rocket.

Norway

Kongsberg Penguin Mk III

Type: air-launched medium-range tactical anti-ship missile
Dimensions: diameter 280 mm (11.02 in); length 3170 mm (124.8 in); span 1000 mm (39.37 in)
Weights: total round 372 kg (820.1 lb); warhead 121 kg (267 lb) impact delay-fused semi-armour-piercing blast fragmentation
Propulsion: one Raufoss Ammunisjons/Atlantic Research Corporation solid-propellant rocket delivering unrevealed thrust
Performance: speed 1100 km/h (684 mph); range 60 km (37.3 miles) from an air launch
Guidance: Kongsberg Vapenfabrikk inertial for midcourse phase, and Kongsberg Vapenfabrikk infra-red for terminal phase

Variants and applications

Penguin Mk II Mod 7: The Penguin was the Western world's first anti-ship missile, the original Penguin Mk I having been conceived in the early 1960s to enter service in 1972 as part of Norway's defence against maritime invasion. This initial model was optimized for good performance in the country's peculiar coastal waters after launch from fast attack craft and other small naval platforms. The result is a missile with infra-red terminal homing, treated as a round of ammunition and launched on information supplied by the launch platform's sensors and fire-control system. This ship-launched version has a two-stage rocket and height control by an interesting pulsed-laser altimeter.

The missile's launch weight is 340 kg (749.6 lb) including the powerful warhead essentially identical with that of the US AGM-12B Bullpup air-to-surface missile, the dimensions include a length of 2960 mm (116.54 in) and span of 1400 mm (55.1 in), and the performance figures include a speed of 980 km/h (609 mph) and range of 20 km (12.4 miles). The missiles still in service are being upgraded to Penguin Mk I Mod 7 standard with the seeker of the Penguin Mk II Mod 3. The Penguin Mk II is an improved Mk I with range boosted to 30 km (18.6 miles) and weight increased to 340 kg (749.6 lb) in the ship-launched Penguin Mk II Mod 3 variant (also used by Sweden with the designation RBS 12). This has a length of 2950 mm (116.14 in) and span of 1400 mm (55.1 in). The type entered service in the early 1980s, and surviving rounds are being upgraded to Penguin Mk II Mod 5 standard with enhanced seeker performance.

The basic type has been further developed for helicopter launch as the Penguin Mk II Mod 7 (US designation **AGM-119B** for use with helicopters such as the Kaman SH-2F Seasprite and Sikorsky SH-60B Seahawk) with a weight of 385 kg (848.8 lb) and a length of 3020 mm (118.9 in). This variant includes a number of penguin Mk III improvements (notably in the seeker and signal processor), and has a new two-stage rocket and fully digital electronics, unlike the earlier Mk II variants which have analog electronics.
Penguin Mk III: This is an air-launched development (also capable of ship launch) intended primarily for the Royal Norwegian air force's General Dynamics F-16 Fighting

A Penguin III anti-ship missile, complete with protective cover over its unique infra-red homing head, awaits trials under a General Dynamics F-16A fighter at Eglin AFB in Florida.

Falcon multi-role fighters. The variant, which possesses the US designation **AGM-119A,** has a longer body, shorter-span wings, a single-stage rocket and a radar altimeter. The Penguin Mk III is a highly capable weapon programmed to fly a circuitous approach to the target via one or more waypoints after launch on the basis of radar information from the launch aircraft, or of visual sighting with data entered into the missile via the pilot's head-up display or optical sight. As with the earlier versions of the missile, the use of infra-red terminal homing (activated only when the missile has reached the target's anticipated position) gives the target vessel virtually no warning of the missile's imminent arrival, so reducing the time available for countermeasures.

Saab RBS 15F

Sweden

Type: air-launched medium/long-range tactical anti-ship missile
Dimensions: diameter 500 mm (19.69 in); length 4350 mm (171.26 in); span 1400 mm (55.1 in)
Weights: total round 598 kg (1,318.3 lb); warhead unrevealed blast fragmentation
Propulsion: on Microturbo TRI 60-2 Model 077 turbojet delivering 377 kg (831 lb) thrust
Performance: speed high subsonic; range 150 km (93.2 miles)
Guidance: preprogrammed autopilot plus radar altimeter for midcourse phase, and Philips Elektronikindustrier active radar for terminal phase

Variant and application

RBS 15F: This highly capable weapon type is due to enter service in 1988 in its 780 kg (1,719.6 lb) RBS 15M variant as a potent ship-launched anti-ship missile. After onboard programming using data derived from the ship's sensors and fire-control system, the missile is fired with the aid of two jettisonable solid-propellant rockets to fly at medium or low altitude (the former capability being essential in the Swedish archipelago) to the point at which the active seeker is turned on, whereupon the missile

Modern anti-ship missiles such as this Saab RBS 15 use air-breathing propulsion for maximum range on a given fuel weight.

drops to sea-skimming height for the attack over a maximum range of 100 km (62.1 miles). Development from 1982 has evolved the RBS 15F air-launched version, which has no requirement for the launch boosters, and is thus lighter and longer-ranged. The particularly notable feature of the RBS 15F is its extremely advanced Philips seeker, which is frequency-agile, and has digital signal processing, a choice of search modes and patterns, variable electronic counter-countermeasures, and target-choice logic to allow the missile to choose the most important of several possible targets.

USSR

AS-2 'Kipper'

Type: air-launched long-range tactical/operational anti-ship missile
Dimensions: diameter 900 mm (35.4 in); length 10000 mm (393.7 in); span 4880 mm (192.13 in)
Weights: total round 4200 kg (9,259.3 lb); warhead 1000 kg (2,204.6 lb) blast fragmentation
Propulsion: one turbojet (possibly Lyul'ka AL-5) delivering 5500 kg (12,125 lb) thrust
Performance: speed Mach 1.2; range 185 km (115 miles) from a high-altitude launch
Guidance: autopilot with radio update capacity for midcourse phase, and active radar for terminal phase

Variant and application
AS-2 'Kipper': Introduced in 1960 as the primary anti-ship weapon of the Tupolev Tu-16 'Badger-C', the AS-2 homes on targets with a large radar signature, and is obsoles-

cent by modern standards because of its size and lack of electronic sophistication. The type flies to the anticipated target position under command of an autopilot, which can be overridden by data-linked commands from midcourse guidance helicopter or aircraft, and then homes on information provided by its onboard active radar seeker. The primary target for such a massive missile is a high-value type such as an aircraft carrier. It is worth noting that range estimates for the missile vary very considerably from a minimum of 185 km (115 miles) to a maximum of 560 km (348 miles), though it is likely that the comparative inaccuracy of the guidance system and the desire to maximize the warhead has probably dictated a comparatively small fuel capacity.

AS-3 'Kangaroo' USSR

Type: air-launched long-range tactical/operational anti-ship and strategic area-attack missile
Dimensions: diameter 1850 mm (72.8 in); length 14900 mm (586.61 in); span 9150 mm (360.24 in)
Weights: total round 11000 kg (24,250 lb); warhead 800 kiloton/1 megaton thermonuclear or (possibly) 2300 kg (5,070.5 lb) blast fragmentation
Propulsion: one turbojet (probably Tumanskii R-11 or R-13) delivering 5100 or 6800 kg (11,243 or 14,991 lb) thrust
Performance: speed Mach 1.8; range 650 km (405 miles) from a high-altitude launch
Guidance: autopilot and radio command, though the range of the missile suggests that inertial or preprogrammed guidance may have been retrofitted

Variant and application
AS-3 'Kangaroo': This massive weapon entered service in 1961 as a strategic weapon carried by Tupolev Tu-95 'Bear-B' aircraft, and was intended as an area-attack weapon without any form of terminal guidance, though it is hard to see how any useful degree of accuracy could have been maintained beyond a range of about 275 km (170 miles) with such a system. The warhead also makes the weapon useful against tactical/operational naval targets such as carrier battle groups. There are indications that the missile may have been derived from the Mikoyan-Gurevich E-2A experimental fighter (with features of the same bureau's E-5 prototype), and the size of the whole vehicle makes it feasible for some type of terminal guidance package to have been retrofitted for continued capability up to the present.

AS-4 'Kitchen' USSR

Type: air-launched long-range tactical/operational anti-ship and strategic area-attack missile
Dimensions: diameter 900 mm (35.4 in); length 11300 mm (444.88 in); span 3000 mm (118.11 in)
Weights: total round 5900 kg (13,007 lb); warhead 200/350 kiloton thermonuclear or 1000 kg (2,204.6 lb) blast fragmentation
propulsion: one liquid-propellant rocket delivering unrevealed thrust

Performance: speed Mach 3.5; range 460 km (286 miles) from a high-altitude launch reducing to 300 km (186 miles) from a low-altitude launch
Guidance: inertial for midcourse phase, and active radar or passive infra-red (see below) for terminal phase

Variant and application
AS-4 'Kitchen': Introduced in 1965 and used by Tupolev Tu-22 'Blinder' and Tu-26 'Backfire' supersonic bombers, the AS-4 is a truly formidable weapon system. The type was the USSR's first genuine multi-role missile, being available with no terminal homing and a thermonuclear warhead as a strategic area attack weapon, or with a choice between active radar, passive infra-red and passive radiation terminal homing as an anti-ship weapon with either a conventional or thermonuclear warhead. The missile's flight profile in the anti-ship role is a high altitude cruise followed by a devastating dive in which the combination of hypersonic speed and small frontal area would make it a difficult target for the defences to counter effectively. The type is now tasked generally with operations against US surface battle groups, which are operationally and strategically important targets with the high electro-magnetic and thermal signatures best suited to the AS-4's terminal guidance system.

USSR

AS-5 'Kelt'

Type: air-launched long-range tactical anti-ship missile
Dimensions: diameter 900 mm (35.43 in); length 8600 mm (338.58 in); span 4600 mm (181.1 in)
Weights: total round 3400 kg (7,496 lb); warhead 1000 kg (2,204.6 lb) blast fragmentation
Propulsion: one liquid-propellant rocket delivering unrevealed thrust
Performance: speed Mach 1.2: range 230 km (143 miles) from a high-altitude launch reducing to 180 km (112 miles) from a low-altitude launch
Guidance: autopilot for midcourse phase, and active or passive radar for terminal phase

Variant and application
AS-5 'Kelt': Introduced in 1966 as replacement for the AS-1 'Kennel' in the air-launched role (and possibly using the same basic airframe), the AS-5 is used exclusively by the Tupolev Tu-16 in its 'Badger-G' variant. The AS-5 has the same basic aeroplane configuration as its predecessor, an HE warhead and more advanced avionics including two terminal homing options. These last are fitted in a nose resembling that of the SS-N-2 'Styx', and it has sensibly been suggested that in addition to its active/passive radar homing system the AS-5 also possesses a passive infra-red homing system like that of the surface-launched SS-N-2.

USSR

AS-6 'Kingfish'

Type: air-launched long-range tactical/operational anti-ship and strategic area-attack missile

Still the USSR's most important missile carrier type is the Tupolev Tu-16 'Badger', this being a 'Badger-G' carrying the multi-capable AS-6 'Kingfish' air-to-surface and anti-ship missile.

Dimensions: diameter 800 mm (31.5 in); length 10000 m (393.7 in); span 2500 mm (98.43 in)
Weights: total round 5000 kg (11,023 lb); warhead 350 kiloton nuclear or 1000 kg (2,204.6 lb) blast fragmentation
Propulsion: on storable liquid-propellant (or possibly solid-propellant) rocket delivering unrevealed thrust
Performance: speed Mach 3; range 560 km (348 miles) from a high-altitude launch reducing to 250 km (155 miles) from a low-altitude launch
Guidance: inertial for midcourse phase, and active radar or passive radiation homing for terminal phase

Variant and application
AS-6 'Kingfish': Introduced in the 1970s as a complement to the AS-4 'Kitchen', and generally carried in pairs by Tupolev Tu-16 'Badger-C (Mod)', Tu-16 'Badger-G (Mod)' and Tu-26 'Backfire-B' aircraft, the AS-6 has the same targeting and homing methods as the AS-4 (and was initially though to be development of the earlier weapon). But the AS-6 possesses greater range, and cruises at an altitude of 18000 m (59,055 ft) after launch at 11000 m (36,090 ft) before the high-supersonic dive onto the target. It is believed that the AS-6 also has a low-level attack profile and, like the AS-4, is considered a potentially decisive weapon against US carrier battle groups.

BAe Sea Eagle UK

Type: air-launched medium/long-range tactical anti-ship missile
Dimensions: diameter 15.75 in (400 mm); length 163.0 in (4140 mm); span 47.3 in (1201 mm)

Above *A weapon whose development has proceeded with great smoothness is the Sea Eagle air-launched anti-ship missile, of which two can be carried by aircraft such as this BAe Sea Harrier FRS. Mk 1. The structure of the missile is based on that of the Martel, but notably different is the rear end of the missile's body, which accommodates the Microturbo turbojet with its inlet located ventrally between the two lower wings. The nose cone covers the antenna for the very advanced active radar seeker developed by Marconi.*

Right *The Sea Eagle has a Royal Ordnance warhead weighing more than 500 lb (227 kg), and this allows even a single round to disable even major warships, as indicated by the damage caused to the decommissioned destroyer HMS* Devonshire *during Sea Eagle trials.*

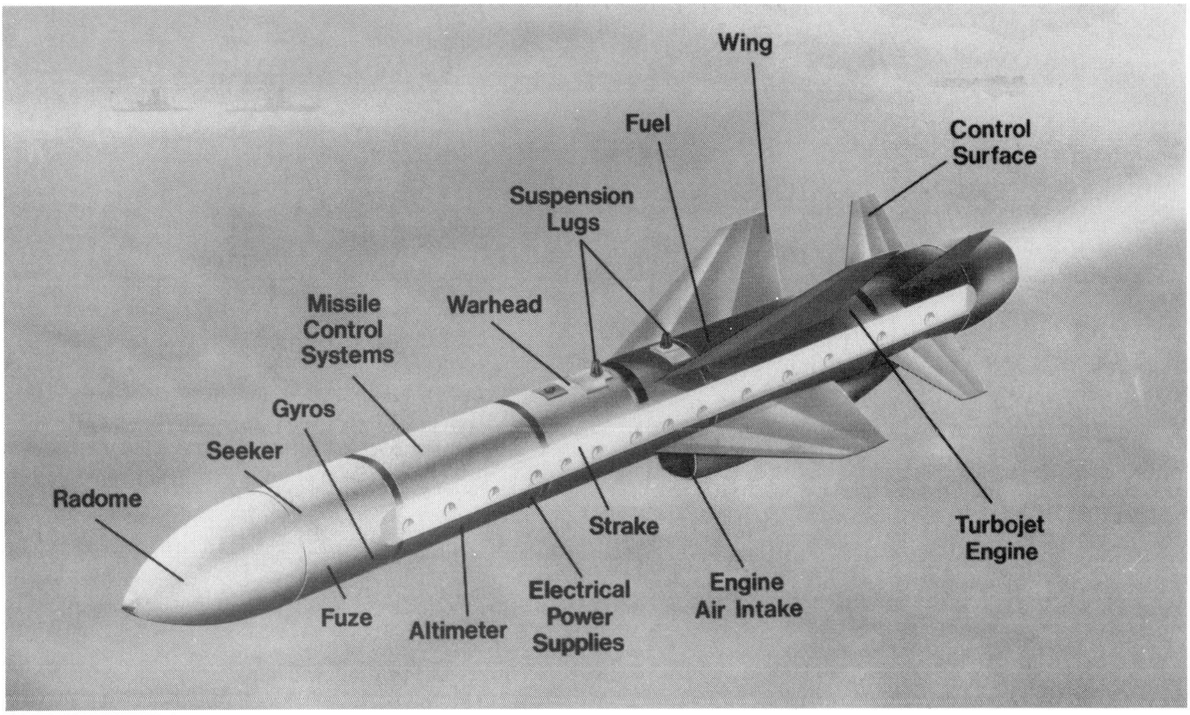

An annotated impression of the Sea Eagle indicates the most important feature of this important weapon. The type has also been developed for launch by the Indian navy's Westland Advanced Sea King Mk 42B helicopters, the lower speed of this launch platform dictating the incorporation of jettisonable booster rockets in this version, which may pave the way for a ship- or coast-launched version.

Weights: total round 1,325 lb (601.0 kg); warhead 500 lb (226.8 kg) penetrating blast fragmentation
Propulsion: one Microturbo TRI 60-1 Model 067 turbojet delivering 367 kg (787 lb) thrust
Performance: speed 685+ mph (1102+ km/h); range 68+ miles (109.4+ km) decreasing with lower launch altitude
Guidance: strapdown inertial plus Plessey radar altimeter for midcourse phase, and Marconi active radar for terminal phase

Variant and applications
Sea Eagle: The Sea Eagle began to enter service in 1986, and was derived from the P3T of 1976. The structural and aerodynamic basis of the missile is the Martel air-to-surface missile, but in this longer-range application it is fitted with a turbojet aspirated through a ventral inlet. The missile was designed mainly for aircraft launch platforms (the BAe Sea Harrier being able to carry two and the BAe Buccaneer four), and is notable for its good speed and range. The Sea Eagle has a thoroughly modern guidance system, the strapdown inertial portion having a microprocessor into which the launch platform loads data on the target's position, bearing, course and speed just before launching the Sea Eagle. The missile then cruises at low altitude to reduce the chance of the target spotting it by electro-magnetical or visual means, and the active seeker is of an advanced type. The large warhead is a potent Royal Ordnance type effective against most naval targets.

The basic missile has also been developed as the Sea Eagle SL (P5T) surface-launched version for use by coastal batteries or warships. In this role it is fitted with a pair of jettisonable solid-propellant booster rockets of the type developed to provide the missile with helicopter-launch capability for the Indian navy's Westland Sea King Mk 42s.

UK

BAe Sea Skua

Type: air-launched short-range tactical anti-ship missile
Dimensions: diameter 9.75 in (248 m); length 98.5 in (2502 mm); span 28.5 in (724 mm)
Weights: total round 325 lb (147.4 kg); warhead 44 lb (20.0 kg) blast fragmentation
Propulsion: one BAJ Vickers solid-propellant booster rocket delivering unrevealed thrust, and one BAJ Vickers solid-propellant sustainer rocket delivering unrevealed thrust
Performance: speed 685+ mph (1102+ km/h); range 22,000 yards (20115 m)
Guidance: strapdown inertial plus BAe/TRT radar altimeter for midcourse phase, and Marconi semi-active radar for terminal phase

Variant and application

Sea Skua: The Sea Skua entered premature service in 1982 as a result of the demands of the Anglo-Argentine Falklands War, and is a helicopter-launched anti-ship missile. Operational experience has combined with combat success to prove the missile an excellent weapon. The Sea Skua is designed to sink small surface units and cripple medium-sized warships. Its light weight and small size mean that it can be carried in useful numbers by helicopters such as the versatile Westland Lynx, which can be fitted with four such missiles.

The autopilot-altimeter cruise phase of the flight, at any one of the four pre-selected

The Sea Skua lightweight anti-ship missile marks a considerable divergence between British and French concepts for such a weapon. Seen here as its small solid-propellant booster rocket ignites after the missile has been dropped from a Westland Lynx HAS.Mk 2, the Sea Skua uses semi-active radar homing dependent on energy from the Lynx's Ferranti Seaspray radar reflected by the target. The French concept, as embodied in the AS.15TT, provides for a cheaper missile using command guidance based on the launcher's Thomson-CSF Agrion 15 radar.

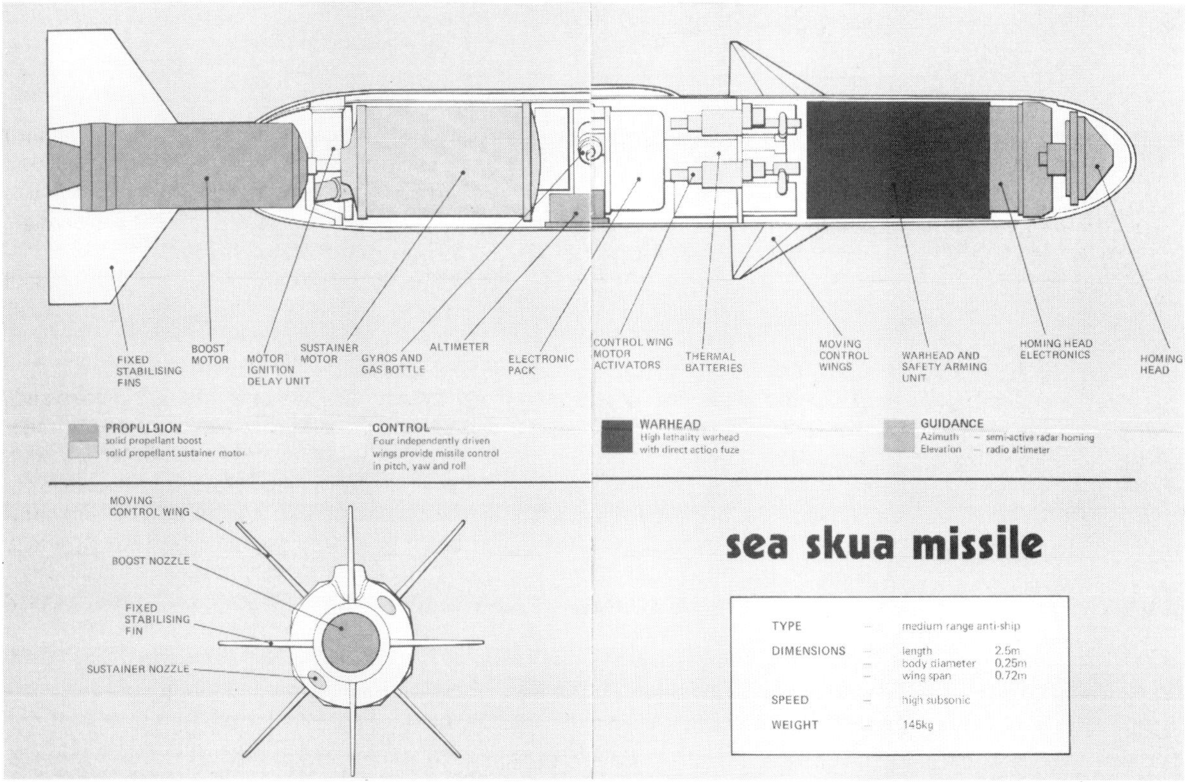

cruise altitudes, is followed by a rapid climb to the altitude at which the semi-active radar homing system can lock onto the reflections of the Ferranti Seaspray radar carried by the Lynx, though the adoption of other types as launch platforms will necessitate alterations to this system. The Sea Skua can also be launched on the basis of target and launcher information provided by electronic support measures and the Decca Tactical Air Navigation System, with the Seaspray illuminating radar activated only at the last minute to provide the target with minimum reaction time.

Manufacturer's artwork shows off the most important features of the Sea Skua lightweight anti-ship missile. An interesting feature is the location of the sustainer motor, which exhausts through laterally inclined nozzles rather than straight to the rear.

McDonnell Douglas AGM-84A Harpoon

USA

Type: air-launched medium-range tactical anti-ship missile
Dimensions: diameter 13.5 in (343 mm); length 151.0 in (3835 mm); span 36.0 in (914 mm)
Weights: total round 1,160 lb (526.2 kg); warhead 488 lb (221.4 kg) proximity and impact delay-fused penetrating blast fragmentation
Propulsion: one Teledyne Continental J402-CA-400 turbojet delivering 680 lb (308 kg) thrust
Performance: speed 645 mph (1038 km/h); range 68 miles (110 km)

Guidance: Lear-Siegler or Northrop strapdown inertial plus Honeywell APN-194 radar altimeter for midcourse phase, and Texas Instruments PR-53/DSQ-58 two-axis active radar for terminal phase

Variant and application

AGM-84A Harpoon: The Western world's most important anti-ship missile, the Harpoon was conceived in the late 1960s as a capable but comparatively cheap weapon emphasizing reliability rather than outright performance in all respects but electronic capability and range, where the use of a turbojet rather than a rocket in the sustainer role pays handsome dividends. The missile was first deployed during the mid-1970s in its RGM-84A ship-launched version with a 14,550 lb (6600 kg) thrust jettisonable Aerojet booster rocket increasing length to 180.75 in (4591 mm) and weight to 1,498 lb (679.5 kg).

This baseline missile can be fired in range and bearing mode, allowing the late activation of the active radar as a means of reducing the chances of the missile being detected through its own emissions. The bearing-only mode for earlier activation of the radar is used where the precise location of the target is not available at missile launch time. If no target is found after the low-level approach, the missile undertakes a preprogrammed search pattern, and acquisition of the target is followed in the 57 mile (91.7 km) range Block I missiles of the initial production batch by a steep pop-up climb and dive onto the target's more vulnerable upper surfaces. Currently in production is the Block II long-range type with range increased to more than 120 miles (193 km), plus variable flight profiles and increased capability against electronic countermeasures. The AGM-84A air-launched version needs no booster and is thus shorter and lighter than the RGM-84A. There is also a UGM-84A Sub-Harpoon version for submarine use, this being launched from a torpedo tube in a watertight capsule, which is discarded on the surface.

AGM-84B Harpoon: This air-launched version has a wholly sea-skimming flight profile without the pop-up feature of the AGM-84A, thus reducing the target's chances of spotting and countering the weapon.

AGM-84C Harpoon: Much improved in comparison with earlier version, this air-launched model incorporates the flight profile capabilities of the AGM-84A (pop-up) and AGM-84B (sea-skimming), and is entering service in the second half of the 1980s. The type is also available in submarine- and ship-launched versions as the UGM-84C Sub-Harpoon and RGM-84C Harpoon respectively.

AGM-84D Harpoon: This is a longer-range version incorporating Black II range capabilities with improved electronic counter-countermeasures, and provision for the preprogramming of dog-leg approach via three waypoints and for the selection of alternative terminal approaches.

SLAM: Due to start trials in 1988, the **Stand-off Land-Attack Missile** is a company-funded derivative of the Harpoon for possible carriage by US Navy and US Marine Corps Grumman A-6 Intruder and McDonnell Douglas F/A-18 Hornet aircraft. The weapon combines the airframe, powerplant and warhead of the Harpoon with the imaging infra-red terminal guidance unit of the AGM-65D Maverick air-to-surface missile, the data-link of the AGM-62 Walleye glide bomb, and a Global Positioning System receiver to create a missile 177.0 in (4496 mm) long and 1,385 lb (628.2 kg) in weight. The SLAM is designed to provide a stand-off capability against high-value targets such as power stations, harbour equipment and bridges, using inertial guidance for the mid-course phase of the flight before the missile is aligned precisely at the target through the GPS link, allowing the operator to acquire the target visually at a maximum range of 10,000 yards (9145 m), lock onto the image and depart the scene as the missile completes its attack.

Texas Instruments Sea Ray

USA

Type: air-launched short-range tactical anti-ship missile
Dimensions: diameter 8.0 in (203 mm); length 117.0 in (2972 mm); span data not available
Weights: total round 400 lb (181.4 kg); warhead 150 lb (68.9 kg) blast fragmentation
Propulsion: one Aerojet Mk 78 solid-propellant rocket delivering unrevealed thrust
Performance: speed supersonic; range 7,650 yards (6995 m)
Guidance: semi-active laser homing

Variant and application
Sea Ray: This interesting weapon is under development as a private venture by Texas Instruments to the order of an undisclosed customer (possibly South Korea) using the powerplant of the AGM-45A Shrike anti-radiation missile and the seeker system of the Paveway III series laser-homing bomb. The Sea Ray can be fired from a ship, aircraft or land launcher, and flies a ballistic trajectory until some 10 seconds from anticipated impact, when the seeker is activated. In its ship-launched version the weapon is associated with the Sea Tiger fire-control system, and enters service in 1987 and 1988. The Sea Ray provides the launch customer and other potential operators with a comparatively light yet effective weapon that can be fitted to many launch platforms, the guidance system offering the possibility of third-party targeting for greater tactical flexibility.

MBB Kormoran

West Germany

Type: air-launched medium-range tactical anti-ship missile
Dimensions: diameter 344 mm (13.54 in); length 4400 mm (173.23 in); span 1000 mm (39.37 in)
Weights: total round 600 kg (1,322.75 lb); warhead 165 kg (363.75 lb) impact delay-fused radial blast fragmentation
Propulsion: two SNPE Prades solid-propellant booster rockets each delivering 2750 kg (6,063 lb) thrust for 1 second, and one SNPE Eole IV solid-propellant sustainer rocket delivering 285 kg (628 lb) thrust for 100 seconds
Performance: speed 1160 km/h (721 mph); range 37 km (23 miles)
Guidance: SFENA-Bodenseewerk strap-down inertial plus TRT radar altimeter for mid-course phase, and Thomson-CSF RE576 active-passive radar for terminal phase

Variants and application
Kormoran: Developed from the Nord AS.34 missile projected in France, the Kormoran entered West German naval service in 1977 as the primary anti-ship weapon of Lockheed F-104G Starfighters operated by the Marineflieger; the weapon is now used on the Panavia Tornadoes operated by the same service and by the Italian air force. The missile cruises at a height of 30 m (100 ft) to the approximate location of the target, then descends to wave-top height for the present passive or active radar attack. The warhead is a particularly impressive feature of the missile, being designed to penetrate up to 90 mm (3.54 in) of metal before the 16 radially-disposed charges detonate to pierce the ship's bottom, decks and internal bulkheads.
Kormoran 2: This improved 630 kg (1,388.9 lb) weapon has a new Thomson-CSF solid-

Though its design concept and general configuration reveals its French origins, the Kormoran was produced in West Germany by MBB as a potent anti-ship missile for carriage by attack aircraft such as this Lockheed F-104G Starfighter.

state radar seeker (with digital signal processing for higher hit probability and enhanced resistance to electronic countermeasures), greater range, and a 220 kg (485.0 lb) warhead of greater penetration capability. The Kormoran 2 is due to enter service on the Tornado in the late 1980s.

TORPEDOES

DCN/ECAN L4

France

Type: air-launched anti-submarine torpedo
Dimensions: diameter 533 mm (21.0 in); length (basic weapon) 3033 mm (119.41 in) or (with parachute stabilizer pack) 3130 mm (123.23 in)
Weights: total round 540 kg (1,190.5 lb); warhead 104 kg (229.3 lb) impact- or proximity-fused HE
Propulsion: one electric motor driving contra-rotating propellers
Performance: speed 30 kts (55.6 km/h; 34.5 mph); range 5500 m (6,015 yards); maximum operational depth 300 m (985 ft)
Guidance: active acoustic homing

Variant and applications

L4: This useful lightweight torpedo was developed by the Direction des Constructions Navales specifically for airborne applications, notably as the primary air-launched anti-submarine weapon of patrol aircraft (typically the Breguet Br.1050 Alizé and the Dassault-Breguet Atlantic as well as its upgraded Atlantique successor) and as the warhead of the Malafon anti-submarine missile launched by surface units of the French navy. The torpedo designed mainly for operations against submerged targets moving at not more than 20 kts (37 km/h; 23 mph), and of the L4's two variants the later type is optimized for anti-submarine engagements in shallow water.

After dropping into the water under the stabilizing and retarding effect of its parachute, the torpedo describes a circular orbit as its seeker looks for a target. As soon as such a target has been detected, the torpedo heads for it and its warhead is finally detonated by an impact or acoustic proximity fuse. The type has been modernized in recent years, and now has a capability against submarines anywhere between periscope depth and 300 m (985 ft). There is also a 570 kg (1,256.6 lb) ship-launched version of the L4, and this is 3300 mm (129.92 in) long.

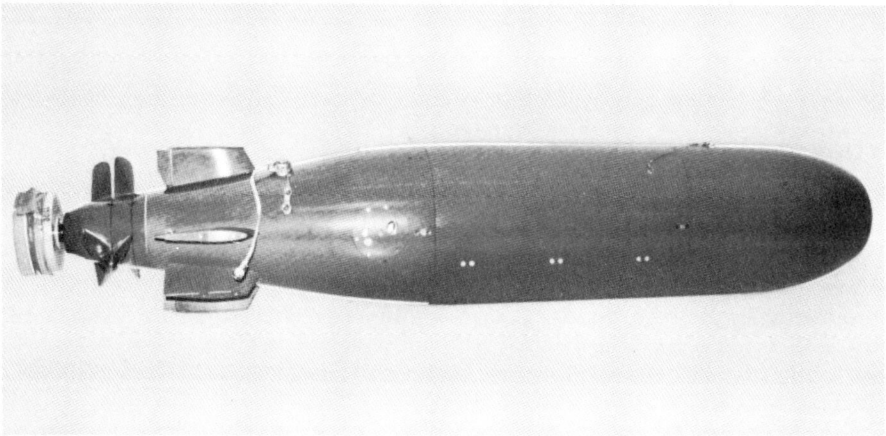

The L4 torpedo is unusual in air-launched weapons of its type in being a full-calibre 533-mm (21-in) weapon, which results in a stubby appearance.

France

DCN/ECAN Murène

Type: air-launched lightweight anti-submarine torpedo
Dimensions: diameter 324 mm (12.75 in); length 2960 mm (116.54 in) with parachute stabilizer pack
Weights: 295 kg (650.4 lb) complete with 15 kg (33.1 lb) parachute stabilizer pack; warhead 59 kg (130.1 lb) shaped-charge HE
Propulsion: one 100 kW (134 hp) electric motor driving contra-rotating propellers
Performance: speed 40 kts (74 km/h; 46 mph) for search and 55 kts (102 km/h; 63 mph) for attack; range 10000 m (10,936 yards) at 55 kts; endurance 12 minutes; operational diving depth 30/1000+ m (100/3,280+ ft).
Guidance: advanced computerized search programme for midcourse phase, and active or passive acoustic terminal phase

Variant and application
Murène: Currently under development for an in-service date of 1991, the Murène is one of the most capable torpedo designs in the world, and has been designed for capability against conventionally or nuclear-powered submarines operating slowly (and thus quietly) over the continental shelf, or at speed (and thus more noisily) in the ocean depths. The electric motor is powered by a silver oxide/aluminium battery which makes use of sea water to dissolve the solid electrolyte. The Thomson-SINTRA homing system occupies the nose of the torpedo and comprises by the three powerful sonar systems (using four transducer arrays in the form of two four-unit arrays disposed laterally, one four-unit array disposed to 'look' obliquely forward and downward, and one 30-unit array 'looking' forward). These listen for echoes returned from ahead, each side and the bottom, the last of accurate determination of depth. Three separate microprocessors compare target indications and choose potentially the most vulnerable, directing the SFIM-designed Capitole guidance system (which uses four separate 16-bit microprocessors) to bring the torpedo to maximum speed and to a course parallel with that of the target before turning in to strike at the optimum right angle to the hull in the area of the target's control room. The shaped-charge HE warhead probably fires a slug of hot metal, and can apparently pierce some 1200 mm (47.24 in) of metal, offering the very real capacity to deal with the most advanced submarines: in trials the warhead has shown the ability, at a 50° impact angle, to penetrate 10 mm (0.79 in) of outer hull, 2.0 m (6.56 ft) of ballast water, 40 mm (1.57 in) of pressur e hull and an additional 300 mm (11.81 in) beyond that. The French design team at the Etablissement des Construction et Armes Navales is already planning an improved version of the Murène with a lithium/thionyl chloride battery.

Italy

Whitehead Motofides A 244/S

Type: air-launched lightweight anti-submarine torpedo
Dimensions: diameter 324 mm (12.75 in); length 2700 mm (106.3 in)
Weights: total round 221 kg (487.2 lb); warhead 34 kg (74.96 lb) shaped-charge HE
Propulsion: one electric motor driving contra-rotating propellers
Performance: speed 30 kts (55.5 km/h; 34.5 mph); range 6500 m (7,110 yards); maximum operational diving depth not revealed
Guidance: active or passive acoustic homing.

Left *The Whitehead A 244/S is an advanced lightweight torpedo specially configured for shallow-water operations against quiet targets. The type is seen here in the classic delivery mode used by the Italian naval air arm, being dropped from an Agusta (Bell) AB.212ASW which has the transducer of its Bendix ASQ-13 sonar 'dunked' to provide up-to-the-second targeting data. It is unlikely that the parachute retarder/stabilizer will deploy fully before the torpedo enters the water, whereupon the parachute is discarded.*

Below *Manufacturer's artwork shows off a typical installation of the Whitehead A 244/S lightweight torpedo on a helicopter or fixed-wing aircraft.*

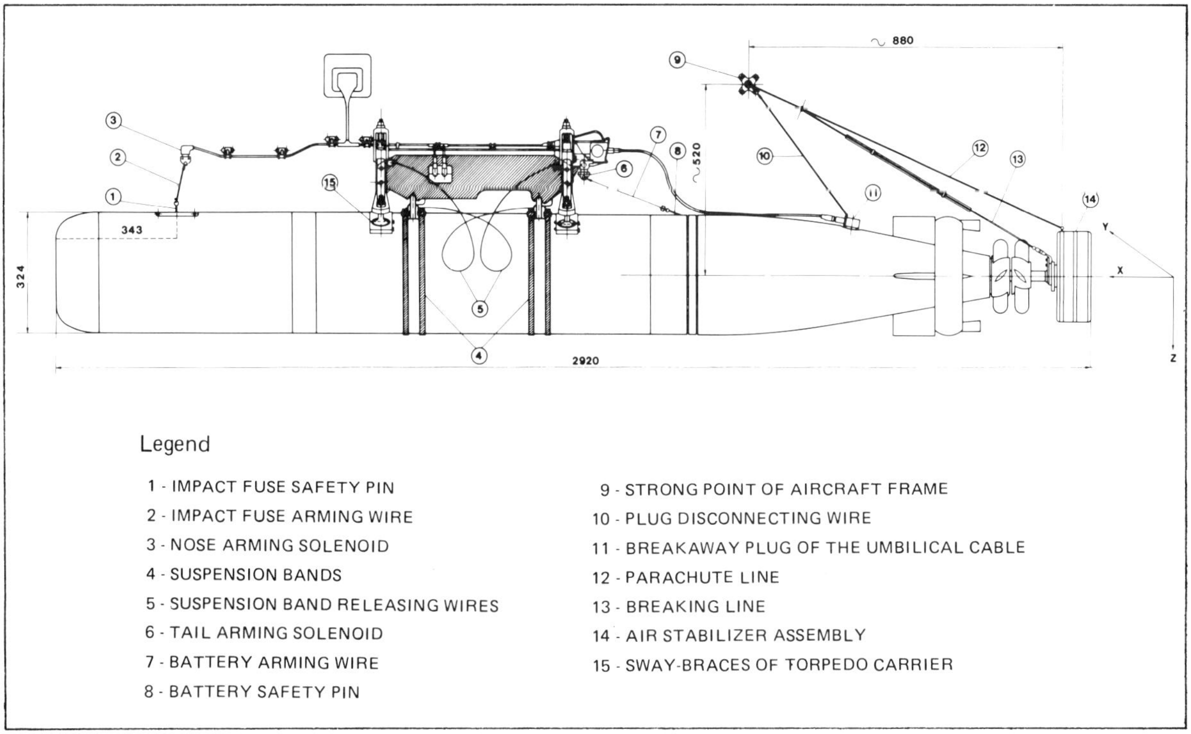

An Agusta (Sikorsky) ASH-3 heavy anti-submarine helicopter drops one of its maximum four Whitehead A 244/S anti-submarine torpedoes.

The moment just after release of a Whitehead A 244/S lightweight anti-submarine torpedo.

The attachment strops fall away and a Whitehead A 244/S torpedo falls free of its launch aircraft, the static line beginning to tighten before it pulls the parachute stabilizer from its pack to the rear of the contra-rotating propellers.

Variants and application

A 244: This lightweight torpedo was designed as successor to the US Mk 44 type, and was thus conceived for air and surface launching with particular emphasis on the type of shallow-water capability required for successful use in the acoustically-difficult waters of the Mediterranean. After being dropped by its launch aircraft (both fixed- and rotary-wing platforms are possible), the torpedo is stabilized in flight by a parachute attached to the tail, and the Selenia AG70 seeker is activated as the torpedo enters the water.

A 244/S: This is an advanced version of the basic torpedo with the Elsag CIACIO-S seeker system, which is capable of active or passive use. Through the use of special signal modulation and processing, the seeker possesses the ability to discriminate between the real target and target-launched decoys even in conditions of heavy acoustic reverberation. The seeker head can search through a large volume of water using pre-formed multiple beams which follow self-adaptive search patterns; other features of the CIACIO-S seeker and associated guidance system are preprogrammed combinations of signal-processing techniques, spatial filtering, the ability to inaugurate tactical manoeuvring to match the torpedo to the changing operational situation, and the ability to select from an assortment of search, attack and second attack courses.

A 290: This is an advanced version of the A 244 currently under development with an upgraded CIACIO-S seeker and greater power for an attack speed of 50 kts (92.5 km/h; 57.5 mph).

FFV Tp 422

Sweden

Type: air-launched lightweight anti-submarine and anti-ship torpedo

Dimensions: diameter 400 mm (15.75 in); length 2440 mm (96.06 in) or 2600 mm (102.36 in) with wire-guidance section

Weights: total round 298 kg (657.0 lb); warhead 45 kg (99.2 lb) shaped-charge HE
Propulsion: electric motor driving contra-rotating propellers
Performance: speed 15 or 25 kts (27.8 or 46.3 km/h; 17.3 or 28.8 mph); range 20000 m (21,875 yards) at 15 kts or 10000 m (10,935 yards) at 25 kts; maximum operational diving depth not revealed
Guidance: external command via trailing wire for midcourse phase, and passive acoustic for terminal phase

Variants and application
Tp 422: This is the baseline model of an important Swedish anti-submarine and anti-ship torpedo family, and entered service in 1983 mainly on the Swedish naval air arm's Kawasaki (Boeing Vertol) 107 twin-rotor helicopters. The torpedo can in theory also be carried by subsonic fixed-wing aircraft, though it is difficult to see how in this application the wire-guidance system could be used to maximize this capable torpedo's attack profile. The weapon is powered by a silver/zinc battery, and is unique in modern Western air-launched torpedoes in its diameter and control system. The former is an intermediate size between that of all other air-launched torpedoes and that of the full-calibre weapons carried by submarines and some surface vessels: the control system employs wire-guidance from the tactical system (sensors and analysis equipment) carried in the parent aircraft. The torpedo switches to a self-adaptive passive acoustic homing for the terminal phase of its attack, or in the event of a break in the guidance wire, this allowing the onboard computer system to make best use of correlative principles for accuracy under difficult sonar conditions. The weapon has two speed options (resulting in reduced range with the higher option), and a change in speed can be made at operator command via the guidance wire, or be pre-set in the onboard guidance system.
Tp 427: This is an export model with changes to the guidance and fusing systems so that Swedish operating frequencies are not compromised.

Sweden

FFV Tp 432

Type: air-launched lightweight anti-submarine and anti-ship torpedo
Dimensions: diameter 400 mm (15.75 in); length 2850 mm (112.2 in)
Weights: total round 280 kg (617.3 lb); warhead 45 kg (99.2 lb) shaped-charge HE
Propulsion: electric motor driving contra-rotating propellers
Performance: speed 15, 25 or 35 kts (27.8, 46.3 or 64.9 km/h; 17.3, 28.8 or 40.3 mph); range 20000 m (21,875 yards) at 15 kts; maximum operational diving depth not revealed
Guidance: external command via trailing wire for midcourse phase, and passive acoustic for terminal phase

Variants and application
Tp 432: Entering service in 1987, the Tp 432 is an improved development of the Tp series with digital microprocessor guidance, modular construction, inbuilt test equipment and special features to make possible attacks on quiet submarines operating in shallow water. Greater tactical flexibility is provided by a choice of three running speeds (including one considerably greater than that attainable by the Tp 42 series) and appropriate ranges. However, the air-launched version has a smaller battery than the 310 kg (683.4 lb) ship-launched version, and cannot therefore match the heavier variant's ranges of 30000, 20000 and 10000 m (32,810, 21,875 and 10,935 yards) at 15, 25 or 35 kts respec-

tively. Should the guidance wire break during the torpedo's mid-course phase, the onboard computer system takes over to calculate the anticipated target position, and then guides the weapon to this location where one of the several possible search patterns is initiated. There is also an exercise version of the torpedo with the same homing system, but fitted with recording equipment and a recovery balloon in place of explosive.
Tp 43XO: This is the export version of the Tp 432, bearing the same relationship to the Swedish version as the Tp 427 to the Tp 422. The Tp 43XO can also be fitted with different propulsion systems to suit customer requirements.

Soviet torpedoes — USSR

Virtually nothing of any real substance has appeared in open literature about the torpedoes used by the Soviets. This stricture covers the whole spectrum of torpedo types, but applies perhaps most strongly to the anti-ship and anti-submarine weapons carried and delivered by Soviet helicopters and maritime patrol aircraft. It is generally believed, however, that these air-launched weapons have a diameter of 406 mm (16 in), and that the torpedoes are produced in two basic lengths. The shorter models are probably medium-range weapons for employment against submarines dived to moderate depths, and use impact fusing for maximum destructive effect. The lengthier models are probably longer-range weapons designed for employment against surface vessels as well as submarines, and use a combination of impact and magnetic-influence fusing, the latter being designed to ensure detonation under the target vessel as a means of breaking its back.

Some Soviet aircraft, including the Mil Mi-14 'Haze' helicopter and fixed-wing machines such as the Beriev M-12 'Mail', Ilyushin Il-38 'May' and Tupolev Tu-142 'Bear', may carry torpedoes of larger diameter and length for use against US Navy carrier battle groups. These more substantial weapons, with a diameter of 660 mm (26 in) and a length of 9144 mm (360.0 in). They are almost certainly capable of carrying a nuclear warhead in place of the conventional types, whereas the smaller weapons are generally reckoned to have conventional warheads only.

The 660 mm torpedo is likely to have an advanced guidance system (the submarine version is thought to have wire guidance with an autonomous terminal phase, probably replaced by a 'fire-and-forget' autonomous type in the air-launched version), and is thought to have range options of 50 km (31.1 miles) at 50 kts (93 km/h); 58 mph) or 100 km (62.1 miles) at 30 kts (56 km/h; 35 mph).

The 406-mm torpedoes carried by the Kamov Ka-25 'Hormone' and Kamov Ka-27 'Helix' anti-submarine helicopters are believed to possess a wire-guidance option, but the standard homing method for Soviet torpedoes is acoustic (active or passive sonar), though another possibility is wake-homing for the larger weapons.

Marconi Sting Ray — UK

Type: air-launched lightweight anti-submarine torpedo
Dimensions: diameter 12.75 in (324 mm); length 102.25 in (2597 mm)
Weights: total round 585.2 lb (265.4 kg); warhead 88 lb (40 kg) shaped-charge HE
Propulsion: one electric motor driving a pump jet

Seen with an armament of two Sting Ray anti-submarine torpedoes (only the starboard weapon being visible) is a Westland Lynx-3 demonstrator. Though this type is not to enter production, its features are to be used in other Lynx variants.

Performance: speed 45 kts (51.8 mph; 83.4 km/h); range 12,150 yards (11110 m): operational diving depths 3,280+ ft (1000+ m)
Guidance: advanced computerized search programme for midcourse phase, and active or passive acoustic terminal phase

Variant and applications

Sting Ray: This is the world's most advanced lightweight torpedo, a highly capable weapon designed for launch from surface ships, aircraft (both fixed and rotary-wing types) and missiles such as the Australian Ikara, and capable of engaging all types of submarine target in shallow or deep waters. The keys to the torpedo's capabilities are high speed (resulting from a combination of powerful motor and pump-jet propulsion), a strong structure for the deep-diving capability required to tackle the latest generation of Soviet titanium-hulled submarines, and an extremely advanced guidance package. In its airborne application, the torpedo is dropped under a retarding parachute only after the onboard tactical system of the launch aircraft has detected and localized the target submarine.

After hitting the water at an oblique angle and discarding the parachute, the Sting Ray initiates a pre-programmed search pattern designed to maximize the acoustic homing system's chances of detecting the target. Once the target has been acquired, the torpedo's fully programmable onboard computer guidance system and multi-mode, multi-beam active/passive sonar make it very difficult for the target to escape: the advanced software of the torpedo's computer system allows it to filter out background and decoy noise, and to make a direct interception (even of a manoeuvring target) rather than adopt an energy-expensive tailchase pursuit. The speed of the torpedo also makes it difficult for a target to outrun the weapon, whose powerful shaped-charge warhead is detonated by an impact fuse to ensure that maximum penetration is achieved.

USA

Mk 44 Mod 1

Type: air-launched lightweight anti-submarine torpedo
Dimensions: diameter 12.75 in (324 mm); length 100.75 in (2559 mm)

Weights: total round 433 lb (196.4 kg); warhead 73 lb (33.1 kg) HE
Propulsion: one electric motor driving contra-rotating propellers
Performance: speed 30 kts (34.5 mph; 55.6 km/h); range 6,000 yards (5485 m); operational diving depths ?/1,000 ft (?/305 m)
Guidance: active sonar homing

Variants and applications
Mk 44 Mod 0: The Mk 44 torpedo was accepted for service in 1956, and was used in three main applications: as an air-launched weapon (from fixed- and rotary-wing launch platforms), as a ship-launched weapon (using the Mk 32 launcher system tied into the parent ship's analog-electronics underwater fire-control system), and as the payload of the RUR-5A ASROC long-range anti-submarine rocket system. The Mk 44 was the West's standard lightweight anti-submarine torpedo throughout the 1960s and well into the 1970s, but is now obsolete against the current generation of high-speed and quiet-running submarines. The weapon is fed with course and depth settings via an umbilical cable as late as possible before launch (to optimize this limited-range weapon's chances of acquiring the target), and its battery is activated by sea water. The acquisition of the homing system is comparatively poor by modern standards, at 650 yards (595 m), and this makes it doubly important that target data are supplied to the torpedo's onboard system as accurately (and late) as possible. The initial Mk 44 Mod 0 has a weight of 425 lb (192.8 kg), a length of 100.0 in (2540 mm) and a 75 lb (34 kg) HE warhead.
Mk 44 Mod 1: This is a slightly improved version with detail modifications in comparison with the Mk 44 Mod 0.

Aerojet Mk 46 Mod 1 USA

Type: air-launched anti submarine torpedo
Dimensions: diameter 12.75 in (324 mm); length 102.0 in (2591 mm)
Weights: total round 508 lb (230.4 kg); warhead 95 lb (43.1 kg) HE
Propulsion: one Otto monopropellant thermo-chemical engine driving contra-rotating propellers
Performance: speed 45 kts (51.8 mph; 83.4 km/h); range 12,000 yards (10975 m) at 50 ft (15 m) depth decreasing to 6,000 yards (5485 m) at 1,500 ft (455 m) depth; operational diving depths 1,500 ft (455 m)
Guidance: active/passive sonar homing

Variants and applications
Mk 46 Mod 0: The Mk 46 programme was initiated during 1960 in response to the perceived obsolescence of the Mk 44 in face of faster, quieter and deeper-diving Soviet submarines. Development of the Mk 46 proceeded quite rapidly to use the same launcher systems as the Mk 44, and the initial Mk 46 Mod 0 began to enter service in 1966, remaining the principal air-launched variant of the weapon right up to the present. The Mk 46 is currently the West's most important lightweight anti-submarine torpedo in both numerical and tactical terms. The initial Mk 46 Mod 0 differs from all others of the series in using a solid-propellant motor. This variant is 105.0 in (2667 mm) long and weighs 568 lb (257.6 kg), but is otherwise comparable to its successor variants. On entering the water it discards its retarding parachute, and then starts a helical search pattern. After acquiring the target at a maximum range estimated at 650 yards (595 m)

it closes in for the kill. The onboard guidance system is capable enough for the weapon to make repeated attacks should the first not lead to warhead detonation.
Mk 46 Mod 1: This is an improved version introduced in 1967 with an Otto mono-propellant engine after difficulties had been experienced with the servicing of the solid-fuel engine of the Mk 46 Mod 0.
Mk 46 Mod 2: This is an improved derivative of the Mk 46 Mod 1 introduced in 1972. The subsequent Mk 46 Mod 4 is basically the same as the Mk 46 Mods 1 and 2, but designed for use in the Mk 50 CAPTOR (enCAPsulated TORpedo) mine system, comprising a Mk 46 torpedo in an environmentally-controlled capsule and moored on the seabed in submarine chokepoints for triggering as a hostile submarine passes overhead.
Mk 46 Mod 5 NEARTIP: This is a modified version designed under the leadership of Honeywell to maintain a capability (albeit a limited capability) against Soviet submarines covered with 'Clusterguard' anechoic coatings to reduce their acoustic reflectivity by some 33%, and thus effectively reducing the Mk 46 torpedo's target-acquisition range by the same factor. NEARTIP (NEAR-Term Improvement Program) torpedoes have been produced as new-built and converted weapons, and feature improved motors and control units, upgraded guidance systems for repeated attacks, and more sensitive sonar transducers.

USA

Honeywell Mk 50 Barracuda

Type: air-launched lightweight anti-submarine torpedo
Dimensions: diameter 12.75 in (324 mm); length 114.0 in (2896 mm)
Weights: total round 800 lb (362.9 kg); warhead 100 lb (45.4 kg) shaped-charge HE
Propulsion: one stored chemical energy propulsion system with a closed-cycle steam turbine driving a pump jet
Performance: speed 55+ kts (63.3+ mph; 101.9+ km/h); range 15,000 yards (13715 m); operational diving depths 1,970 ft (600 m)
Guidance: advanced computerized search programme for midcourse phase, and active or passive acoustic terminal phase

Variant and applications
Mk 50 Barracuda: This is an advanced weapon in the same cybernetic category as the British Sting Ray torpedo, and its development was begun in 1975 as the Advanced Light-Weight Torpedo to supersede the Mk 46 even in the latter's most capable Mk 46 Mod 5 NEARTIP form. Features required of the new torpedo were clearly better range, speed and maximum depth (all needed for successful engagements with the latest types of Soviet submarine), as well as a highly capable computer guidance system and improved sonar capability, the latter in both active and passive modes, with emphasis on the acquisition of targets at the maximum possible range. The result is expected to enter service in about 1990, and particular features are considerable diving depth (though still not to the depths of which the latest Soviet submarines are thought capable) without degradation of range performance, and a self-adaptive guidance system (based on the AYK-14 digital computer) to make possible multiple attacks and energy-conserving interceptions rather than tailchase pursuits.

AIR-LAID SEA MINES

Thomson-SINTRA sea mines

France

Based at Brest, Thomson-SINTRA is primarily responsible for the development and production of the sea mines used by the French forces. The French navy is a keen exponent of submarine offensive minelaying (using the 850 kg/1,873.9 lb multi-fused TSM 3510 [otherwise known as the MCC 23] dispersed through standard torpedo tubes) and surface ship defensive minelaying (using the 1500 kg/3,306.9 lb TSM 3530 [otherwise known as the MCT 15] dispensed from stern rails). A certain amount of air minelaying capability is also envisaged; however, the primary platform for this role being the Dassault-Breguet Atlantic 1, to be supplemented and eventually replaced by the Atlantique 2 (sic), which can each carry nine 250 kg (551.1 lb) seabed mines in the lower-fuselage weapons bay.

Italian sea mines

Italy

Italy has a long and enviable record as an exponent of the sea mine, and this reputation is still deserved today as a result of the number and quality of sea mine types currently produced by the Italian Valtec Italiana and MISAR firms.

The main product of the first organization is the **Valtec VS SM600**, an extremely advanced multi-sensor offensive bottom mine that can be laid from aircraft, ships and submarines with submarines and medium/large-tonnage surface vessels as its main targets. The weapon is a cylinder 533 mm (21 in) in diameter and 2750 mm (108.27 in) long, weighing a total of 780 kg (1,719.6 lb) including 600 kg (1,322.75 lb) of explosive. The weapon can be laid in water between 10 and 150 m (33 and 492 ft) deep, with five safety devices and a delayed-arming system preset for any interval between 1 and 32 hours. The mine has one pressure, one magnetic and two acoustic (low- and audio-frequency) sensors, and its sophisticated processor system allows the mine to deal with a specific type of vessel or with any large target of approximately the right type. The VS SM600 has automatic 'sterilization' (self-neutralization) and advanced anti-removal systems.

MISAR is the other main Italian producer of sea mines, its range including the excellent **MISAR MR-80.** This is again a cylindrical bottom mine designed for offensive use after being laid by aircraft, ships or submarines. The weapon has a diameter of 533 mm (21 in) and a length varying from 2750 mm (108.27 in) in the **MR-80 modello A** to 2096 mm (82.52 in) in the **MR-80 modello B** and 1646 mm (64.8 in) in the **MR-80 modello C.** These three variants weigh 1070, 820 and 655 kg (2,358.9, 1,807.8 and 1,444.0 lb), and contain 870, 630 and 460 kg (1,918.0, 1,388.9 and 1,014.1 lb) of HBX-3 or similar explosive respectively. The mine has a shelf life of 30 years (with 5-year maintenances) and an operational life of 999 days, after which an automatic 'sterilization' programme is initiated by the quartz timing mechanism. The MR-80 has one magnetic, one pressure and two acoustic (low- and audio-frequency) sensors feeding data to an extremely advanced processing and 'final logic' unit able to detonate the mine against the right target while lying on the bottom at any depth between 8 and 300 m (26 and 985 ft).

The MR-80 was introduced in the early 1980s, and since that time the company has introduced two considerably more sophisticated versions, the **MP80** and **MRP** to counter the efforts of more capable sweeping technology. Both the variants retain the mechanical features of the MR-80, but have an updated firing mechanism and, most significantly, an improved processor section using four digital microprocessors to compare the signatures of potential targets with an onboard threat library to allow the selection of the most important target. The MP80 is designed for use by NATO navies, while the MRP is the export model with features that do not compromise the NATO model. The MRP can be laid at any depth between 5 and 300 m (16 and 985 ft), is 2096 mm (82.52 in) long, and weighs in the order of 780 kg (1,719.6 lb).

The same company also produces a lightweight shallow-water mine designed for use against landing craft and the like. This is the **MISAR Manta,** which can be laid by light surface vessels, frogmen and helicopters. The design is similar to that of land mines in that it is a comparatively small conoidal design, but it contains 140 kg (308.6 lb) of TNT or 170 kg (374.8 lb) of HBX-3 in an overall weight of 220 or 240 kg (485.0 or 529.1 lb). The mine is 470 mm (18.5 in) high and 980 mm (38.58 in) in diameter, and has an operational life of more than a year at depths between 2.5 and 100 m (8.2 and 330 ft). A delay circuit allows the mine to be activated anything up to 63 days after being laid, and an acoustic/magnetic firing mechanism is included.

USSR

Soviet sea mines

There is little doubt that the USSR is currently the world's most avid exponent of sea mine warfare, and is now believed to have in its inventory more than 400,000 weapons of this type for laying by surface vessels, by submarines (both conventionally- and nuclear-powered types) and by aircraft of Soviet naval aviation. There are four basic categories of Soviet mine: defensive moored mines, offensive bottom mines, offensive moored anti-submarine mines, and nuclear mines. All can be laid by aircraft.

The defensive moored mine comes in a number of small, medium and large sizes all characterized by an inertial firing mechanism using contact-electrical, contact-mechanical or influence activating systems. The two smallest mines of this type are known as the **YaRM** and **YaM:** these are both spherical contact mines intended mainly for lake, river and coastal use, the former type containing 3 kg (6.6 lb) of explosive and the latter 20 kg (44.1 lb) of explosive. In the medium category there is the spherical contact **M08/39,** an elderly but still useful coastal barrier type with some 120 kg (264.6 lb) of explosive. In the large category there are three cylindrical mines each containing more than 200 kg (440.9 lb) of explosive: the **KB** is a contact type, the **KRAB** is an influence type with a 30 m (98.5 ft) target detection range, and the **MAG** is an antenna type with 30 m (98.5 ft) antennae above and below the weapon itself. The KRAB and MAG are designed largely for the creation of coastal anti-submarine barriers, the former in normal waters and the latter in waters with strong tides or currents.

The offensive bottom mine is particularly important in Soviet military thinking, and comes in two basic forms as the **AMD-500** and **AMD-1000,** the numerical suffix indicating the variants' weights in kilogrammes. The AMD-500 contains about 300 kg (661.4 lb) of explosive, and the AMD-1000 some 700 kg (1,543.2 lb) of explosive. Both types can be laid by aircraft, there being suggestions that the variants for these roles are designated **KMD-500** and **KMD-1000.** The mines are generally deposited in water between 4 and 70 m (13 and 230 ft) deep, but the AMD-1000 is probably effective in the anti-submarine role down to a laying depth of 200 m (656 ft). Both sizes are available in four models: the first has a magnetic influence fuse working on either the intensity of the horizontal or vertical component of the target's magnetic field, or on the rate of change of the target's magnetic field; the second has an acoustic influence fuse working on either or both low- and high-noise; the third has a pressure influence fuse working on the pressure reduction of the column of water next to the mine caused by the passage of a potential target; and the fourth has a combination fuse using two or all three of the above (magnetic/pressure or acoustic/pressure being the most common combinations) and a ship-counter mechanism.

The offensive moored anti-submarine mine is available in three basic variants: two of these can be laid by aircraft, ship or submarine (and are known as rising mines), while the third can be laid only by submarine or ship (and is known as the underwater electric potential mine). All three models were developed for the creation of operational/strategic anti-submarine barriers in choke points or on main transit routes, and possess a powerful secondary anti-ship capability. Of the two types that can be laid by aircraft, the NATO-designated **'Cluster Bay'** is intended for use on the continental shelf and the NATO-designated **'Cluster Gulf'** in the deeper waters of the continental ledge. It is thought that both types are moored torpedo-type weapons with rocket propulsion and a passive/active acoustic sensor system, the passive component for long-range target detection and the active component for medium/short-range target localization and weapon initiation. The version designed for laying in water up to 600 m (1,970 ft) deep is believed to weigh about 1200 kg (2,645 lb) and possess a 225 kg (496 lb) HE warhead.

The USSR has also developed an air-layable chokepoint weapon system comparable to the US Mk 60 CAPTOR. This is thought to weigh 1500 kg (3,306.9 lb) and be usable in water up to 750 m (2,460 ft) deep. As in its American counterpart, the Soviet weapon carries as payload a lightweight acoustic-homing torpedo initiated by the computer-controlled sensor system in the environmental capsule.

The nuclear mine is designed for the engagement of high-value targets such as aircraft-carriers or nuclear-powered ballistic missile submarines. Soviet nuclear mines have yields between 5 and 20 kilotons, and it is believed that while the primary laying platforms are conventional and nuclear-powered submarines, aircraft could be used for emergency deployment or the mining of major bases.

British sea mines and depth bombs

UK

The Royal Navy and Royal Air Force are both involved in British sea mine warfare, and still hold in their inventories a number of old (and probably obsolete) sea mines such as the **Mk 12A** ground mine (for delivery by fixed- or rotary-wing aircraft). In recent years, however, considerable effort has been expended on the development of new weapons, and these are now available in useful (but still inadequate) numbers.

The **British Aerospace Sea Urchin** is a bottom mine that can be laid by surface ships, submarines and aircraft. The weapon is effective in water depths from 15 ft (4.6 m) down to 300 ft (91.4 m), and can be used against the whole range of surface and sub-surface targets. The weapon comprises a warhead section, which can be used in any of three possible sizes (250, 500 or 750 kg/551, 1,102 or 1,653 lb), and a sensor/processor unit containing the microprocessor, electronics, battery, sensors (depth, acoustic, pressure, and magnetic), and safety/arming unit. The sensor/processor unit is a very advanced package capable of being programmed for variable mission profiles (including single- or multi-influence actuation), selectable delay, multiple ship count, selectable 'sterilization' delay, and automatic 'sterilization' in the event of any malfunction. The Sea Urchin has a diameter of 533 mm (21 in), and its lengths are 1.0 m (3.28 ft) with the 250 kg warhead, 1.5 m (4.92 ft) with the 500 kg warhead, and 2.2 m (7.22 ft) with the 750 kg warhead.

The **Marconi Underwater Systems Stonefish** is another modular type of bottom mine, the system allowing the creation of individual weapons tailored to the particular mission requirements. Like the Sea Urchin, the Stonefish can be laid in all its variants by aircraft, ships and submarines. The weapon has a shelf life of 20 years, and an operational life of 400 days, the advanced sensor/processor unit ensuring that the weapon remains perfectly adjusted to its operating conditions. The sensor section contains acoustic, magnetic and pressure units feeding data to analog conditioning and digital processing systems associated with a microprocessor-based control system into which can be entered other factors such as thresholds and 'sterilization' periods. The Stonefish can be used against surface targets in water between 30 and 300 ft (9.1 and 91.4 m), and against submarine targets in water down to 650 ft (198 m). The weapon has a diameter of 533 mm (21 in), and weighs between 205 and 900 kg (451.9 and 1,984.1 lb) depending on warhead type and the launch kit fitted.

Another product of Marconi Underwater Systems is the **Marconi Dragonfish** light bottom mine. This is designed to counter invasion forces, and can be laid from vehicles as diverse as light surface craft and aircraft. The mine is based on the most advanced of modern electronics to allow it a capability to select between a number of possible targets and engage the nearest, and to adjust itself to its environmental conditions. The mine has an extremely low acoustic signature, and detonates only when the separate acoustic and magnetic sensors confirm the target's vulnerability. The Dragonfish has a no-maintenance shelf life of 20 years and a minimum laid life of 200 days.

In 1987 this manufacturer announced the existence of a new sea mine, the **Marconi Underwater Systems Hammerhead,** developed jointly with Loral in the USA. This mine is designed for deployment to a predetermined depth (beyond the continental shelf if required), where it waits until the computer-controlled onboard sensor system detects and identifies one of a specific range of naval targets, whereupon a homing mine is launched to sink the intended submarine or ship. This is a singularly 'smart' weapon, and is designed for flexible tactical use as it can be laid by aircraft, ship or submarine. The Hammerhead is scheduled to enter service with the Royal and US Navies in the 1990s.

It is worth noting in this context that many countries still preserve stocks of obsolescent and indeed obsolete sea mines. These are still capable weapons so far as their mechanical and explosive features are concerned, but woefully lacking in their 'intelligence' by comparison with modern sea mines. There is thus a potential market for the upgrading of such weapons (the British Mk 12 is a good example) with a new sensor/processing package, British Aerospace has addressed itself to this need and produced a modular update package aimed mainly at mines produced during the 1950s using operational experience of mine warfare in World War 2. BAe has based its package on development of the Sea Urchin bottom mine and VEMS training mine. The result is a compact package that can be fitted into the volume previously occupied by less advanced control equipment of Western and Soviet mines. The basic system has a high-power battery of compact size, a microprocessor-controlled sensor and processor unit, and a combined safety and arming unit.

The only conventional depth bomb in service with the British forces is the **BAe Mk 11.** This was developed in its basic form during World War 2, and comprises of a cylindrical depth charge fitted with a ring-tail stabilizer that breaks off as the weapon hits the water. The weapon's dimensions include an overall length of 38.2 in (970 mm) and diameter of 11 in (279 mm). It weighs 320 lb (145.2 kg) complete with 176 lb (80 kg) of explosive detonated by a hydropneumatic arming/fusing system. The weapon can be used on targets down to a depth of 300 ft (91 m). While the older **Mk 11 Mod 2** version is now used mainly for training, the first-line variant is the **Mk 11 Mod 3** with a strengthened casing and revised fuse to mitigate the vibration effects of helicopter carriage.

American sea mines and depth bombs

USA

In common with other exponents of sea mine warfare, the Americans employ a number of such weapons for laying by aircraft, ship or submarine. The main types of standard sea mine in the current US Navy inventory are the Mk 52, Mk 53, Mk 55, Mk 56 and Mk 57, of which the first four can be laid by aircraft but the last only by ships or submarines. The Mk 53 is carried primarily by the Lockheed P-3 Orion long-range maritime patrol/anti-submarine aircraft, but no details of this weapon have been released.

The **Mk 52** is designed for use in shallow waters against submarines and ships, and can be laid by submarines as well as aircraft. The weapon is a bottom mine, and its dimensions include a diameter of 33.25 in (845 mm) and a length of 89 in (2261 mm). The warhead comprises 595 lb (269.9 kg) of HBX-1 explosive, and the overall weights of the variants are 1,195 lb (542.1 kg) for the acoustic **Mk 52 Mod 1,** 1,250 lb (567.0 kg) for the magnetic **Mk 52 Mod 2,** 1,260 lb (571.5 kg) for the magnetic/pressure **Mk 52 Mod 3**, 1,255 lb (569.3 kg) for the acoustic/magnetic **Mk 52 Mod 5**, and 1,240 lb (562.5 kg) for the acoustic/magnetic/pressure **Mk 52 Mod 6.** With the exception of the Mk 52 Mod 2, which can be used in water as deep as 600 ft (183 m), the maximum depth of the Mk 52 series is 150 ft (45.7 m).

The **Mk 55** is designed for use in shallow waters against submarines, and can be laid by submarines as well as aircraft. The weapon is a bottom mine, and its dimensions include a diameter of 40.5 in (1029 mm) and a length of 114 in (2896 mm). The warhead comprises 1,290 lb (585.1 kg) of HBX-1 explosive, and the overall weights of the variants are 2,040 lb (925.3 kg) for the magnetic **Mk 55 Mod 1,** 2,110 lb (957.1 kg) for the magnetic **Mk 55 Mod 2,** 2,120 lb (961.6 kg) for the magnetic/pressure **Mk 55 Mod 3,** 2,120 lb (961.1 kg) for the acoustic/magnetic **Mk 55 Mod 5,** 2,195 lb (995.7 kg) for the acoustic/magnetic/pressure **Mk 55 Mod 6,** and 2,115 lb (959.4 kg) for the dual-channel magnetic **Mk 55 Mod 7.** With the exception of the Mk 55 Mods 2 and 3, which can be used in water as deep as 600 ft (183 m), the maximum depth of the Mk 55 series is 150 ft (45.7 m).

The **Mk 56** (produced in only one variant as the **Mk 56 Mod 0**) is designed for use in medium-depth waters against submarines, and can only be laid by aircraft. The weapon is a moored mine, and its dimensions include a diameter of 41.75 in (1060 mm) and a length of 121 in (3073 mm). The warhead comprises 350 lb (158.8 kg) of HBX-3 explosive, and the overall weight of the mine is 2,225 lb (1009.3 kg). Activation is entrusted to a dual-channel total field magnetic sensor, and the weapon can be used in water as deep as 1,200 ft (365.8 m). The submarine-laid version of the Mk 56 is the 2,055 lb (932.1 kg) Mk 57 with a fibreglass case and a cast steel anchor.

Also used as sea mines are a number of air-dropped general-purpose bombs in the Mk 80 series. These can be dropped without a parachute retarder, and are essentially similar to the GP series other than the addition of a Mk 75 Mod 0 Destructor Modification Kit to create the Service Destructor series. The use of this kit turns the 500 lb (227 kg) Mk 82 bomb into the Mk 36 destructor, the 1,000 lb (454 kg) Mk 83 bomb into the Mk 40 destructor, and the 2,000 lb (907 kg) Mk 84 bomb into the Mk 41 destructor.

The **Mk 36** is designed for use in shallow waters against ships, and can be laid only by aircraft. The weapon is a bottom mine, and its dimensions include a diameter of 15.75 in (400 mm) and a length of 88.5 in (2248 mm). The warhead comprises 192 lb (87.1 kg) of H-6 explosive, and the overall weights of the variants are 529 lb (240.0 kg) with the fixed conical fin or 575 lb (260.8 kg) with the retarder tail. The **Mk 36 Mod 0** and **Mk 36 Mod 3** have magnetic activation, while the **Mk 36 Mod 4** and **Mk 36 Mod 5** have combined magnetic/seismic activation. The series can be used in water as deep as 300 ft (91.4 m).

The **Mk 40** is designed for use in shallow waters against ships, and can be laid only by aircraft. The weapon is a bottom mine, and its dimensions include a diameter of 22.45 in (570 mm) and a length of 112.5 in (2858 mm). The warhead comprises 450 lb (204.1 kg) of H-6 explosive, and the overall weight of the variants are 985 lb (446.8 kg) with the fixed conical fin or 1,060 lb (480.8 kg) with the retarder tail. The **Mk 40 Mod 0** and **Mk 40 Mod 3** have magnetic activation, while the **Mk 40 Mod 4** and **Mk 40 Mod 5** have combined magnetic/seismic activation. The series can be used in water as deep as 300 ft (91.4 m).

The **Mk 41** is designed for use in shallow waters against ships, and can be laid only by aircraft. The weapon is a bottom mine, and its dimensions include a diameter of 24.8 in (630 mm) and a length of 150.75 in (3829 mm). The warhead comprises an unspecified weight of H-6 explosive, and the overall weights of the variants are 2,040 lb (925.3 kg) for the magnetic **Mk 41 Mod 0** and **Mk 41 Mod 3,** or 2,030 lb (920.8 kg) for the magnetic/seismic **Mk 41 Mod 4** and **Mk 41 Mod 5.**

The same three GP bombs are also used in the US Navy's modern 'Quickstrike' programme for the development of mines using the same TDD-57 target detection and classification package attached to weapon casings of different sizes and weights to create the 500-lb (227-kg). **'Quickstrike' Mk 62**, 1,000 lb (454 kg) **'Quickstrike' Mk 63** and 2,000 lb (907 kg) **'Quickstrike' Mk 64**. 'Quickstrike' weapons are designed for deployment mainly by aircraft, though ships and submarines can also be used to lay the mines of this series. The Mk 60 series can be used in water up to 330 ft (100 m) deep, and requires the surface ship target to approach quite close to the weapon before its sensor system is activated.

Another mine in the 'Quickstrike' series is the **'Quickstrike' Mk 65 Mod 0,** though this is a more effective weapon than its companions as it has a purpose-designed thin-wall casing and the more effective PBX explosive. The weapon has a total weight of 2,390 lb (1084.1 kg), and its dimensions include a diameter of 21 in (533 mm), a length of 128 in (3251 mm) and a fin span of 28.9 in (734 mm). The standard delivery method for this weapon is under a retarding/stabilizing parachute after aerial release: after entering the water the weapon sheds its tail section (aerodynamic stabilizer and parachute stowage/attachment unit) and sinks to the bottom, where the rear-mounted sensor/fuse assembly is activated. The Mk 65 is made by Aerojet General, and the weapon is also being developed for export sale with the target detection device evolved by MISAR for the Italian MP-80 and MRP versions of the MR-80 mine (see Italian section).

The **DST 115A** falls into a different category, being an aircraft-laid surface mine. The weapon weighs 135 lb (61.2 kg) including 53 lb (24.0 kg) of HBX-3 explosive, and its primary dimensions include a diameter of 24.4 in (620 mm) and a length of 17.75 in (451 mm). The weapon has magnetic/seismic (acoustic) activation.

Also falling into another quite different category, though still capable of being air-laid, is the **Mk 60 CAPTOR** (enCAPsulated TORpedo). In its air-launched version this is a parachute-retarded environmental capsule complete with seabed mooring weight, Mk 46 Mod 4 or Mk 46 Mod 5 NEARTIP torpedo (see Torpedo section) as payload, and a DCU (Detection and Control Unit, comprising an passive-active sensor system and associated processor system). the weapon is of the 'rising mine' type and has an operational life of about six months before the onboard self-neutralization system is activated. The CAPTOR is laid in deep water (down to 3,000 ft/914.4 m) in anticipated submarine choke points, whereupon its sensor system (gated to ignore surface vessels) listens for any submarine target within a range estimated at 3,250 ft (990.6 m) and, at the appropriate moment, fires the Mk 46 torpedo to make an autonomous attack or attacks. The whole system has a diameter of 21 in (533 mm), a length of 145 in (3680 mm) and a loaded weight of 2,160 lb (1183.9 kg).

Air-dropped depth bombs were widely developed and used in World War 2 as the primary anti-submarine weapons carried by Allied maritime patrol aircraft involved in the campaign against German, Italian and Japanese submarines. In recent years, however, the lightweight acoustic-homing torpedo has assumed the primary role amongst air-launched anti-submarine weapons. The USA still deploys two anti-submarine depth bombs fitted with hydrostatic pistol detonators, however, in the form of the **Mk 41** and **Mk 54.** The Mk 41 is an obsolescent weapon now being phased out of service, while the Mk 54 continues in service in a number of forms. The weapon is based on a thin casing of sheet steel, and carries either 226 lb (102.5 kg) of TNT or 248 lb (112.5 kg) of HBX for a maximum weight of 324 lb (147 kg) or 346 lb (156.9 kg) respectively.

West German sea mines

West Germany

West Germany is again beginning to move to the front of sea mine technology, notably with the **Dornier SGM-80** multi-influence bottom mine. No details of this weapon have been released, though it is known that four companies (AEG, Dornier, Krupp-Atlas and MBB) are bidding to produce the type for the West German and Danish navies. It is likely that the mine is optimized for defensive purposes in the Baltic, where the two navies have an important NATO task in bottling up the Soviet Baltic Fleet.

Index

210